T0337858

Fundamentals of the Finite Element Method for Heat and Mass Transfer

WILEY SERIES IN COMPUTATIONAL MECHANICS

Series Advisors:

René de Borst
Perumal Nithiarasu
Tayfun E. Tezduyar
Genki Yagawa
Tarek Zohdi

Fundamentals of the Finite Element Method for Heat and Mass Transfer	Nithiarasu, Lewis and Seetharamu	January 2016
Introduction to Computational Contact Mechanics: A Geometrical Approach	Konyukhov	April 2015
Extended Finite Element Method: Theory and Applications	Khoei	December 2014
Computational Fluid-Structure Interaction: Methods and Applications	Bazilevs, Takizawa and Tezduyar	January 2013
Introduction to Finite Strain Theory for Continuum Elasto-Plasticity	Hashiguchi and Yamakawa	November 2012
Nonlinear Finite Element Analysis of Solids and Structures, Second Edition	De Borst, Crisfield, Remmers and Verhoosel	August 2012
An Introduction to Mathematical Modeling: A Course in Mechanics	Oden	November 2011
Computational Mechanics of Discontinua	Munjiza, Knight and Rougier	November 2011
Introduction to Finite Element Analysis: Formulation, Verification and Validation	Szabó and Babuška	March 2011

Fundamentals of the Finite Element Method for Heat and Mass Transfer

Second Edition

P. Nithiarasu
Zienkiewicz Centre for Computational Engineering
College of Engineering, Swansea University, UK

R. W. Lewis
Zienkiewicz Centre for Computational Engineering
College of Engineering, Swansea University, UK

K. N. Seetharamu
Department of Mechanical Engineering
PESIT, Bangalore, Karnataka, India

Library of Congress Cataloging-in-Publication Data

Names: Nithiarasu, Perumal. | Lewis, R. W. (Roland Wynne) | Seetharamu, K. N.
  | Lewis, R. W. (Roland Wynne). Fundamentals of the finite element method for heat and fluid flow.
Title: Fundamentals of the finite element method for heat and mass transfer.
Description: Second edition / P. Nithiarasu, R.W. Lewis, K.N. Seetharamu. | Chichester, West Sussex :
  John Wiley & Sons, Inc., 2016. | First edition: Fundamentals of the finite element method for heat and fluid flow /
  Roland W. Lewis, Perumal Nithiarasu, Kankanhalli N. Seetharamu (Hoboken, NJ : Wiley, 2004). | Includes
  bibliographical references and index.
Identifiers: LCCN 2015034600 | ISBN 9780470756256 (cloth : alk. paper)
Subjects: LCSH: Finite element method. | Heat equation. | Heat–Transmission.
  | Fluid dynamics. | Mass transfer.
Classification: LCC QC20.7.F56 L49 2016 | DDC 530.15/5353–dc23 LC record available at
http://lccn.loc.gov/2015034600

A catalogue record for this book is available from the British Library.

Set in 10/12.5pt Times by Aptara Inc., New Delhi, India.

Cover images: courtesy of the authors.

Printed in the UK

Contents

Preface to the Second Edition   **xii**

Series Editor's Preface   **xiv**

1 Introduction   **1**
    1.1   Importance of Heat and Mass Transfer 1
    1.2   Heat Transfer Modes 2
    1.3   The Laws of Heat Transfer 3
    1.4   Mathematical Formulation of Some Heat Transfer Problems 5
         1.4.1   Heat Transfer from a Plate Exposed to Solar Heat Flux 5
         1.4.2   Incandescent Lamp 7
         1.4.3   Systems with a Relative Motion and Internal Heat Generation 8
    1.5   Heat Conduction Equation 10
    1.6   Mass Transfer 13
    1.7   Boundary and Initial Conditions 13
    1.8   Solution Methodology 15
    1.9   Summary 15
    1.10   Exercises 16
    References 17

2 Some Basic Discrete Systems   **19**
    2.1   Introduction 19
    2.2   Steady-state Problems 20
         2.2.1   Heat Flow in a Composite Slab 20
         2.2.2   Fluid Flow Network 23
         2.2.3   Heat Transfer in Heat Sinks 26
    2.3   Transient Heat Transfer Problem 28
    2.4   Summary 31
    2.5   Exercises 31
    References 36

3 The Finite Element Method   **39**
    3.1   Introduction 39
    3.2   Elements and Shape Functions 42

   3.2.1 One-dimensional Linear Element 43
   3.2.2 One-dimensional Quadratic Element 46
   3.2.3 Two-dimensional Linear Triangular Element 49
   3.2.4 Area Coordinates 53
   3.2.5 Quadratic Triangular Element 55
   3.2.6 Two-dimensional Quadrilateral Elements 58
   3.2.7 Isoparametric Elements 63
   3.2.8 Three-dimensional Elements 72
  3.3 Formulation (Element Characteristics) 76
   3.3.1 Ritz Method (Heat Balance Integral Method – Goodman's Method) 78
   3.3.2 Rayleigh–Ritz Method (Variational Method) 79
   3.3.3 The Method of Weighted Residuals 82
   3.3.4 Galerkin Finite Element Method 86
  3.4 Formulation for the Heat Conduction Equation 89
   3.4.1 Variational Approach 90
   3.4.2 The Galerkin Method 93
  3.5 Requirements for Interpolation Functions 94
  3.6 Summary 100
  3.7 Exercises 100
  References 102

4 Steady-State Heat Conduction in One-dimension   **105**
  4.1 Introduction 105
  4.2 Plane Walls 105
   4.2.1 Homogeneous Wall 105
   4.2.2 Composite Wall 107
   4.2.3 Finite Element Discretization 108
   4.2.4 Wall with Varying Cross-sectional Area 110
   4.2.5 Plane Wall with a Heat Source: Solution by Linear Elements 112
   4.2.6 Plane Wall with Heat Source: Solution by Quadratic Elements 115
   4.2.7 Plane Wall with a Heat Source: Solution by Modified Quadratic
      Equations (Static Condensation) 117
  4.3 Radial Heat Conduction in a Cylinder Wall 118
  4.4 Solid Cylinder with Heat Source 120
  4.5 Conduction – Convection Systems 123
  4.6 Summary 126
  4.7 Exercises 127
  References 129

5 Steady-state Heat Conduction in Multi-dimensions   **131**
  5.1 Introduction 131
  5.2 Two-dimensional Plane Problems 132
   5.2.1 Triangular Elements 132
  5.3 Rectangular Elements 142

5.4   Plate with Variable Thickness   145
5.5   Three-dimensional Problems   146
5.6   Axisymmetric Problems   148
      5.6.1   Galerkin Method for Linear Triangular Axisymmetric Elements . . .   150
5.7   Summary   153
5.8   Exercises   153
References   155

6   Transient Heat Conduction Analysis   157
6.1   Introduction   157
6.2   Lumped Heat Capacity System   157
6.3   Numerical Solution   159
      6.3.1   Transient Governing Equations and Boundary and Initial
               Conditions   159
      6.3.2   The Galerkin Method   160
6.4   One-dimensional Transient State Problem   162
      6.4.1   Time Discretization-Finite Difference Method (FDM)   163
      6.4.2   Time Discretization-Finite Element Method (FEM)   168
6.5   Stability   169
6.6   Multi-dimensional Transient Heat Conduction   169
6.7   Summary   171
6.8   Exercises   171
References   173

7   Laminar Convection Heat Transfer   175
7.1   Introduction   175
      7.1.1   Types of Fluid Motion Assisted Heat Transport   176
7.2   Navier-Stokes Equations   177
      7.2.1   Conservation of Mass or Continuity Equation   177
      7.2.2   Conservation of Momentum   179
      7.2.3   Energy Equation   183
7.3   Nondimensional Form of the Governing Equations   184
7.4   The Transient Convection-Diffusion Problem   188
      7.4.1   Finite Element Solution to the Convection-Diffusion Equation   189
      7.4.2   A Simple Characteristic Galerkin Method for Convection-Diffusion
               Equation   191
      7.4.3   Extension to Multi-dimensions   197
7.5   Stability Conditions   202
7.6   Characteristic Based Split (CBS) Scheme   202
      7.6.1   Spatial Discretization   208
      7.6.2   Time-step Calculation   211
      7.6.3   Boundary and Initial Conditions   211
      7.6.4   Steady and Transient Solution Methods   213
7.7   Artificial Compressibility Scheme   214

7.8   Nusselt Number, Drag and Stream Function 215
        7.8.1   Nusselt Number 215
        7.8.2   Drag Calculation 216
        7.8.3   Stream Function 217
7.9   Mesh Convergence 218
7.10   Laminar Isothermal Flow 219
7.11   Laminar Nonisothermal Flow 231
        7.11.1 Forced Convection Heat Transfer 232
        7.11.2 Buoyancy-driven Convection Heat Transfer 238
        7.11.3 Mixed Convection Heat Transfer 240
7.12   Extension to Axisymmetric Problems 243
7.13   Summary 246
7.14   Exercises 247
References 249

8   Turbulent Flow and Heat Transfer   253
8.1   Introduction 253
        8.1.1   Time Averaging 254
        8.1.2   Relationship between κ, ϵ, ν_T and α_T 256
8.2   Treatment of Turbulent Flows 257
        8.2.1   Reynolds Averaged Navier-Stokes (RANS) 257
        8.2.2   One-equation Models 258
        8.2.3   Two-equation Models 259
        8.2.4   Nondimensional Form of the Governing Equations 260
8.3   Solution Procedure 262
8.4   Forced Convective Flow and Heat Transfer 263
8.5   Buoyancy-driven Flow 272
8.6   Other Methods for Turbulence 275
        8.6.1   Large Eddy Simulation (LES) 275
8.7   Detached Eddy Simulation (DES) and Monotonically Integrated LES
        (MILES) 278
8.8   Direct Numerical Simulation (DNS) 278
8.9   Summary 279
References 279

9   Heat Exchangers   281
9.1   Introduction 281
9.2   LMTD and Effectiveness-NTU Methods 283
        9.2.1   LMTD Method 283
        9.2.2   Effectiveness – NTU Method 285
9.3   Computational Approaches 286
        9.3.1   System Analysis 286
        9.3.2   Finite Element Solution to Differential Equations 289
9.4   Analysis of Heat Exchanger Passages 289

	9.5	Challenges	297
	9.6	Summary	299
	References		299

10 Mass Transfer   **301**
	10.1	Introduction	301
	10.2	Conservation of Species	302
		10.2.1 Nondimensional Form	304
		10.2.2 Buoyancy-driven Mass Transfer	305
		10.2.3 Double-diffusive Natural Convection	306
	10.3	Numerical Solution	307
	10.4	Turbulent Mass Transport	317
	10.5	Summary	319
	References		319

11 Convection Heat and Mass Transfer in Porous Media   **321**
	11.1	Introduction	321
	11.2	Generalized Porous Medium Flow Approach	324
		11.2.1 Nondimensional Scales	327
		11.2.2 Limiting Cases	329
	11.3	Discretization Procedure	329
		11.3.1 Temporal Discretization	330
		11.3.2 Spatial Discretization	331
		11.3.3 Semi- and Quasi-Implicit Forms	332
	11.4	Nonisothermal Flows	333
	11.5	Porous Medium-Fluid Interface	342
	11.6	Double-diffusive Convection	347
	11.7	Summary	349
	References		349

12 Solidification   **353**
	12.1	Introduction	353
	12.2	Solidification via Heat Conduction	354
		12.2.1 The Governing Equations	354
		12.2.2 Enthalpy Formulation	354
	12.3	Convection During Solidification	356
		12.3.1 Governing Equations and Discretization	358
	12.4	Summary	363
	References		364

13 Heat and Mass Transfer in Fuel Cells   **365**
	13.1	Introduction	365
		13.1.1 Fuel Cell Types	367
	13.2	Mathematical Model	368

	13.2.1	Anodic and Cathodic Compartments	371
	13.2.2	Electrolyte Compartment	373
13.3	Numerical Solution Algorithms		373
	13.3.1	Finite Element Modeling of SOFC	374
13.4	Summary		378
	References		378

14 An Introduction to Mesh Generation and Adaptive Finite Element Methods 379

	14.1	Introduction	379
	14.2	Mesh Generation	380
		14.2.1	Advancing Front Technique (AFT)	381
		14.2.2	Delaunay Triangulation	382
		14.2.3	Mesh Cosmetics	387
	14.3	Boundary Grid Generation	390
		14.3.1	Boundary Grid for a Planar Domain	390
		14.3.2	NURBS Patches	391
	14.4	Adaptive Refinement Methods	392
	14.5	Simple Error Estimation and Mesh Refinement	393
		14.5.1	Heat Conduction	394
	14.6	Interpolation Error Based Refinement	397
		14.6.1	Anisotropic Adaptive Procedure	398
		14.6.2	Choice of Variables and Adaptivity	399
	14.7	Summary	401
	References		402

15 Implementation of Computer Code 405

	15.1	Introduction	405
	15.2	Preprocessing	406
		15.2.1	Mesh Generation	406
		15.2.2	Linear Triangular Element Data	408
		15.2.3	Element Area Calculation	409
		15.2.4	Shape Functions and Their Derivatives	410
		15.2.5	Boundary Normal Calculation	411
		15.2.6	Mass Matrix and Mass Lumping	412
		15.2.7	Implicit Pressure or Heat Conduction Matrix	414
	15.3	Main Unit	416
		15.3.1	Time-step Calculation	416
		15.3.2	Element Loop and Assembly	419
		15.3.3	Updating Solution	420
		15.3.4	Boundary Conditions	421
		15.3.5	Monitoring Steady State	422
	15.4	Postprocessing	423
		15.4.1	Interpolation of Data	424
	15.5	Summary	424
	References		424

A   Gaussian Elimination   **425**
    Reference   426

B   Green's Lemma   **427**

C   Integration Formulae   **429**
    C.1   Linear Triangles   429
    C.2   Linear Tetrahedron   429

D   Finite Element Assembly Procedure   **431**

E   Simplified Form of the Navier–Stokes Equations   **435**

F   Calculating Nodal Values of Second Derivatives   **437**

Index   **439**

Preface to the Second Edition

In this second and enhanced edition of the book, we provide the readers with a detailed step-by-step application of the finite element method to heat and mass transfer problems. In addition to the fundamentals of the finite element method and heat and mass transfer, we have attempted to take the readers through some advanced topics of heat and mass transfer. The first edition of the book covered only the application of the finite element method to heat conduction and flow aided laminar heat convection. The second edition of the book has been enhanced further with turbulent flow and heat transfer, and mass transfer, in addition to advanced topics such as fuel cells. We believe that the second edition provides a comprehensive text for students, engineers and scientists who would like to pursue a finite element based heat transfer analysis. This textbook is suitable for beginners, senior undergraduate students, postgraduate students, engineers and early career researchers.

The first three chapters of the book deal with the essential fundamentals of both the heat conduction and the finite element method. In the first chapter, the fundamentals of energy balance and the standard derivations of relevant equations for the heat conduction analysis are discussed. Chapter 2 deals with the basic discrete systems which provide a basis for the finite element method formulations in the following chapters. The discrete system analysis is demonstrated through a variety of simple heat transfer and fluid flow problems. The third chapter gives a comprehensive account of the finite element method formulations and relevant history. Several examples and exercises included in Chapter 3 give the readers a complete overview of the theory and practice associated with the finite element method.

The application of the finite element method to heat conduction problems are discussed in detail in Chapters 4, 5 and 6. The conduction analysis starts with a simple one-dimensional steady-state heat conduction in Chapter 4 and is extended to multi-dimensions in Chapter 5. Chapter 6 gives the transient solution procedures for heat conduction problems.

Chapters 7, 8 and 9 deal with heat transfer by convection. In Chapter 7, heat transfer aided by the laminar motion of a single phase flow is discussed in detail. All the relevant differential equations are derived from first principles. All the three types of convection modes; forced, mixed and natural convection, are discussed in detail. Several examples and comparisons are provided to support the accuracy and flexibility of the finite element procedures discussed. In Chapter 8 the turbulent flow and heat transfer are discussed in some detail. Some examples and comparisons provide the readers a chance to assess the accuracy of the methods employed. Chapter 9 utilizes the finite element method developed in Chapters 1, 7 and 8 to provide a solution approach to flow and heat transfer in compact heat exchangers. Chapter 10 provides an introduction to the application of the finite element to problems of mass transfer. A detailed description of heat and mass transfer in porous media is then provide in Chapter 11. Two important applications of the finite element method for heat and mass transfer are explained in Chapters 12 and 13. Chapter 12 briefly introduces solidification problems using both heat conduction and convection approaches. Simple examples of solidification in this chapter may serve as a reference for students and researchers working in the area of solidification. In Chapter 13, we introduced a finite element solution approach to studying heat and mass transfer in fuel cells. Although the approach is only explained for solid oxide fuel cells, the method can be easily generalized to other types of fuel cells. Chapter 14 gives the reader sufficient information to understand the process of mesh generation. The main focus of this chapter is automatic and unstructured mesh generation. Some aspects of the adaptive mesh generation are also covered in this chapter. Finally, Chapter 15 briefly introduces the topic of computer implementation. The readers will be able to download the two-dimensional source codes and documentations from the website: **www.zetacomp.com**

Many people have assisted the authors either directly or indirectly during the preparation of this textbook. In particular, the authors wish to thank Dr Alessandro Mauro, Universitá degli Studi di Napoli Parthenope, for proofreading Chapter 13 and Dr Igor Sazonov, Swansea University, for helping the authors to put together part of Chapter 14. We would also like thank all our students, postdoctoral researchers and colleagues for providing help and support.

P. Nithiarasu, Swansea
R. W. Lewis, Swansea
K. N. Seetharamu, Bangalore

Series Editor's Preface

It is known that heat transfer provides a good context for teaching finite element methods and other computational mechanics topics. Fundamental concepts can be explained with such simple examples as heat conduction in 1D, then in 2D and 3D, and convective terms can be added to describe the special methods needed to deal with that class of partial differential equations. This book in our series does that, and with its distinguished, experienced authors, does it well. It not only teaches how to solve heat and mass transfer problems with finite element methods, but it also serves the purpose of teaching many different concepts in finite element methods. Readers from very diverse backgrounds will be able to benefit from this book. The book can be used by engineering undergraduate students to learn the fundamentals of heat and mass transfer and numerical methods, by graduate students in engineering and sciences to learn the advanced topics they need to know, and by practicing engineers and scientists as a good source and guide for research and development work in heat and mass transfer.

1

Introduction

1.1 Importance of Heat and Mass Transfer

The subject of heat and mass transfer is of fundamental importance in many branches of engineering. A *mechanical engineer* may be interested to know the mechanisms of heat transfer involved in the operation of equipment, for example, boilers, condensers, air preheaters, economizers etc., in a thermal power plant in order to improve their performance. Nuclear power plants require precise information on heat transfer as safe operation is an important factor in their design. Refrigeration and air-conditioning systems also involve heat-exchanging devices, which need careful design. *Electrical engineers* are keen to avoid material damage in electric motors, generators and transformers due to hot spots, developed by improper heat transfer design. An *electronic engineer* is interested in knowing efficient methods of heat dissipation from chips and semi-conductor devices so that they function within safe operating temperatures. A *computer hardware engineer* is interested to know the cooling requirements of circuit-boards, as the miniaturization of computing devices is advancing at a rapid rate. *Chemical engineers* are interested in heat and mass transfer processes in various chemical reactions. A *metallurgical engineer* would be interested in knowing the rate of heat transfer required for a particular heat treatment process, e.g. the rate of cooling in a casting process has a profound influence on the quality of the final product. *Aeronautical engineers* are interested in knowing the heat transfer rate in rocket nozzles and in heat shields used in re-entry vehicles. An *agricultural engineer* would be interested in the drying of food grains, food processing and preservation. A *civil engineer* would need to be aware of the thermal stresses developed in quick setting concrete, the influence of heat and mass transfer on building and building materials as well as the effect of heat on nuclear containment and buildings etc. An *environmental engineer* is concerned with the effect of heat on dispersion of pollutants in air, transport of pollutants in soils, lakes and seas and their impact on life. A *bioengineer* is often interested in the heat and

Fundamentals of the Finite Element Method for Heat and Mass Transfer, Second Edition.
P. Nithiarasu, R. W. Lewis, and K. N. Seetharamu.
© 2016 John Wiley & Sons, Ltd. Published 2016 by John Wiley & Sons, Ltd.

mass transfer processes, such as hypothermia and hyperthermia associated with the human body.

The above-mentioned applications are only a sample of heat and mass transfer applications. The solar system and the associated energy transfer from the sun are the principal factors for existence of life on Earth. It is not untrue to say that it is extremely difficult, often impossible, to avoid some form of heat transfer in any process on Earth.

The study of heat and mass transfer provides economical and efficient solutions for many critical problems encountered in diverse engineering items of equipment. For example, we can consider the development of heat pipes which can transport heat at a much greater rate than that of copper or silver rods of the same dimensions and even at almost isothermal conditions. The development of present-day gas turbine blades, where the gas temperature exceeds the melting point of the blade material, is possible by providing efficient cooling systems. This is another example of the success of heat transfer design methods. The design of computer chips, which encounter heat flux of the order occurring in re-entry vehicles, especially when the surface temperature of the chips is limited to less than 100 °C, is again a success story of heat transfer design.

Although there are many successful heat transfer designs, further developments on heat and mass transfer studies are necessary in order to increase the life span and efficiency of the many devices discussed previously, which can lead to many more new inventions. Also, if we are to protect our environment, it is essential to understand the many heat and mass transfer processes involved and if necessary to take appropriate action.

1.2   Heat Transfer Modes

Heat transfer is that section of engineering science that studies the energy transport between material bodies due to temperature difference (Bejan 1993; Holman 1989; Incropera and Dewitt 1990; Sukhatme 1992). The three modes of heat transfer are:

 (a) conduction

 (b) convection and

 (c) radiation.

The conduction mode of heat transport occurs either because of an exchange of energy from one molecule to another without actual motion of the molecules, or is due to the motion of free electrons if they are present. Therefore, this form of heat transport depends heavily on the properties of the medium and takes place in solids, liquids and gases if a difference in temperature exists.

Molecules present in liquids and gases have freedom of motion and by moving from a hot to a cold region, they carry energy with them. The transfer of heat from one region to another due to such macroscopic motion in a liquid or gas, added to the energy transfer by conduction within the fluid, is called heat transfer by convection. Convection may be either free, forced or mixed. When fluid motion occurs due to a density variation caused by temperature differences, the situation is said to be a free or natural convection. When the fluid motion is caused by an external force, such as pumping or blowing, the state is defined as being forced convection.

A mixed convection state is one in which both natural and forced convection are present. Convection heat transfer also occurs in boiling and condensation processes.

All bodies emit thermal radiation at all temperatures. This is the only mode which does not require a material medium for heat transfer to occur. The nature of thermal radiation is such that a propagation of energy, carried by *electromagnetic waves*, is emitted from the surface of the body. When these electromagnetic waves strike other body surfaces, a part is reflected, a part transmitted and the remaining part is absorbed.

All modes of heat transfer are generally present in varying degrees in a real physical problem. The important aspects in solving heat transfer problems are to identify the significant modes and to decide whether the heat transferred by other modes can be neglected.

1.3   The Laws of Heat Transfer

It is important to quantify the amount of energy being transferred per unit time and for that we require the use of rate equations. For heat conduction, the rate equation is known as *Fourier's law* (Fourier 1955) which is expressed for one dimension, as

$$q_x = -k\frac{dT}{dx}, \qquad (1.1)$$

where q_x is the heat flux in the x direction (W/m^2); k is the thermal conductivity (W/mK, a property of the material, see Table 1.1) and dT/dx the temperature gradient (K/m).

Table 1.1   Typical values of thermal conductivity of some materials in W/mK at 20 °C.

Material	Thermal conductivity, k
Metals:	
Pure silver	410
Pure copper	385
Pure aluminium	200
Pure iron	73
Alloys:	
Stainless steel (18% Cr, 8% Ni)	16
Aluminium alloy (4.5% Cr)	168
Non metals:	
Plastics	0.6
Wood	0.2
Liquid:	
Water	0.6
Gasses:	
Dry air	0.025 (at atmospheric pressure)

Table 1.2   Typical values of heat transfer coefficient in W/m^2K

Gases (stagnant)	15
Gases (flowing)	15–250
Liquids (stagnant)	100
Liquids (flowing)	100–2000
Boiling liquids	2000–35 000
Condensing vapors	2000–25 000

For convective heat transfer, the rate equation is given by *Newton's law* of cooling (Whewell 1866) as

$$q = h(T_w - T_a), \tag{1.2}$$

where q is the convective heat flux; (W/m^2); $(T_w - T_a)$ the temperature difference between the wall and the fluid and h is the convection heat transfer coefficient (W/m^2K) (or film coefficient, see Table 1.2).

The convection heat transfer coefficient frequently appears as a boundary condition in the solution of heat conduction through solids, where h is often known (Table 1.2).

The maximum flux that can be emitted by radiation from a black surface is given by the *Stefan-Boltzmann Law* (Boltzmann 1884; Stefan 1879), that is,

$$q = \sigma T_w{}^4, \tag{1.3}$$

where q is the radiative heat flux (W/m^2); σ is the Stefan-Boltzmann constant (5.669×10^{-8}), in W/m^2K^4 and T_w is the surface temperature (K).

The heat flux emitted by a real surface is less than that of a black surface and is given by

$$q = \epsilon \sigma T_w{}^4, \tag{1.4}$$

where ϵ is the radiative property of the surface and is referred to as the emissivity. The net radiant energy exchange between any two surfaces 1 and 2 is given by

$$Q = F_e F_G \sigma A_1 (T_1^4 - T_2^4), \tag{1.5}$$

where F_e is a factor which takes into account the nature of the two radiating surfaces; F_G a factor which takes into account the geometric orientation of the two radiating surfaces and A_1 is the area of surface 1.

When a heat transfer surface, at temperature T_1, is completely enclosed by a much larger surface at temperature T_2, the net radiant exchange can be calculated by

$$Q = qA_1 = \epsilon_1 \sigma A_1 (T_1^4 - T_2^4). \tag{1.6}$$

With respect to the laws of thermodynamics, only the first law (Clausius 1850) is of interest in heat transfer problems. The increase of energy in a system is equal to the difference between the energy transfer by heat to the system and the energy transfer by work done on the surroundings by the system, that is,

$$dE = dQ - dW, \tag{1.7}$$

where Q is the total heat entering the system and W is the work done by the system on the surroundings. Since we are interested in the rate of energy transfer in heat transfer processes, we can restate the first law of thermodynamics as:

"The rate of increase of the energy of the system is equal to the difference between the rate at which energy enters the system and the rate at which the system does work on the surroundings," that is,

$$\frac{dE}{dt} = \frac{dQ}{dt} - \frac{dW}{dt}, \tag{1.8}$$

where t is the time.

1.4 Mathematical Formulation of Some Heat Transfer Problems

In analyzing a thermal system, the engineer should be able to identify the relevant heat transfer processes and only then can the system behavior be quantified properly. In this section, some typical heat transfer problems are formulated by identifying the appropriate heat transfer mechanisms.

1.4.1 Heat Transfer from a Plate Exposed to Solar Heat Flux

Consider a plate of size L x B x d exposed to the solar flux of intensity q_s as shown in Figure 1.1. In many solar applications, such as a solar water heater, solar cooker etc., the temperature of the plate is a function of time. The plate loses heat by convection and radiation to the ambient air, which is at temperature T_a. Some heat flows through the plate and is convected

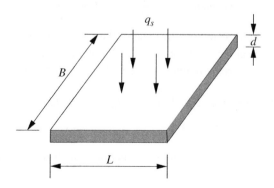

Figure 1.1   Heat transfer from a plate subjected to solar heat flux.

to the atmosphere from the bottom side. We shall apply the law of conservation of energy to derive an appropriate equation, the solution of which gives the temperature of the plate with respect to time.

Heat entering the top surface of the plate:

$$q_s A_T. \tag{1.9}$$

Heat loss from the plate to the surroundings:
Top surface:

$$h A_T (T - T_a) + \epsilon \sigma A_T (T^4 - T_a^4), \tag{1.10}$$

Side surface:

$$h A_S (T - T_a) + \epsilon \sigma A_S (T^4 - T_a^4), \tag{1.11}$$

Bottom surface:

$$h A_B (T - T_a) + \epsilon \sigma A_B (T^4 - T_a^4), \tag{1.12}$$

where the subscripts T, S and B refer respectively to the top, side and bottom surface areas. The topic of radiation exchange between a gas and a solid surface is not simple. Readers are referred to appropriate texts for details (Holman 1989; Siegel and Howell 1992). Under steady-state conditions, the heat received by the plate is lost to the surroundings, thus

$$q_s A_T = h A_T \left(T - T_a \right) + \epsilon \sigma A_T \left(T^4 - T_a^4 \right) + h A_S \left(T - T_a \right)$$
$$+ \epsilon \sigma A_S \left(T^4 - T_a^4 \right) + h A_B \left(T - T_a \right) + \epsilon \sigma A_B \left(T^4 - T_a^4 \right). \tag{1.13}$$

This is a nonlinear algebraic equation because of the presence of the T^4 term. The solution of this equation results in the steady-state temperature of the plate. If we want to calculate the temperature of the plate as a function of time, t, then we have to consider the rate of rise in the internal energy of the plate. Substituting $E = volume \times \rho \times c_p \times T$ into the LHS of the Equation (1.8) gives

$$(volume) \times \rho c_p \frac{dT}{dt} = (LBd) \rho c_p \frac{dT}{dt}, \tag{1.14}$$

where ρ is the density and c_p is the specific heat of the plate. Thus, at any instant of time, the difference between the heat received and lost (work done on the surroundings) by the plate will be equal to the rate of change in internal energy heat stored (Equation (1.8)). Thus,

$$(LBd) \rho c_p \frac{dT}{dt} = q_s A_T - \left[h A_T (T - T_a) + \epsilon \sigma A_T \left(T^4 - T_a^4 \right) + \right.$$
$$\left. \epsilon \sigma A_S \left(T^4 - T_a^4 \right) + h A_B (T - T_a) + \epsilon \sigma A_B \left(T^4 - T_a^4 \right) \right]. \tag{1.15}$$

This is a first-order nonlinear differential equation, which requires an initial condition, viz.,

$$\text{at} \quad t = 0, T = T_a. \tag{1.16}$$

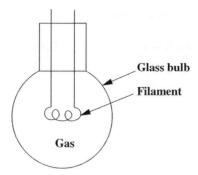

Figure 1.2   Energy balance in an incandescent light source.

The solution is determined iteratively because of the nonlinearity of the problem. Equation (1.15) can be simplified by substituting relations for the surface areas. It should be noted, however, that this is a general equation, which can be used for similar systems.

It is important to note that the spatial variation of temperature within the plate is neglected here. However, this variation can be included via Fourier's law of heat conduction, Equation (1.1). Such a variation is necessary if the plate is not thin enough to reach equilibrium instantly (Section 1.5).

1.4.2   Incandescent Lamp

Figure 1.2 shows an idealized incandescent lamp. The filament is heated to a temperature T_f by an electric current. Heat is convected to the surrounding gas and is radiated to the wall, which also receives heat from the gas by convection. The wall in turn convects and radiates heat to the ambient at T_a. A formulation of equations, based on energy balance, is necessary in order to determine the temperature of the gas and the wall with respect to time.

1.4.2.1   Gas

Rise in internal energy of the gas:

$$\rho_g c_{pg} \frac{dT_g}{dt}. \tag{1.17}$$

Convection from the filament to the gas:

$$h_f A_f (T_f - T_g). \tag{1.18}$$

Convection from the gas to the wall:

$$h_g A_g (T_g - T_w). \tag{1.19}$$

Radiation from the filament to the gas:

$$\epsilon_f A_f \sigma \left(T_f^4 - T_g^4 \right). \tag{1.20}$$

Now, the energy balance for the gas gives

$$\rho_g c_{pg} \frac{dT_g}{dT} = h_f A_f (T_f - T_g) - h_g A_g (T_g - T_w) + \epsilon_f A_f \sigma \left(T_f^4 - T_g^4 \right). \tag{1.21}$$

1.4.2.2   Wall

Rise in internal energy of the wall:

$$\rho_w c_{pw} \frac{dT_w}{dt}. \tag{1.22}$$

Radiation from the filament to the wall:

$$\epsilon_f \sigma A_f \left(T_f^4 - T_w^4 \right). \tag{1.23}$$

Convection from the wall to ambient:

$$h_w A_w (T_w - T_a). \tag{1.24}$$

Radiation from the wall to ambient:

$$\epsilon_w \sigma A_w \left(T_w^4 - T_a^4 \right). \tag{1.25}$$

Energy balance for the wall gives

$$\rho_w c_{pw} \frac{dT_w}{dt} = h_g A_g (T_g - T_w) + \epsilon_f \sigma A_f \left(T_f^4 - T_w^4 \right) - h_w A_w (T_w - T_a) - \epsilon_w \sigma A_w \left(T_w^4 - T_a^4 \right), \tag{1.26}$$

where ρ_g is the density of the gas in the bulb; c_{pg} the specific heat of the gas; ρ_w the density of the wall of the bulb; c_{pw} the specific heat of the wall; h_f the heat transfer coefficient between filament and gas; h_g the heat transfer coefficient between gas and wall; h_w the heat transfer coefficient between wall and ambient and ϵ the emissivity. The subscripts f, w, g and a respectively indicate the filament, wall, gas and ambient.

Equations (1.21) and (1.26) are first-order nonlinear differential equations. The initial conditions required are

At $t = 0$,

$$T_g = T_a \quad \text{and} \quad T_w = T_a. \tag{1.27}$$

The simultaneous solution of Equations (1.21) and (1.26), along with the above initial condition, results in the temperatures of the gas and the wall as functions of time.

1.4.3   Systems with a Relative Motion and Internal Heat Generation

The extrusion of plastics, drawing of wires and artificial fiber (optical fiber), suspended electrical conductors of various shapes, continuous casting etc. can be treated alike.

In order to derive an energy balance for such a system, we consider a small differential control volume of length, Δx, as shown in Figure 1.3. In this problem, the heat lost to the environment by radiation is assumed to be negligibly small. The energy is conducted, convected

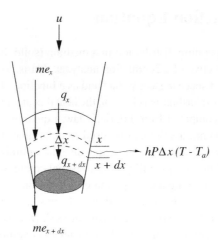

Figure 1.3   Conservation of energy in a moving body.

and transported with the material in motion. With reference to Figure 1.3, we can write the following equations of conservation of energy, that is,

$$Q_x + me_x + GA\Delta x = Q_{x+dx} + me_{x+dx} + hP\Delta x(T - T_a), \qquad (1.28)$$

where $Q = Aq$ is the total heat; m is the mass flow ρAu and is assumed to be constant; e_x is the specific energy; ρ the density of the material; A the cross-sectional area; P the perimeter of the control volume; G is the heat generated per unit volume and u is the velocity at which the material is moving. Using the Taylor series of expansion we obtain

$$m(e_x - e_{x+dx}) = -m\frac{de_x}{dx}\Delta x = -mc_p\frac{dT}{dx}\Delta x. \qquad (1.29)$$

Note that $de_x = c_p dT$ at constant pressure. Similarly, using Fourier's law (Equation (1.1)),

$$Q_x - Q_{x+dx} = -\frac{dQ_x}{dx} = \frac{d}{dx}\left[kA\frac{dT}{dx}\right]. \qquad (1.30)$$

On substituting Equations (1.29) and (1.30) into Equation (1.28), we obtain the following conservation equation,

$$\frac{d}{dx}\left[kA\frac{dT}{dx}\right] - hP(T - T_a) - \rho c_p Au\frac{dT}{dx} + GA = 0. \qquad (1.31)$$

In the above equation, the first term is derived from the heat diffusion (conduction) within the material, the second term is due to convection from the material surface to ambient, the third term represents the heat transport due to the motion of the material, and finally the last term is added to account for heat generation within the body.

1.5   Heat Conduction Equation

The determination of temperature distribution in a medium (solid, liquid, gas or combination of phases) is the main objective of a conduction analysis, that is, to know the temperature in the medium as a function of space at steady state and as a function of time during the transient state. Once this temperature distribution is known, the heat flux at any point within the medium, or on its surface, may be computed from Fourier's law, Equation (1.1). A knowledge of the temperature distribution within a solid can be used to determine the structural integrity via a determination of the thermal stresses and distortion. The optimization of the thickness of an insulating material and the compatibility of any special coatings or adhesives used on the material can be studied by knowing the temperature distribution.

We shall now derive the conduction equation in Cartesian coordinates by applying the energy conservation law to a differential control volume as shown in Figure 1.4. The solution of the resulting differential equation, with prescribed boundary conditions, gives the temperature distribution in the medium.

The Taylor series expansion gives:

$$Q_{x+dx} = Q_x + \frac{\partial Q_x}{\partial x} \Delta x$$

$$Q_{y+dy} = Q_y + \frac{\partial Q_y}{\partial y} \Delta y$$

$$Q_{z+dz} = Q_z + \frac{\partial Q_z}{\partial z} \Delta z. \tag{1.32}$$

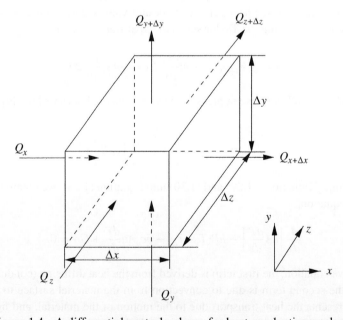

Figure 1.4   A differential control volume for heat conduction analysis.

Note that second and higher order terms are neglected in the above equation. The heat generated in the control volume is $G\Delta x\Delta y\Delta z$ and the rate of change in energy storage is given as

$$\rho c_p(\Delta x\Delta y\Delta z)\frac{\partial T}{\partial t}.$$ (1.33)

Now, with reference to Figure 1.4, we can write the energy balance as

"energy inlet + energy generated = energy stored + energy exit"

that is:

$$(Q_x + Q_y + Q_z) + G(\Delta x\Delta y\Delta z) = \rho(\Delta x\Delta y\Delta z)\frac{\partial T}{\partial t} + Q_{x+dx} + Q_{y+dy} + Q_{z+dz}.$$ (1.34)

Substituting Equation (1.32) into the previous equation and rearranging results in;

$$-\frac{\partial Q_x}{\partial x}\Delta x - \frac{\partial Q_y}{\partial y}\Delta y - \frac{\partial Q_z}{\partial z}\Delta z + G(\Delta x\Delta y\Delta z) = \rho c_p(\Delta x\Delta y\Delta z)\frac{\partial T}{\partial t}.$$ (1.35)

The total heat transfer Q in each direction can be expressed as (area perpendicular to heat flux direction × heat flux):

$$Q_x = (\Delta y\Delta z)q_x = -k_x(\Delta y\Delta z)\frac{\partial T}{\partial x}$$
$$Q_y = (\Delta x\Delta z)q_y = -k_y(\Delta x\Delta z)\frac{\partial T}{\partial y}$$
$$Q_z = (\Delta x\Delta y)q_z = -k_z(\Delta x\Delta y)\frac{\partial T}{\partial z}.$$ (1.36)

Substituting Equation (1.36) into Equation (1.35) and dividing by the volume, $\Delta x\Delta y\Delta z$, we get

$$\frac{\partial}{\partial x}\left[k_x\frac{\partial T}{\partial x}\right] + \frac{\partial}{\partial y}\left[k_y\frac{\partial T}{\partial y}\right] + \frac{\partial}{\partial z}\left[k_z\frac{\partial T}{\partial z}\right] + G = \rho c_p\frac{\partial T}{\partial t}.$$ (1.37)

Equation (1.37) is the transient heat conduction equation for a stationary system expressed in Cartesian coordinates. The thermal conductivity, k, in the above equation is a vector. In its most general form, the thermal conductivity can be expressed as a tensor, that is,

$$\mathbf{k} = \begin{bmatrix} k_{xx} & k_{xy} & k_{xz} \\ k_{yx} & k_{yy} & k_{yz} \\ k_{zx} & k_{zy} & k_{zz} \end{bmatrix}.$$ (1.38)

The preceding Equations (1.37) and (1.38) are valid for solving heat conduction problems in anisotropic materials with directional variation in thermal conductivities. In many situations, however, thermal conductivity can be taken as a nondirectional property, that is, the material is isotropic in nature. In such materials, the heat conduction equation is written as (constant thermal conductivity):

$$\frac{\partial^2 T}{\partial x^2} + \frac{\partial^2 T}{\partial y^2} + \frac{\partial^2 T}{\partial z^2} + \frac{G}{k} = \frac{1}{\alpha}\frac{\partial T}{\partial t}, \tag{1.39}$$

where $\alpha = k/\rho c_p$ is the *thermal diffusivity*, which is an important parameter in transient heat conduction analyses. If the analysis is restricted only to steady-state heat conduction without heat generation, the equation is reduced to

$$\frac{\partial^2 T}{\partial x^2} + \frac{\partial^2 T}{\partial y^2} + \frac{\partial^2 T}{\partial z^2} = 0. \tag{1.40}$$

For a one-dimensional case, the steady-state heat conduction equation is further reduced to

$$\frac{d}{dx}\left(k\frac{dT}{dx}\right) = 0. \tag{1.41}$$

The heat conduction equation for a cylindrical coordinate system is given by

$$\frac{1}{r}\frac{\partial}{\partial r}\left[k_r r\frac{\partial T}{\partial r}\right] + \frac{1}{r^2}\frac{\partial}{\partial \phi}\left[k_\phi \frac{\partial T}{\partial \phi}\right] + \frac{\partial}{\partial z}\left[k_z \frac{\partial T}{\partial z}\right] + G = \rho c_p \frac{\partial T}{\partial t}. \tag{1.42}$$

In cylindrical coordinates, the heat fluxes can be expressed as

$$q_r = -k_r \frac{\partial T}{\partial r}$$

$$q_\phi = -\frac{k_\phi}{r}\frac{\partial T}{\partial \phi}$$

$$q_z = -k_z \frac{\partial T}{\partial z}, \tag{1.43}$$

where r, ϕ and z are the cylindrical coordinate directions. The heat conduction equation for a spherical coordinate system is given by

$$\frac{1}{r^2}\frac{\partial}{\partial r}\left[k_r r^2 \frac{\partial T}{\partial r}\right] + \left(\frac{1}{r^2 sin^2\theta}\right)\frac{\partial}{\partial \phi}\left[k_\phi \frac{\partial T}{\partial \phi}\right] +$$

$$\left(\frac{1}{r^2 sin\theta}\right)\frac{\partial}{\partial \theta}\left[k_\theta sin\theta \frac{\partial T}{\partial \theta}\right] + G = \rho c_p \frac{\partial T}{\partial t}. \tag{1.44}$$

The heat fluxes in a spherical coordinate system can be expressed as

$$q_r = -k_r \frac{\partial T}{\partial r}$$

$$q_\phi = -\frac{k_\phi}{r sin\theta}\frac{\partial T}{\partial \phi}$$

$$q_\theta = -\frac{k_\theta}{r}\frac{\partial T}{\partial \theta}, \tag{1.45}$$

where r, ϕ and θ are the spherical coordinate directions. It should be noted that for both cylindrical and spherical coordinate systems (Equations (1.42) and (1.44)) can be derived in a similar fashion as for Cartesian coordinates by considering the appropriate differential control volumes.

1.6   Mass Transfer

When a concentration gradient exists in a fluid mixture, mass transfer takes place from a higher concentration to a lower concentration location. Such mass transport often takes place at the molecular level in the form of mass diffusion. The mass transport at the macroscopic level is referred to as mass convection. Thus, the modes of mass transfer are very similar to the first two modes of heat transfer, that is, conduction (diffusion) and convection. Mass diffusion is often described using Fick's law of mass transport (Fick 1855). This states that the mass flux of a constituent per unit area is proportional to the concentration gradient, that is,

$$J_A = \frac{\dot{m}_A}{A} = -D_{AB}\frac{dC_A}{dx}, \tag{1.46}$$

where $\dot{m}_A$ is the mass flux per unit time, D_{AB} is the diffusion coefficient and C_A is the mass concentration of the component A. As seen, this expression is very similar to Fourier's law of heat conduction (Equation (1.1)). The convective mass flux per unit area may be defined as

$$\frac{\dot{m}_A}{A} = h_A(C_A - C_{A\infty}), \tag{1.47}$$

where h_A is the mass transfer coefficient and $C_A - C_{A\infty}$ is the concentration difference through which mass transfer occurs. Equation (1.47) is analogous to the Newton's law of cooling for heat transfer (Equation (1.2)). Further details on mass transfer are given in Chapter 10.

1.7   Boundary and Initial Conditions

The heat conduction equations discussed in Section 1.5 will be complete for any problem only if the appropriate boundary and initial conditions are stated. With the necessary boundary and initial conditions, a solution to the heat conduction equation is possible. The boundary conditions for the conduction equation can be of two types or a combination of these: the *Dirichlet* condition, in which the temperature on the boundaries is known and/or the *Neumann* condition, in which the heat flux is imposed, that is (see Figure 1.5):

   Dirichlet condition:

$$T = T_o \quad \text{on} \quad \Gamma_T. \tag{1.48}$$

   Neumann condition:

$$q = -k\frac{\partial T}{\partial n} = \bar{q} \quad \text{on} \quad \Gamma_{qf}. \tag{1.49}$$

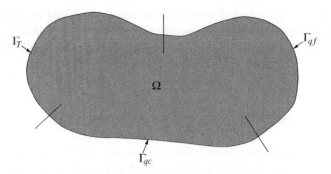

Figure 1.5   Boundary conditions.

In the above equations (Equations (1.48) and (1.49)), T_o is the prescribed temperature; Γ the boundary surface; n is the outward direction normal to the surface and $\bar{q}$ is the constant flux given. The insulated, or adiabatic, condition can be obtained by substituting $\bar{q} = 0$. The convective heat transfer boundary condition also falls into the *Neumann* category and can be expressed as

$$-k\frac{\partial T}{\partial n} = h(T_w - T_a) \quad \text{on} \quad \Gamma_{qc}. \tag{1.50}$$

It should be observed that the heat conduction equation has second-order terms and hence faces two types of boundary conditions. Since the time appears as a first-order term, at least one initial value (i.e., at some instant of time all temperatures must be known) is to be specified for the entire body, that is,

$$T = T_0 \quad \text{all over the domain} \quad \Omega \quad \text{at} \quad t = t_0, \tag{1.51}$$

where t_0 is a reference time.

The constant or variable temperature conditions are generally easy to implement as temperature is a scalar. However, the implementation of surface fluxes is not as straightforward. Equation (1.49) can be rewritten with direction cosines of the outward normals as

$$-\left(k_x\frac{\partial T}{\partial x}\tilde{l} + k_y\frac{\partial T}{\partial y}\tilde{m} + k_z\frac{\partial T}{\partial z}\tilde{n}\right) = \bar{q} \quad \text{on} \quad \Gamma_{qf}. \tag{1.52}$$

Similarly, Equation (1.50) can be rewritten as

$$-\left(k_x\frac{\partial T}{\partial x}\tilde{l} + k_y\frac{\partial T}{\partial y}\tilde{m} + k_z\frac{\partial T}{\partial z}\tilde{n}\right) = h(T - T_a) \quad \text{on} \quad \Gamma_{qc}, \tag{1.53}$$

where $\tilde{l}$, $\tilde{m}$ and $\tilde{n}$ are the direction cosines of the appropriate outward surface normals.

In many industrial applications, for example, wire drawing, crystal growth, continuous casting, etc., the material will have a motion in space and this motion may be restricted to one direction, as in the example (Section 1.4.3) cited previously. The general energy equation for heat conduction, taking into account the spatial motion of the body, is given by

$$\frac{\partial}{\partial x}\left(k_x\frac{\partial T}{\partial x}\right) + \frac{\partial}{\partial y}\left(k_y\frac{\partial T}{\partial y}\right) + \frac{\partial}{\partial z}\left(k_z\frac{\partial T}{\partial z}\right) + G = \rho c_p\left(\frac{\partial T}{\partial t} + u\frac{\partial T}{\partial x} + v\frac{\partial T}{\partial y} + w\frac{\partial T}{\partial z}\right), \quad (1.54)$$

where u, v and w are the components of the velocity in the three directions, x, y and z respectively.

The governing equations for convection heat transfer are very similar to the above and will be discussed in Chapter 7.

1.8   Solution Methodology

Although a number of analytical solutions for conduction heat transfer problems are available (Carslaw and Jaeger 1959; Ozisik 1968), in many practical situations, the geometry and the boundary conditions are so complex that an analytical solution is not possible. Even if one could develop analytical relations for such complicated cases, these will invariably involve complex series solutions and would thus be practically difficult to implement. In such situations, conduction heat transfer problems do need a numerical solution. Some commonly employed numerical methods are the Finite Difference (Ozisik and Czisik 1994), Finite Volume (Patankar 1980), Finite Element and Boundary Element (Ibanez and Power 2002) techniques. This text will address issues related to the Finite Element Method (FEM) only (Comini *et al.* 1994; Huang and Usmani 1994; Lewis *et al.* 1996, 2004; Reddy and Gartling 2000).

In contrast to an analytical solution, which allows for the temperature determination at any point in the medium, a numerical solution enables the determination of temperature only at discrete points. The first step in any numerical analysis must therefore be to select these points. This is done by dividing the region of interest into a number of smaller regions. These regions are bounded by points. These reference points are termed nodal points and their assembly results in a grid or mesh. It is important to note that each node represents a certain region surrounding it and its temperature is a measure of the temperature distribution in that region. The numerical accuracy of these calculations depends strongly on the number of designated nodal points, which control the number of elements generated. The accuracy approaches an exact value as the mesh size (region size) approaches zero.

Further details on the numerical methods, mesh generation, accuracy and error are discussed in later chapters.

1.9   Summary

In this chapter, the subject of heat transfer was introduced and various modes of heat transport were discussed. The fundamentals of energy conservation principles and the application of such principles to some selected problems were also presented. Finally, the general heat conduction equations in multi-dimensions were derived and the appropriate boundary and initial conditions were given. Although this chapter has been brief, we trust that it has given the reader the essential fundamental concepts involved in heat transfer in general and some detailed understanding of conduction heat transfer in particular.

1.10   Exercises

Exercise 1.10.1   *Derive the energy balance equation for a rectangular fin of variable cross-section as shown in Figure 1.6. The fin is stationary and is attached to a hot heat source. (Hint: This is similar to the problem given in Section 1.4.3. but without relative motion.)*

Exercise 1.10.2   *The inner body temperature of a healthy person remains constant at 37°C while the temperature and humidity of the environment change. Explain, via heat transfer mechanisms between the human body and the environment, how the human body keeps itself cooler in summer and warmer in winter.*

Exercise 1.10.3   *Discuss the modes of heat transfer that determine the equilibrium temperature of a space shuttle when it is in orbit. What happens when it reenters the Earth's atmosphere?*

Exercise 1.10.4   *A closed plastic container used to serve coffee in a seminar room is made of two layers with an air gap placed between them. List all heat transfer processes associated with the cooling of the coffee in the inner plastic vessel. What steps do you consider for a better container design so as to reduce the heat loss to the ambient?*

Exercise 1.10.5   *A square chip of size 8 mm is mounted on a substrate with the top surface being exposed to a coolant flow at 20°C. All other surfaces of the chip are insulated. The chip temperature must not exceed 80°C in order for the chip to function properly. Determine the maximum allowable power, which can be applied to the chip if the coolant is air with a heat*

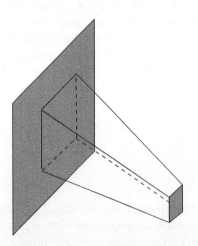

Figure 1.6   Rectangular fin.

transfer coefficient of 250 W/m²K. If the coolant is a dielectric liquid with a heat transfer coefficient of 2500 W/m²K, how much additional power can be dissipated compared to air cooling?

Exercise 1.10.6 *Consider a person standing in a room, which is at a temperature of 21°C. Determine the total heat rate from this person if the exposed surface area of the person is 1.6 m² and the average outer surface temperature of the person is 30°C. The convection coefficient from the surface of the person is 5 W/m² °C. What is the effect of radiation if the emissivity of the surface of the person is 0.90?*

Exercise 1.10.7 *A thin metal plate has one large insulated bottom surface and the top large surface exposed to solar radiation at a rate of 600 W/m². The surrounding air temperature is 20°C. Determine the equilibrium surface temperature of the plate if the convection heat transfer coefficient from the plate surface is 20 W/m²K and the emissivity of the top surface of the plate is 0.8.*

Exercise 1.10.8 *A long, thin copper wire of radius r and length L has an electrical resistance of ρ per unit length. The wire is initially kept at a room temperature of T_a and subjected to an electric current flow of I. The heat generation due to the current flow is simultaneously lost to the ambient by convection. Set up an equation to determine the temperature of the wire as a function of time. Mention the assumptions made in the derivation of the equation.*

Exercise 1.10.9 *In a continuous casting machine, the billet moves at a rate of u m/s. The hot billet is exposed to an ambient temperature of T_a. Set up an equation to find the temperature of the billet as a function of time in terms of pertinent parameters. Assume that radiation also plays a role in the dissipation of heat to ambient.*

Exercise 1.10.10 *In a double pipe heat exchanger, hot fluid (mass flow M kg/s and specific heat C kJ/kg °C) flows inside a pipe and cold fluid (mass flow m kg/s and specific heat c kJ/kg °C) flows outside in the annular space. The hot fluid enters the heat exchanger at T_{h1} and leaves at T_{h2} whereas the cold fluid enters at T_{c1} and leaves at T_{c2}. Set up the differential equation to determine the temperature variation (along the heat exchanger) for hot and cold fluids.*

References

Bejan A (1993) *Heat Transfer.* John Wiley & Sons, Inc., New York.

Boltzmann L (1884) Ableitung des Stefan'schen Gesetzes, betreffend die Abhngigkeit der Wrmestrahlung von der Temperatur aus der electromagnetischen Lichttheorie. *Annalen der Physik und Chemie*, Bd. 22, S. 291–294.

Carslaw HS and Jaeger JC (1959) *Conduction of Heat in Solids,* 2nd Edition. Oxford University Press, Fairlawn, N.J.

Clausius R (1850) Ueber die bewegende Kraft der Wrme und die Gesetze, welche sich daraus für die Wrmelehre selbst ableiten lassen, *Annalen der Physik und Chemie* (Poggendorff, Leipzig), 155, 368–394.

Comini G, del Guidice S and Nonino C (1994) *Finite Element Analysis in Heat Transfer Basic.* Formulation and Linear Problems Series in Computational and Physical Processes in Mechanics and Thermal Sciences. Taylor & Francis, Bristol, PA.

Fick A (1855) On liquid diffusion. *Philosophical Magazine,* 10, 30.

Fourier J (1955) *The Analytical Theory of Heat.* Dover Publications, Inc., New York.

Holman JP (1989) *Heat Transfer.* McGraw Hill, New York.

Huang H-C and Usmani AS (1994) *Finite Element Analysis for Heat Transfer.* Springer-Verlag, London.

Ibanez MT and Power H (2002) *Advanced Boundary Elements for Heat Transfer (Topics in Engineering).* WIT Press, Southampton, UK.

Incropera FP and Dewitt DP (1990) *Fundamentals of Heat and Mass Transfer.* John Wiley & Sons, Inc., New York.

Lewis RW, Morgan K, Thomas HR and Seetharamu KN (1996) *Finite Element Methods in Heat Transfer Analysis.* John Wiley & Sons, Inc., New York.

Lewis RW, Nithiarasu P and Seetharamu KN (2004) *Fundamentals of the Finite Element Method for Heat and Fluid Flow.* John Wiley & Sons, Inc., New York.

Ozisik MN (1968) *Boundary Value Problems of Heat Conduction.* International Text Book Company, Scranton, Pa.

Ozisik MN and Czisik MN (1994) *Finite Difference Methods in Heat Transfer.* CRC Press, London.

Patankar SV (1980) *Numerical Heat Transfer and Fluid Flow,* Hemisphere, Arlington, VA.

Reddy JN and Gartling GK (2000), *The Finite Element Method in Heat Transfer and Fluid Dynamics,* 2nd Edition. CRC Press, London.

Siegel R and Howell JR (1992) *Thermal Radiation Heat Transfer,* 3rd Edition. Hemisphere, Arlington, VA.

Stefan J (1879) Über die Beziehung zwischen der Wrmestrahlung und der Temperatur. *Sitzungsberichte der mathematisch-naturwissenschaftlichen Classe der kaiserlichen Akademie der Wissenschaften.* Bd. 79 (Wien 1879), S. 391–428.

Sukhatme SP (1992) *A Text Book on Heat Transfer,* 3rd Edition. Orient Longman Publishers, Hyderabad, India.

Whewell W (1866). Correction of Newton's law of cooling. In History of the Inductive Sciences, pp. 149–150. Appleton, New York.

2

Some Basic Discrete Systems

2.1 Introduction

Many engineering systems may be simplified by subdividing them into components or elements. These elements can readily be analyzed from first principles and, by assembling these together, the analysis of a full original system can be reconstructed. We refer to such systems as *discrete systems*. In a large number of situations a reasonably adequate model can be obtained using a finite number of well-defined components. This chapter discusses the application of such techniques for the formulation of certain heat and fluid flow problems. The problems presented here provide a valuable basis for the discussion of the finite element method (Bathe 1982; Huebner and Thornton 1982; Hughes 2000; Lewis *et al.* 2004; Reddy 1993; Segerlind 1984; Zienkiewicz *et al.* 2013a,b), which is presented in subsequent chapters.

In the analysis of a discrete system, the actual system response is described directly by the solution of a finite number of unknowns. However, a continuous system is the one in which a continuum is described by complex differential equations. In other words, the system response is described by an infinite number of unknowns. It is often difficult to obtain an exact solution for a continuum problem and therefore appropriate numerical methods are required.

If the characteristics of a problem can be represented by relatively simplified equations, it may be analyzed by employing a finite number of components and simple matrices as shown in the following sections of this chapter. Such procedures reduce the continuous system to an idealization that can be analyzed as a discrete physical system. In reality, an important preliminary study to be made by the engineer is whether an engineering system can be treated as discrete or continuous.

If a system is to be analyzed using complex governing differential equations, then one has to make a decision on how these equations can be discretized by an appropriate numerical method. Such a system is a refined version of discrete systems and the solution accuracy can be controlled by changing the number of unknowns and elements. The importance of the

Fundamentals of the Finite Element Method for Heat and Mass Transfer, Second Edition.
P. Nithiarasu, R. W. Lewis, and K. N. Seetharamu.
© 2016 John Wiley & Sons, Ltd. Published 2016 by John Wiley & Sons, Ltd.

finite element method finds a place here, that is, finite element techniques, in conjunction with the digital computer, have enabled the numerical approximation and solution of continuous systems in a systematic manner. This in effect has made possible the practical extension and application of classical procedures to very complex engineering systems.

In this chapter we deal with some basic discrete or lumped-parameter systems, that is, systems with a finite number of degrees of freedom. The steps in the analysis of a discrete system are as follows:

- *Step1: Approximation of system:* System is idealized as an assembly of elements.

- *Step2: Element characteristics:* The characteristics of each element, or component, is found in terms of the primitive variables.

- *Step3: Assembly:* A set of simultaneous equations is defined via the assembly of element characteristics for the unknown state variables.

- *Step4: Solution of equations:* The simultaneous equations are solved to determine all the primitive variables on a selected number of points.

We consider in the following sections some heat transfer and fluid flow problems. The same procedure can be extended to structural, electrical and other problems.

2.2   Steady-state Problems

2.2.1   Heat Flow in a Composite Slab

Consider the heat flow through a composite slab under steady-state conditions as shown in Figure 2.1. The problem is similar to that of a roof slab subjected to a solar flux on the left hand face. This is subjected to a constant flux of q W/m^2 and the right hand face is subjected to a convection environment. We are interested in determining the temperatures T_1, T_2 and T_3 at nodes 1, 2 and 3 respectively.

The steady-state heat conduction equation for a one-dimensional slab with a constant thermal conductivity is given by Equation (1.41), that is,

$$\frac{d^2T}{dx^2} = 0. \tag{2.1}$$

Integration of the above equation yields the following temperature gradient and temperature distribution

$$\frac{dT}{dx} = a \tag{2.2}$$

and

$$T = ax + b. \tag{2.3}$$

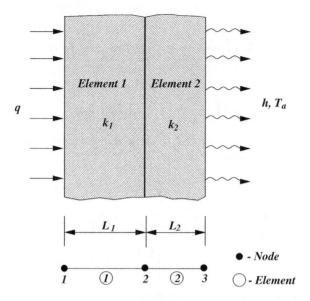

Figure 2.1   Heat transfer through a composite slab.

To determine the constants a and b, consider a homogeneous slab of thickness, L, with the following boundary conditions (in one dimension):

$$\text{At} \quad x = 0, T = T_0 \quad \text{and} \quad \text{At} \quad x = L, T = T_L. \tag{2.4}$$

The substitution of these boundary conditions into Equation (2.3) results in

$$b = T_0 \quad \text{and} \quad a = \frac{T_L - T_0}{L}. \tag{2.5}$$

The heat flux can be calculated from Equations (2.3) and (2.5) as

$$q = -\frac{dT}{dx} = -k \left(\frac{T_L - T_0}{L} \right); \tag{2.6}$$

or, the total heat flow is expressed as

$$Q = qA = -kA \left(\frac{T_L - T_0}{L} \right), \tag{2.7}$$

where A is the area perpendicular to the direction of heat flow.

The total heat flow is constant at any section perpendicular to the heat flow direction (conservation of energy). Applying the above principle to the composite slab shown in Figure 2.1, we have the following heat balance equations at different nodes:

At node 1

$$qA = k_1 A \left(\frac{T_1 - T_2}{L_1} \right). \tag{2.8}$$

At node 2

$$k_1 A \left(\frac{T_1 - T_2}{L_1} \right) = k_2 A \left(\frac{T_2 - T_3}{L_2} \right).$$ (2.9)

At node 3

$$k_2 A \left(\frac{T_2 - T_3}{L_2} \right) = hA(T_3 - T_a),$$ (2.10)

where h is the heat transfer coefficient and T_a is the ambient temperature. We can rearrange the previous three nodal equations as follows:

$$\frac{k_1 A}{L_1} T_1 - \frac{k_1 A}{L_1} T_2 = qA$$

$$-\frac{k_1 A}{L_1} T_1 + \left[\frac{k_1 A}{L_1} + \frac{k_2 A}{L_2} \right] T_2 - \frac{k_2 A}{L_2} T_3 = 0$$

$$-\frac{k_2 A}{L_2} T_2 + \left[\frac{k_2 A}{L_2} + hA \right] T_3 = hAT_a.$$ (2.11)

The above equation can be rewritten in matrix form as

$$\begin{bmatrix} \frac{k_1 A}{L_1} & \frac{-k_1 A}{L_1} & 0 \\ \frac{-k_1 A}{L_1} & \left[\frac{k_1 A}{L_1} + \frac{k_2 A}{L_2} \right] & \frac{-k_2 A}{L_2} \\ 0 & \frac{-k_2 A}{L_2} & \left[\frac{k_2 A}{L_2} + hA \right] \end{bmatrix} \begin{Bmatrix} T_1 \\ T_2 \\ T_3 \end{Bmatrix} = \begin{Bmatrix} qA \\ 0 \\ hAT_a \end{Bmatrix}$$ (2.12)

or

$$[\mathbf{K}]\{\mathbf{T}\} = \{\mathbf{f}\},$$ (2.13)

where

$$[\mathbf{K}] = \begin{bmatrix} \frac{k_1 A}{L_1} & \frac{-k_1 A}{L_1} & 0 \\ \frac{-k_1 A}{L_1} & \left[\frac{k_1 A}{L_1} + \frac{k_2 A}{L_2} \right] & \frac{-k_2 A}{L_2} \\ 0 & \frac{-k_2 A}{L_2} & \left[\frac{k_2 A}{L_2} + hA \right] \end{bmatrix}; \{T\} = \begin{Bmatrix} T_1 \\ T_2 \\ T_3 \end{Bmatrix} \quad \text{and} \quad \{\mathbf{f}\} = \begin{Bmatrix} qA \\ 0 \\ hAT_a \end{Bmatrix}.$$ (2.14)

The solution of Equation (2.13) gives the unknown temperatures T_1, T_2 and T_3. In the case of heat conduction there is only one degree of freedom at each node as temperature is a scalar. The following important features of Equation (2.13) should be observed.

- The characteristics of each layer of the slab for heat conduction can be written as

$$\frac{kA}{L} \begin{bmatrix} 1 & -1 \\ -1 & 1 \end{bmatrix} \begin{Bmatrix} T_i \\ T_j \end{Bmatrix} = \begin{Bmatrix} Q \\ -Q \end{Bmatrix}.$$ (2.15)

where Q is the total heat flow and is constant and i and j are the two nodes on both ends of an element (see Figure 2.1).

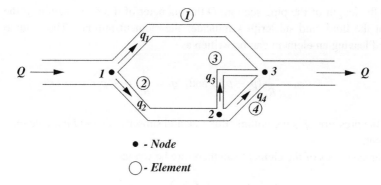

● - *Node*

◯ - *Element*

Figure 2.2   Fluid flow network.

- The global stiffness matrix [**K**] can be obtained by assembling the stiffness matrices of each layer and the result is a symmetric and positive definite matrix.

- The effect of the heat flux boundary condition appears only in the loading terms {**f**}.

- The convective heat transfer effect appears both in the stiffness matrix and the loading vector.

- The thermal force vector consists of known values. The method of assembly can be extended to more than two layers of the slab.

In summary, if [**K**] and {**f**} can be formed, the temperature distribution can be determined by any standard solution procedure to a set of simultaneous equations.

2.2.2   Fluid Flow Network

Many practical problems require a knowledge of the flow in various circuits, for example water distribution systems, ventilation ducts in electrical machines (including transformers), electronic cooling systems, internal passages in gas turbine blades etc. In order to illustrate the flow calculations in each circuit, laminar incompressible flow is considered in the network of circular pipes[1] as shown in Figure 2.2. If a quantity $Q\ m^3/s$ of fluid enters and leaves the pipe network, it is possible to compute the fluid nodal pressures and the volume flow rate in each pipe. We shall make use of a four-element and three-node model as shown in Figure 2.2.

The fluid resistance for an element is written as (Poiseuille flow (Shames 1982))

$$R_k = \frac{\Delta p}{Q} = \frac{128L\mu}{\pi D^4},\tag{2.16}$$

---

[1]It should be noted that we use the notation Q for both total heat flow and fluid flow rate.

where L is the length of the pipe section; D the diameter of the pipe section, μ the dynamic viscosity of the fluid, and subscript k indicates the element number. The volume flux rate entering and leaving an element can be written as

$$q_i = \frac{1}{R_k}(p_i - p_j) \quad \text{and} \quad q_j = \frac{1}{R_k}(p_j - p_i), \tag{2.17}$$

where p is the pressure, q is the volume flux rate and subscripts i and j indicate the two nodes of an element.

The characteristics of the element can therefore be written as

$$\frac{1}{R_k}\begin{bmatrix} 1 & -1 \\ -1 & 1 \end{bmatrix}\begin{Bmatrix} p_i \\ p_j \end{Bmatrix} = \begin{Bmatrix} q_i \\ q_j \end{Bmatrix}. \tag{2.18}$$

Similarly, we can construct the characteristics of each element in Figure 2.2 as:

Element 1

$$\frac{1}{R_1}\begin{bmatrix} 1 & -1 \\ -1 & 1 \end{bmatrix}\begin{Bmatrix} p_1 \\ p_3 \end{Bmatrix} = \begin{Bmatrix} q_1 \\ -q_1 \end{Bmatrix}. \tag{2.19}$$

Note that the mass flux rate entering an element is positive and leaving an element is negative.

Element 2

$$\frac{1}{R_2}\begin{bmatrix} 1 & -1 \\ -1 & 1 \end{bmatrix}\begin{Bmatrix} p_1 \\ p_2 \end{Bmatrix} = \begin{Bmatrix} q_2 \\ -q_2 \end{Bmatrix}. \tag{2.20}$$

Element 3

$$\frac{1}{R_3}\begin{bmatrix} 1 & -1 \\ -1 & 1 \end{bmatrix}\begin{Bmatrix} p_2 \\ p_3 \end{Bmatrix} = \begin{Bmatrix} q_3 \\ -q_3 \end{Bmatrix}. \tag{2.21}$$

Element 4

$$\frac{1}{R_4}\begin{bmatrix} 1 & -1 \\ -1 & 1 \end{bmatrix}\begin{Bmatrix} p_2 \\ p_3 \end{Bmatrix} = \begin{Bmatrix} q_4 \\ -q_4 \end{Bmatrix}. \tag{2.22}$$

From the above element equations, it is possible to write the following nodal equations:

$$\left[\frac{1}{R_1} + \frac{1}{R_2}\right]p_1 - \frac{1}{R_2}p_2 - \frac{1}{R_1}p_3 = q_1 + q_2 = Q$$

$$-\frac{1}{R_2}p_1 + \left[\frac{1}{R_2} + \frac{1}{R_3} + \frac{1}{R_4}\right]p_2 - \left[\frac{1}{R_3} + \frac{1}{R_4}\right]p_3 = q_3 + q_4 - q_2 = 0$$

$$-\frac{1}{R_1}p_1 - \left[\frac{1}{R_3} + \frac{1}{R_4}\right]p_2 + \left[\frac{1}{R_1} + \frac{1}{R_3} + \frac{1}{R_4}\right]p_3 = -q_1 - q_3 - q_4 = -Q. \tag{2.23}$$

Table 2.1 Details of pipe network

Component number	Diameter, cm	Length, m
1	2.50	30.00
2	2.00	20.00
3	2.00	25.00
4	1.25	20.00

Now the following matrix form can be written from the above equation.

$$
\begin{bmatrix}
\left[\frac{1}{R_1}+\frac{1}{R_2}\right] & -\frac{1}{R_2} & -\frac{1}{R_1} \\
-\frac{1}{R_2} & \left[\frac{1}{R_2}+\frac{1}{R_3}+\frac{1}{R_4}\right] & -\left[\frac{1}{R_3}+\frac{1}{R_4}\right] \\
-\frac{1}{R_1} & -\left[\frac{1}{R_3}+\frac{1}{R_4}\right] & \left[\frac{1}{R_1}+\frac{1}{R_3}+\frac{1}{R_4}\right]
\end{bmatrix}
\begin{Bmatrix} p_1 \\ p_2 \\ p_3 \end{Bmatrix} =
$$

$$
\begin{Bmatrix} q_1 + q_2 \\ -q_2 + q_3 + q_4 \\ -q_1 - q_3 - q_4 \end{Bmatrix} = \begin{Bmatrix} Q \\ 0 \\ -Q \end{Bmatrix}. \tag{2.24}
$$

Note that $q_1 + q_2 = Q$ and $q_2 = q_3 + q_4$.

In this fashion we can solve such problems as electric networks, radiation networks etc. Let us consider a numerical example to illustrate the above.

Example 2.2.1 *In a pipe network as shown in Figure 2.2, water enters the network at a rate of 0.1 m^3/s with a viscosity of 0.96×10^{-3} Ns/m^2. The component details are given in Table 2.1. Determine the pressure and flow distributions.*

On substitution of the various values, we get the following resistances from Equation (2.16):

$$
\begin{aligned}
R_1 &= 0.3 \times 10^7 \\
R_2 &= 0.5 \times 10^7 \\
R_3 &= 0.6 \times 10^7 \\
R_4 &= 3.2 \times 10^7.
\end{aligned}
\quad \left(\frac{N}{M^2} \quad \frac{S}{M^3}\right)
$$

Now Equation (2.24) can be formulated as

$$
10^{-7} \begin{bmatrix} 5.33 & -2.00 & -3.33 \\ -2.00 & 3.98 & -1.98 \\ -3.33 & -1.98 & 5.31 \end{bmatrix} \begin{Bmatrix} p_1 \\ p_2 \\ p_3 \end{Bmatrix} = \begin{Bmatrix} 0.1 \\ 0.0 \\ -0.1 \end{Bmatrix}. \tag{2.25}
$$

The reduction of the above simultaneous system of equations with $p_3 = 0.0$ (assumed as reference atmospheric pressure) results in

$$
10^{-7} \begin{bmatrix} 5.33 & -2.00 \\ -2.00 & 3.98 \end{bmatrix} \begin{Bmatrix} p_1 \\ p_2 \end{Bmatrix} = \begin{Bmatrix} 0.1 \\ 0.0 \end{Bmatrix}. \tag{2.26}
$$

The solution to the above set of simultaneous equations (see Appendix A) gives the relative pressure (actual minus atmospheric pressure) as

$$p_1 = 0.236 \times 10^6 \; N/m^2$$
$$p_2 = 0.110 \times 10^6 \; N/m^2$$

From Equations (2.19), (2.20), (2.21) and (2.22) we can calculate the flow quantities as

$$q_1 = \frac{p_1 - p_2}{R_1} = 0.042 \; m^3/s$$

$$q_2 = \frac{p_1 - p_3}{R_2} = 0.0472 \; m^3/s$$

$$q_3 = \frac{p_2 - p_3}{R_3} = 0.0183 \; m^3/s$$

$$q_4 = \frac{p_2 - p_3}{R_4} = 0.0034 \; m^3/s. \tag{2.27}$$

It is possible to take into account the entrance loss, exit loss, bend loss etc. in the calculation of nodal pressures and flows in each circuit. If the fluid flow in the network is turbulent, it is still possible to define an element but the element equations are no longer linear as can be seen from an empirical relation governing fully developed turbulent pipe flow (Darcy-Weisbach formula (Shames 1982)):

$$p_1 - p_2 = \frac{8fLQ^2\rho}{\pi^2 D^5}. \tag{2.28}$$

where f is the Moody friction factor, which is a function of the Reynolds number and the pipe roughness. The fluidity matrix would contain known functions of the flow rate Q instead of constants. Hence, the problem becomes nonlinear.

2.2.3 Heat Transfer in Heat Sinks

In order to increase the heat dissipation by convection from a given primary surface, additional surfaces may be added. The additional material added is either referred to as an "Extended Surface" or a "Fin." A familiar example is in motor cycles where fins extend from the outer surface of the engine to dissipate more heat by convection. A schematic diagram of such a fin array is shown in Figure 2.3. This is a good example of a heat sink.

We shall assume for simplicity that there is no variation in temperature along the thickness and width of the fins. We will also assume that the temperature varies only along the longitudinal direction of the fin and the height direction of the hot body to which the fin is attached. We can then derive a simplified model as shown in Figure 2.4. A typical element in the fin array is shown in Figure 2.5.

Assuming that the surface temperature is average temperature two nodes,

$$Q_i = \frac{kA}{L}(T_i - T_j) + \frac{hPL}{2}\left[\frac{T_i + T_j}{2} - T_a\right] \tag{2.29}$$

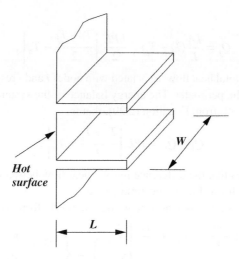

Figure 2.3   Array of thin rectangular fins.

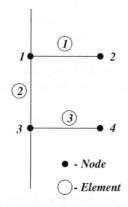

Figure 2.4   A simplified model of the rectangular fins of Figure 2.3.

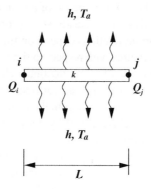

Figure 2.5   A typical component from the rectangular fin arrangement and conductive–convective heat transfer mechanism. k – thermal conductivity, h – heat transfer coefficient, T_a – atmospheric temperature.

and

$$Q_j = \frac{kA}{L}(T_i - T_j) - \frac{hPL}{2}\left[\frac{T_i + T_j}{2} - T_a\right], \tag{2.30}$$

where Q_i and Q_j are the total heat flow associated with nodes i and j respectively, A is the cross sectional area and P is the perimeter. The energy balance of the system can easily verified by subtracting Equation (2.30) from Equation (2.29), that is,

$$Q_i - Q_j = hPL\left[\frac{T_i + T_j}{2} - T_a\right]. \tag{2.31}$$

This clearly indicates that the difference in total heat flow between nodes i and j is equal to the heat lost from the fin surface to the atmosphere.

Equations (2.29) and (2.30) may now be written in matrix form as

$$\begin{bmatrix} \frac{kA}{L} + \frac{hPL}{4} & -\frac{kA}{L} + \frac{hPL}{4} \\ -\frac{kA}{L} + \frac{hPL}{4} & \frac{kA}{L} + \frac{hPL}{4} \end{bmatrix}\left\{\begin{matrix} T_i \\ T_j \end{matrix}\right\} = \left\{\begin{matrix} Q_i + \frac{hPLT_a}{2} \\ -Q_j + \frac{hPLT_a}{2} \end{matrix}\right\}. \tag{2.32}$$

In the above equation, either Q_i or T_i is often known and quantities such as T_a, h, k, L and P are also generally known *a priori*. The above problem is therefore reduced to finding three unknowns Q_i or Q_j and T_i, T_j. In addition to the above two equations, an additional equation relating Q_i and Q_j as given by Equation (2.31) may also be used.

It is now possible to solve the system to find the unknowns. If there is more than one element, then an assembly procedure is necessary as discussed in the previous section. Equation (2.32) is reduced to (2.15) if surface convection is absent. Also, if the terms $(T_i + T_j)/2$ in Equation (2.32) is replaced by $(2T_i + T_j)/3$, then we obtain the standard Galerkin weighted residual form discussed in Chapter 3.

2.3   Transient Heat Transfer Problem

In a transient or propagation problem the response of a system changes with time. The same methodology as used in the analysis of a steady-state problem is employed here, but the temperature and element equilibrium relations depend on time. The objective of the transient analysis is to calculate the temperatures with respect to time.

Figure 2.6 shows an idealized case of a heat treatment chamber. A metallic part is heated to an initial temperature, T_p, and is placed in a heat treatment chamber where an inert gas, such as nitrogen, is present. Heat is transferred from the metallic part to the gas by convection. The gas in turn loses heat to the enclosure wall by convection and radiation. The wall also receives heat by direct radiation from the metallic part as the gas is assumed to be transparent to radiation. The wall loses heat to the atmosphere by radiation and convection.

The unknown variables in the present analysis are the temperature of the metallic part T_p; the temperature of the gas T_g; and the temperature of the enclosure wall T_w. For simplicity, we use a lumped parameter approach, that is, the temperature variations within the metal, gas and wall are ignored.

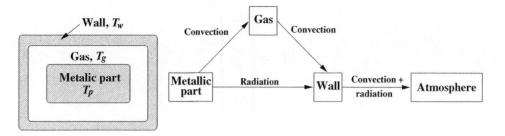

Figure 2.6   Heat treatment chamber and associated heat transfer processes.

Let C_p, C_g and C_w be the heat capacities of the metallic part, the gas and the wall respectively. Neglecting radiation from and to the gas, the time-dependent heat balance equations may be derived as follows.

For the metallic part:

$$C_p \frac{dT_p}{dt} = -\left\{ h_p A_p (T_p - T_g) + \epsilon_p \sigma A_p \left(T_p^4 - T_w^4 \right) \right\}. \tag{2.33}$$

For the gas component:

$$C_g \frac{dT_g}{dt} = h_p A_p (T_p - T_g) - h_g A_g (T_g - T_w). \tag{2.34}$$

For the furnace wall:

$$C_w \frac{dT_w}{dt} = \epsilon_p \sigma A_p \left(T_p^4 - T_w^4 \right) + h_g A_g (T_g - T_w)$$
$$- h_w A_w (T_w - T_a) - \epsilon_w \sigma A_w \left(T_w^4 - T_a^4 \right). \tag{2.35}$$

The heat balance for the whole system may be obtained by adding all three energy balance equations. This results in the total rate of change of energy within the whole system being equal to the heat exchanged by the external wall to the atmosphere, proving that the system is in dynamic equilibrium with the atmosphere.

Equations (2.33), (2.34) and (2.35) can be recast in matrix form as

$$[C] \left\{ \dot{T} \right\} + [K] \{ T \} = \{ f \}, \tag{2.36}$$

where

$$[C] = \begin{bmatrix} C_p & 0.0 & 0.0 \\ 0.0 & C_g & 0.0 \\ 0.0 & 0.0 & C_w \end{bmatrix} \tag{2.37}$$

$$\{\dot{\mathbf{T}}\} = \left\{ \begin{array}{c} \dfrac{dT_p}{dt} \\[2mm] \dfrac{dT_g}{dt} \\[2mm] \dfrac{dT_w}{dt} \end{array} \right\} \tag{2.38}$$

$$\{\mathbf{T}\} = \left\{ \begin{array}{c} T_p \\ T_g \\ T_w \end{array} \right\} \tag{2.39}$$

$$[\mathbf{K}] = \begin{bmatrix} h_p A_p & -h_p A_p & 0.0 \\[2mm] -h_p A_p & h_p A_p + h_g A_g & -h_g A_g \\[2mm] 0.0 & -h_g A_g & h_g A_g + h_w A_w \end{bmatrix} \tag{2.40}$$

and

$$\{\mathbf{f}\} = \left\{ \begin{array}{c} -\epsilon_p \sigma A_p (T_p^4 - T_w^4) \\[3mm] 0 \\[3mm] h_w A_w T_a + \epsilon_p \sigma A_p \left(T_p^4 - T_w^4\right) - \epsilon_w \sigma A_w \left(T_w^4 - T_a^4\right) \end{array} \right\}. \tag{2.41}$$

where h_p is the heat transfer coefficient from the metallic part to the gas; A_p the surface area of the metallic part in contact with the gas; h_g the heat transfer coefficient of the gas to the wall; A_g the surface area of the gas in contact with the wall; h_w is the heat transfer coefficient from wall to atmosphere; A_w is the wall area in contact with atmosphere; ϵ_p and ϵ_w emissivity of the metallic part and the wall respectively and σ the Stefan-Boltzmann constant (Chapter 1).

Although we follow the SI system of units, it is essential to reiterate here that the temperatures T_p, T_g, T_w and T_a should be used in K (Kelvin) as radiation heat transfer is involved in the given problem. In view of the radiation terms appearing in the governing equations (i.e., temperature to the power of 4), the problem is highly nonlinear and an iterative procedure is necessary. An initial guess of the unknown temperature values is also essential to start any iterative procedure.

In this example, if the time terms are neglected, we can recover the steady-state formulation. However, the time-dependent load terms are necessary to carry out any form of transient analysis. In practice, the reduction of an appropriate discrete system that contains all the important characteristics of the actual physical system is usually not straightforward. In general a different discrete model should be chosen for a transient response prediction than that chosen for a steady-state analysis.

The time derivative terms used in the above formulation have to be approximated in order to obtain a temperature distribution. As discussed in later chapters, approximations such as backward Euler, central difference etc. may well be employed.

2.4 Summary

In this chapter, we have discussed some basic discrete system analyses. The application of such an analysis to a heat exchanger problem is provided in Chapter 9. It is important to reiterate that this chapter gives only a brief discussion of the system analysis. We believe that the material provided in this chapter is sufficient to give the reader a starting point. It should be noted that the system analysis is straight forward and works for many simple heat transfer problems. However for complex continuum problems, a standard discretization of the governing equations and solution methodology is essential. We will discuss these problems in detail in the following chapters.

2.5 Exercises

Exercise 2.5.1   *Use the system analysis procedure described in this chapter and construct the discrete system for heat conduction through the composite wall shown in Figure 2.7. Also, from the following data, calculate the temperature distribution in the composite wall.*
   Areas: $A_1 = 2.0\,m^2, A_2 = 1.0\,m^2$ and $A_3 = 1.0\,m^2$.
   Thermal conductivity: $k_1 = 2.00\,W/mK, k_2 = 2.5\,W/mK$ and $k_3 = 1.5\,W/mK$.
   Heat transfer coefficient: $h = 0.1\,W/m^2K$.
   Atmospheric temperature: $T_a = 30\,°C$.
   Temperature at the left face of wall: $T_1 = 75.0\,°C$.

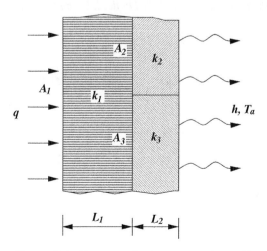

Figure 2.7   Heat transfer in a composite wall.

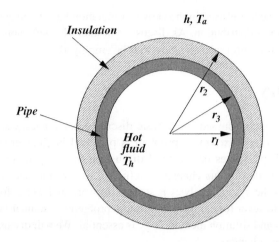

Figure 2.8   Heat transfer through an insulating material.

Exercise 2.5.2   *The cross-section of an insulated pipe carrying a hot fluid is shown in Figure 2.8. The inner and outer radii of the pipe are r_1 and r_2 respectively. The thickness of the insulating material is $r_3 - r_2$. Assume appropriate conditions and form the discrete system equations.*

Exercise 2.5.3   *The pipe network used to circulate hot water in a domestic central heating arrangement is shown in Figure 2.9. The flow rate at the entrance is Qm^3/s. Neglecting any loss of mass, construct a system of simultaneous equations to calculate the pressure distribution at selected points using a discrete system analysis. Assume laminar flow occurs in the system.*

Exercise 2.5.4   *A schematic diagram of a counter flow heat exchanger is shown in Figure 2.10. The hot fluid enters the central, circular pipe from the left and exits at the right. The cooling fluid is circulated around the inner tube to cool the hot fluid. Using the principles of heat exchanger*

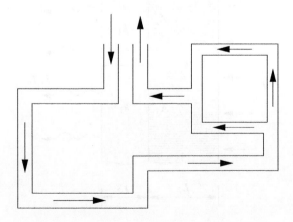

Figure 2.9   Pipe network for central heating.

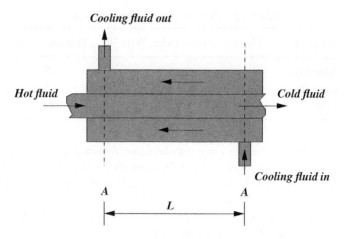

Figure 2.10   Counter flow heat exchanger.

system discussed in chapter 9, construct a discrete system to determine the temperature distribution.

Exercise 2.5.5   *A transient analysis is very important in the casting industry. In Figure 2.11, a simplified casting arrangement is shown (without a runner or raiser). The molten metal is poured into the mold and the metal loses heat to the mold and solidifies. It is often possible to have a small air gap between the metal and mold. The figure shows an idealized system which has a uniform gap all around the metal. Assume that heat is transferred from the metal to the mold via radiation and conduction. Then heat is conducted through the mold and convected to the atmosphere. Stating all assumptions, derive a system of equations to carry out a transient analysis.*

Exercise 2.5.6   *Consider a 0.6 m high and 2 m wide double-pane window consisting of two 4 mm thick layers of glass ($k = 0.80$ W/m °C) separated by a 8 mm wide stagnant air space*

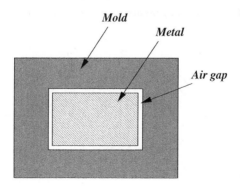

Figure 2.11   Counter flow heat exchanger.

Table 2.2   Details of the composite wall

Material	Thermal conductivity, W/m °C	Thickness, cm
Aluminium	200	5
Copper	400	15
Steel	50	20

(k = 0.025 W/m °C). Determine the steady-state heat transfer through the double-pane window and the temperature of the inner surface for a day when the outside air temperature is −15 °C and the room temperature is 20 °C. The heat transfer coefficient on the inner and outer surface of the window to be 10 W/m² °C and 40 W/m²° C respectively. Note that these heat transfer coefficients include the effect of radiation. If the air gap is not provided, what is the temperature of the glass inside the room? Comment on the result.

Exercise 2.5.7   *A simplified model can be applied to describe the steady-state temperature distribution through the core region, muscle region and skin region of the human body. The core region temperature T_c, is the mean operating temperature of the internal organs. The muscle temperature, T_m is the operating temperature of the muscle layer of the human body. Muscle is a shell tissue and can be either resting or actively working. The skin temperature, T_s, is the operating temperature of the surface region of the body consisting of a subcutaneous fat layer, the dermal layer and finally epidermal layer. If the metabolic heat rate of a common man is 45 W/m² and the skin temperature is 32.6 °C, calculate the core region temperature if the thermal conductivity of the core, muscle and skin are 0.48 W/m °C and the thickness of the layers are 4 cm, 2 cm and 1 cm respectively. Also calculate the muscle temperature.*

Exercise 2.5.8   *A composite wall consists of layers of aluminum, copper and steel. The steel external surface is 350 °C and the external surface of the aluminum is exposed to an ambient of 25 °C with a heat transfer coefficient of 5 W/m² °C. Calculate the heat loss and interfacial temperature using a three-element model using the data given in Table 2.2.*

Exercise 2.5.9   *An incompressible fluid flows through a pipe network of circular pipes as shown in Figure 2.12. If 0.1 m³/s of fluid enters and leaves the piping network, using a 4-node 5-element model, calculate the nodal pressure and the volume flow in each pipe. If the nodes*

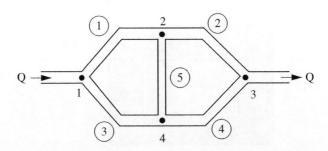

Figure 2.12   Incompressible flow through a pipe network.

Table 2.3   Pipe network element details

Element number	Nodes	Diameter, D(cm)	Length, L(m)
1	1,2	5	25
2	2,3	5	25
2	1,4	5	25
4	4,3	5	25
5	2,4	10	90

1 and 3 are directly connected, in addition to the existing arrangement, what change takes place in the nodal pressure and volume fluid in each pipe? The viscosity of the fluid is 1×10^{-2} N s/m^2. For laminar flow, the resistance for the flow is given by $128\,\mu L/\pi D^4$. The details of the elements are given in Table 2.3.

Exercise 2.5.10   *Figure 2.13 shows a direct current circuit. The voltage at the output terminals are also shown in Figure 2.13. Calculate the voltage at each node and the current in each of the branches using the finite element method.*

Exercise 2.5.11   *A cross-section of a heat sink used in electronic cooling is shown in Figure 2.14. All the fins are of the same size. Calculate the heat dissipating capacity of the heat sink per unit length of heat sink.*

Exercise 2.5.12   *The details of a double pipe heat exchanger are given as: (a) cold fluid heat capacity rate $C_1 = 1100$ W/$^\circ$C; (b) Hot fluid heat capacity rate $C_2 = 734$ W/$^\circ$C; (c) overall heat transfer coefficient $U = 600$ W/m$^{2\circ}$C (d) heat exchanger area $A = 4\,m^2$ (e) cold fluid entry temperature $T_{ci} = 20\,^\circ$C (f) hot fluid entry temperature $T_{hi} = 80\,^\circ$C. Set up the stiffness matrix and them solve for the outlet temperature and the effectiveness of the heat exchanger by using 1 element, 2 elements and 4 elements for the heat exchanger. Also determine the minimum number of elements required for converged solution (refer to Chapter 9).*

Exercise 2.5.13   *Figure 2.15 shows an arrangement for cooling of an electronic equipment consisting of a number of printed circuit boards (PCB) enclosed in a box. Air is forced through the box by a fan. Select a typical element and write down the stiffness matrix and show that this*

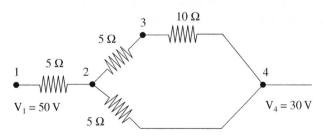

Figure 2.13   A direct current circuit.

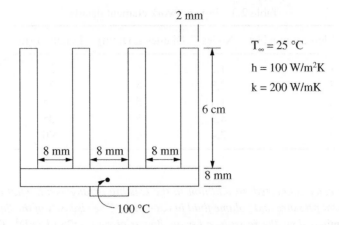

Figure 2.14   A heat sink.

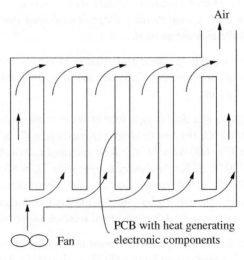

Figure 2.15   A printed circuit board.

method can take care of nonuniform flow (by using the methodology similar to the problem 4, the nonuniform flow in each channel can be determined) and nonuniform heat generation in individual PCB.

References

Bathe KJ (1982) *Finite Element Procedures in Engineering Analysis*. Prentice Hill Inc., Englewood
    Cliffs, New Jersey.
Holman JP (1989) *Heat Transfer*. McGraw-Hill, New York.

Huebner K and Thornton EA (1982) *The Finite Element Method for Engineers*, 2nd Edition. John Wiley & Sons, Inc., New York.

Hughes TJR (2000), *The Finite Element Method: Linear Static and Dynamic Finite Element Analysis*. Dover Publications, New York.

Incropera FP and Dewitt DP (1990) *Fundamentals of Heat and Mass Transfer*. John Wiley & Sons, Inc., New York.

Lewis RW, Nithiarasu P and Seetharamu KN (2004) *Fundamentals of the Finite Element Method for Heat and Fluid Flow*. John Wiley & Sons, Inc., New York.

Reddy JN (1993) *An Introduction to Finite Element Method*, 2nd Edition. McGraw-Hill, New York.

Segerlind LJ (1984) *Applied Finite Element Analysis*, 2nd Edition. John Wiley & Sons, Inc., New York.

Shames IH (1982) *Mechanics of Fluids*. McGraw-Hill, Singapore.

Zienkiewicz OC, Taylor RL and Nithiarasu P (2013) *The Finite Element Method. Vol. 3. Fluid Dynamics*, 7th Edition. Elsevier, Amsterdam.

Zienkiewicz OC, Taylor RL and Zhu JZ (2013) *The Finite Element Method. Vol. 1. The Basis*. 7th Edition, Elsevier, Amsterdam.

3

The Finite Element Method

3.1 Introduction

The finite element method is a numerical tool for determining approximate solutions to a large class of engineering problems. The method was originally developed to study the stresses in complex air-frame structures (Clough 1960) and was later extended to the general field of continuum mechanics (Zienkiewicz and Cheung 1965). There have been many articles on the history of finite elements written by numerous authors with conflicting opinions on the origins of the technique (Gupta and Meek 1996; Oden 1996; Zienkiewicz 1996).

The finite element method has received considerable attention in engineering education and in industry because of its diversity and flexibility as an analysis tool. It is often necessary to obtain approximate numerical solutions for complex industrial problems where exact closed-form solutions are difficult to obtain. An example of such a complex situation can be found in the cooling of electronic equipment (or chips). Also, the dispersion of pollutants during nonuniform atmospheric conditions, metal wall temperatures in the case of gas turbine blades where the inlet gas temperatures exceed the melting point of the material of the blade, cooling problems in electrical motors, various phase change problems etc., are many other examples of such complex situations. Although it is possible to derive the governing equations and boundary conditions from first principles, it is often difficult to obtain any form of analytical solution to such problems. This is due to the fact that either the geometry is irregular or boundary condition is complex.

Among the various numerical methods that have evolved over the years, the most commonly used techniques are the finite difference, finite volume and finite element methods. The finite difference is a well-established and conceptually simple method, which requires a pointwise approximation to the governing equations. The numerical model, formed by writing

Fundamentals of the Finite Element Method for Heat and Mass Transfer, Second Edition.
P. Nithiarasu, R. W. Lewis, and K. N. Seetharamu.
© 2016 John Wiley & Sons, Ltd. Published 2016 by John Wiley & Sons, Ltd.

the difference equations for an array of grid points, can be improved by increasing the number of points. Although many heat transfer problems may be solved using the finite difference method (Ozisik and Czisik 1994), as soon as irregular geometries or an unusual specification of boundary conditions are encountered, then the finite difference technique becomes difficult to use.

The finite volume method is a further refined version of the finite difference method and has become popular in computational fluid dynamics (Patankar 1980). The vertex centered finite volume technique is very similar to the linear finite element method (Malan *et al.* 2002; Zienkiewicz *et al.* 2013).

The finite element method (Baker 1985; Bathe 1982; Chandrupatla and Belegundu 1991; Huebner and Thornton 1982; Hughes 2000; Lewis et al. 1996; Rao 1989; Segerlind 1984; Zienkiewicz and Morgan 1983; Zienkiewicz and Taylor 2000; Zienkiewicz et al. 2013a,b) considers that the solution region comprises many small, interconnected, subregions or elements and gives a piecewise approximation to the governing equations, that is, the complex partial differential equations are reduced to a set of either linear or nonlinear simultaneous equations. Thus, the finite element discretization procedure (i.e., dividing the domain into a number of smaller regions) reduces the continuum problem which has an infinite number of unknowns, to one with a finite number of unknowns at specified points, referred to as nodes. Since the finite element method allows us to form the elements, or subregions, in an arbitrary sense, a close representation of boundaries of complicated domains is possible.

Most of the finite difference schemes used in fluid dynamics and heat transfer can be viewed as special cases within a weighted-residual framework. For weighted residual procedures, the error in the approximate solution of the conservation equations is not set to zero, but instead its integral, with respect to selected "weights," is required to vanish. Within this family, the collocation method reproduces the classical finite difference equations, whereas the finite volume algorithm is obtained by using constant weights.

In this book we intend to present a step-by-step procedure of the finite element method as applied to heat transfer problems. In doing so, we intend to present the topic in as simplified a form as possible so that both students and practicing engineers can benefit.

A numerical model for a heat transfer problem starts with the physical problem, an example of which is shown in Figure 3.1. As can be seen one part of the model deals with the discretization of the domain and the other carries out the discrete approximation of the partial differential equations. Finally, by combining both, the numerical solution to the problem is achieved.

The solution of a continuum problem by the finite element method is approximated by the following step-by-step process.

1. *Discretization of the continuum*

   Divide the solution region into nonoverlapping elements or subregions. The finite element discretization allows a variety of element shapes, for example, triangles or quadrilaterals. Each element is formed by the connection of a certain number of nodes (Figure 3.2). The number of nodes employed to form an element depends on the type of element (or interpolation function).

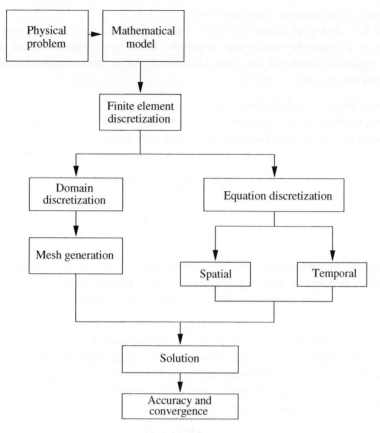

Figure 3.1   Numerical model for heat transfer calculations.

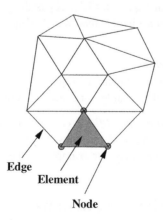

Figure 3.2   Typical finite element mesh. Elements, nodes and edges.

2. *Select interpolation or shape functions*

   The next step is to choose the type of interpolation function, which represents the variation of the field variable over an element. The number of nodes to form an element, the nature and number of unknowns at each node decide the variation of a field variable within the element.

3. *Form element equations (formulation)*

   Next, we have to determine the matrix equations, which express the properties of the individual elements by forming an element Left Hand Side (LHS) matrix and load vector. For example, a typical LHS matrix and a load vector can be written as

$$[\mathbf{K_e}] = \frac{Ak}{l} \begin{bmatrix} 1 & -1 \\ -1 & 1 \end{bmatrix} \tag{3.1}$$

$$\{\mathbf{f_e}\} = \begin{Bmatrix} Q_i \\ Q_j \end{Bmatrix}, \tag{3.2}$$

   where the subscript e represents an element; Q is the total heat transferred; k is the thermal conductivity; l is the length of a one-dimensional linear element and i and j represent the nodes forming an element. The unknowns are the temperature values on the nodes.

4. *Assemble the element equations to obtain a system of simultaneous equations*

   To find the properties of the overall system, we must assemble all the individual element equations, that is, combine the matrix equations of each element in an appropriate way such that the resulting matrix represents the behaviour of the entire solution region of the problem. The boundary conditions must be incorporated after the assemblage of the individual element contributions, that is,

$$[\mathbf{K}]\{\mathbf{T}\} = \{\mathbf{f}\}, \tag{3.3}$$

   where $[\mathbf{K}]$ is the global LHS matrix, which is the assemblage of the individual element LHS matrices as given in Equation (3.1), $\{\mathbf{f}\}$ is the global load vector, which is the assemblage of the individual element load vectors (Equation (3.2)) and $\{\mathbf{T}\}$ is the global unknown vector.

5. *Solve the system of equations*

   The resulting set of algebraic equations, Equation (3.3), may now be solved to obtain the nodal values of the field variable, for example, temperature.

6. *Calculation of the secondary quantities (post-processing)*

   From the nodal values of the field variables, for example, temperatures, we can then calculate the secondary quantities, for example, the heat fluxes.

3.2   Elements and Shape Functions

As shown in Figure 3.1, the finite element method involves the discretization of both the domain and governing equations. In this process, the variables are represented in a piecewise

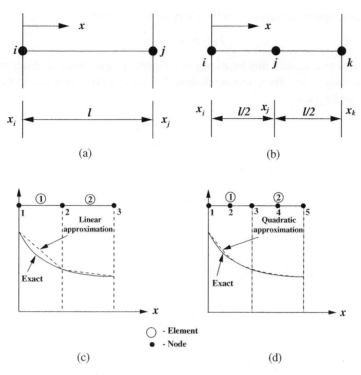

Figure 3.3 One-dimensional finite elements: (a) a linear element; (b) a quadratic element; (c) linear; and (d) quadratic variation of temperature over an element.

manner over the domain. By dividing the solution region into a number of small regions, called "elements" (refer to Chapter 14 for mesh generation), and approximating the solution over each small region, or element, by a suitable known function, a relation between differential equations and elements is established. The functions employed to represent the nature of the solution within each element are called shape functions, or interpolating functions, or basis functions. They are called interpolating functions as they are used to determine the value of the field variable within an element by interpolating the nodal values. They are also known as "basis functions" as they form the basis of the discretization method. Polynomial type functions have been most widely used as they can be integrated, or differentiated, easily and the accuracy of the results can be improved by increasing the order of the polynomial as shown in Figure 3.3.

3.2.1 One-dimensional Linear Element

Many industrial and environmental problems may be approximated using a one-dimensional finite element model. For instance, pipe flow, river flow, heat conduction through a fin etc. can be resolved approximately using a one-dimensional assumption. Figure 3.3 shows the temperature profile in an element as represented by linear and quadratic polynomials. Let us consider a typical linear element with end nodes i and j with the corresponding temperature being denoted by T_i and T_j respectively (Figure 3.3a).

The linear temperature variation in the element is represented by

$$T(x) = \alpha_1 + \alpha_2 x, \tag{3.4}$$

where T is the temperature at any location x and α_1, and α_2 are constants. Since there are two arbitrary constants in the linear representation, it requires only two nodes to determine the values of α_1, and α_2, viz.,

$$T_i = \alpha_1 + \alpha_2 x_i$$
$$T_j = \alpha_1 + \alpha_2 x_j. \tag{3.5}$$

From the above equations, we get

$$\alpha_1 = \frac{T_i x_j - T_j x_i}{x_j - x_i}$$
$$\alpha_2 = \frac{T_j - T_i}{x_j - x_i}. \tag{3.6}$$

On substituting the values of α_1, and α_2 into Equation (3.4) we get

$$T = T_i \left[\frac{x_j - x}{x_j - x_i} \right] + T_j \left[\frac{x - x_i}{x_j - x_i} \right] \tag{3.7}$$

or

$$T = N_i T_i + N_j T_j = \begin{bmatrix} N_i & N_j \end{bmatrix} \begin{Bmatrix} T_i \\ T_j \end{Bmatrix}, \tag{3.8}$$

where N_i and N_j are called "Shape functions" or "Interpolation functions" or "Basis functions" and defined as

$$N_i = \left[\frac{x_j - x}{x_j - x_i} \right]$$
$$N_j = \left[\frac{x - x_i}{x_j - x_i} \right]. \tag{3.9}$$

Equation (3.8) can be rewritten as

$$T = [\mathbf{N}]\{\mathbf{T}\}, \tag{3.10}$$

where

$$[\mathbf{N}] = \begin{bmatrix} N_i & N_j \end{bmatrix} \tag{3.11}$$

is the shape function matrix and

$$\{\mathbf{T}\} = \begin{Bmatrix} T_i \\ T_j \end{Bmatrix} \tag{3.12}$$

is the vector of unknown temperatures

Equation (3.8) shows that the temperature T at any location x can be calculated using the shape functions N_i and N_j evaluated at x. The shape functions at different locations within an element are tabulated in Table 3.1.

Table 3.1 Properties of shape functions within an element

Item	Node, i	Node, j	Arbitrary x
N_i	1	0	$0 \le N_i \le 1$
N_j	0	1	$0 \le N_j \le 1$
$N_i + N_j$	1	1	1

The shape function assumes a value of unity at the designated node and zero at all other nodes. We also see that the sum of all the shape functions in an element is equal to unity anywhere within the element including the boundaries. These are the two essential requirements of the properties of the shape functions of any element in one, two or three dimensions. Figure 3.4 shows the variation of the shape functions and their derivatives within a linear element. A typical linear variation of temperature is also shown in this figure. As seen, the derivatives of the linear shape functions are constant within an element.

From Equation (3.8) the temperature gradient is calculated as

$$\frac{dT}{dx} = \frac{dN_i}{dx}T_i + \frac{dN_j}{dx}T_j = -\left(\frac{1}{x_j - x_i}\right)T_i + \left(\frac{1}{x_j - x_i}\right)T_j, \tag{3.13}$$

or

$$\frac{dT}{dx} = \begin{bmatrix} -\frac{1}{l} & \frac{1}{l} \end{bmatrix} \begin{Bmatrix} T_i \\ T_j \end{Bmatrix}, \tag{3.14}$$

where l is the length of an element equal to $(x_j - x_i)$.

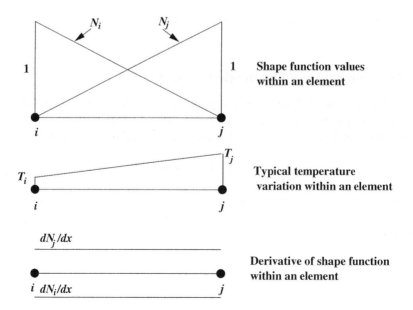

Shape function values within an element

Typical temperature variation within an element

Derivative of shape function within an element

Figure 3.4   Variation of shape functions, temperature and derivatives within a linear element.

Thus, we observe that the temperature gradient is constant within an element as the temperature variation is linear. We can rewrite Equation (3.14) as

$$g = [\mathbf{B}]\{\mathbf{T}\},\tag{3.15}$$

where g is the gradient of the field variable T, $[\mathbf{B}]$ is the derivative matrix, or strain matrix in structural mechanics, which relates the gradient of the field variable to the nodal values and $\{\mathbf{T}\}$ is the temperature vector.

The shape function matrix $[\mathbf{N}]$ and the derivative matrix $[\mathbf{B}]$ are the two important matrices which are used in the determination of the element properties as we shall see later in this chapter.

Example 3.2.1   *Calculate the temperature of an 8 cm long bar at a distance of 5 cm from one end where the temperature is 120 °C with the other end at a temperature of 200 °C. Assume a linear temperature variation between the two end points.*

From Equation (3.8), the temperature distribution over an element can be written as

$$T = N_i T_i + N_j T_j,\tag{3.16}$$

where, at x = 5 m

$$N_i = \frac{x_j - x}{x_j - x_i} = \frac{3}{8}$$

$$N_j = \frac{x - x_i}{x_j - x_i} = \frac{5}{8}.\tag{3.17}$$

Substituting into Equation (3.16), we get $T = 170 °C$. Note that $N_i + N_j = 1$.

3.2.2 One-dimensional Quadratic Element

We can see from Figure 3.3(d) that a better approximation for the temperature profile could be achieved if we use parabolic arcs over each element rather than linear segments. The function $T(x)$ would therefore be quadratic in x within each element and is of the form

$$T(x) = \alpha_1 + \alpha_2 x + \alpha_3 x^2.\tag{3.18}$$

We now have three parameters to determine and hence we need the temperature value at three points. We choose the mid point in addition to the end values to get three equations for the temperature. Assuming x_i is the origin in Figure 3.3(b) we obtain

$$T_i = \alpha_1$$

$$T_j = \alpha_1 + \alpha_2 \frac{l}{2} + \alpha_3 \left(\frac{l}{2}\right)^2$$

$$T_k = \alpha_1 + \alpha_2 l + \alpha_3 l^2.\tag{3.19}$$

From the above three equations, we obtain the following values for the three constants α_1, α_2 and α_3.

$$\alpha_1 = T_i$$

$$\alpha_2 = \frac{1}{l}(-3T_i + 4T_j - T_k)$$

$$\alpha_3 = \frac{2}{l^2}(T_i - 2T_j + T_k). \tag{3.20}$$

Substituting the values of α_1, α_2 and α_3, into Equation (3.18) and collating the coefficients of T_i, T_j and T_k, we get

$$T = T_i\left[1 - \frac{3x}{l} + \frac{2x^2}{l^2}\right] + T_j\left[4\frac{x}{l} - 4\frac{x^2}{l^2}\right] + T_k\left[2\frac{x^2}{l^2} - \frac{x}{l}\right] \tag{3.21}$$

or

$$T = N_iT_i + N_jT_j + N_kT_k. \tag{3.22}$$

Hence, the shape functions for a one-dimensional quadratic element are obtained from Equation (3.21) as follows:

$$N_i = \left[1 - \frac{3x}{l} + \frac{2x^2}{l^2}\right]$$

$$N_j = \left[4\frac{x}{l} - 4\frac{x^2}{l^2}\right]$$

$$N_k = \left[2\frac{x^2}{l^2} - \frac{x}{l}\right]. \tag{3.23}$$

The variation of temperature and shape functions of a typical quadratic element is shown in Figure 3.5. The first derivative of temperature can now be written as

$$\frac{dT}{dx} = \frac{dN_i}{dx}T_i + \frac{dN_j}{dx}T_j + \frac{dN_k}{dx}T_k \tag{3.24}$$

or

$$\frac{dT}{dx} = \left[\frac{4x}{l^2} - \frac{3}{l}\right]T_i + \left[\frac{4}{l} - \frac{8x}{l^2}\right]T_j + \left[\frac{4x}{l^2} - \frac{1}{l}\right]T_k. \tag{3.25}$$

In matrix form

$$g = [\mathbf{B}]\{\mathbf{T}\}. \tag{3.26}$$

The [**B**] matrix is given as

$$[\mathbf{B}] = \left[\left(\frac{4x}{l^2} - \frac{3}{l}\right) \quad \left(\frac{4}{l} - \frac{8x}{l^2}\right) \quad \left(\frac{4x}{l^2} - \frac{1}{l}\right)\right]. \tag{3.27}$$

The above relations show that $N_i = 1$ at i and 0 at j and k, $N_j = 1$ at j and 0 at i and k and $N_k = 1$ at k and 0 at i and j.

It can be verified easily that within an element the summation over the shape functions is equal to unity, that is,

$$\sum_{i=1}^{3} N_i = 1. \tag{3.28}$$

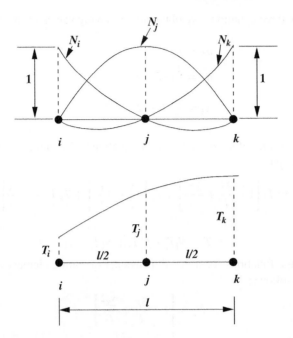

Figure 3.5   Variation of shape functions and their derivatives over a one-dimensional quadratic element.

For example the point at $x = l/4$, the shape function values are

$$N_i = 1 - \frac{3}{4} + \frac{2}{16} = \frac{6}{16}$$

$$N_j = 1 - \frac{4}{16} = \frac{12}{16}$$

$$N_k = \frac{2}{16} - \frac{1}{4} = -\frac{2}{16}, \qquad\qquad (3.29)$$

and it can be easily seen that the sum of the above three shape functions is equal to 1.

It can also be observed that even though the derivatives of the quadratic element are functions of the independent variable x, they will not be continuous at the inter-element nodes. The type of interpolation used here is known as Lagrangian (as they can be generated by Lagrangian interpolation formulae) and it only guarantees the continuity of the function across the inter-element boundaries. These types of elements are known as C° elements where the superscript zero indicates that only the derivatives of zero order are continuous, that is, only the function is continuous. The elements that also assure the continuity of derivatives across inter-element boundaries, in addition to the continuity of functions, are known as C^1 elements and such functions are known as *Hermite polynomials*.

The C° shape functions can be determined in a general way by using Lagrangian polynomial formulae. The one-dimensional $(n - 1)$ th order Lagrange interpolation polynomial is the

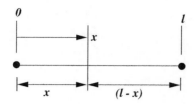

Figure 3.6   A one-dimensional linear element represented by local coordinates.

ratio of two products. For an element with n nodes, $(n - 1)$ order polynomial, the interpolation function is

$$N_k^e(x) = \Pi_{i=1}^{n} \frac{x - x_i}{x_k - x_i}. \tag{3.30}$$

Note that the above equation is only valid when $k \neq i$. For a one-dimensional linear element, the shape functions can be written using Equation (3.30), as $(n = 2)$:

$$N_1 = \frac{x - x_2}{x_1 - x_2} \quad \text{and} \quad N_2 = \frac{x - x_1}{x_2 - x_1}. \tag{3.31}$$

Note that N_1 and N_2 are the shape functions corresponding to the two nodes of a one-dimensional linear element (N_i and N_j). If we use local coordinates, as shown in Figure 3.6, with $x_1 = 0$ and $x_2 = l$, then the shape functions of the first and second node of an element become

$$N_i = \left(1 - \frac{x}{l}\right) = L_i \quad \text{and} \quad N_j = \left(\frac{x}{l}\right) = L_j, \tag{3.32}$$

where L_i and L_j are the shape functions defined by local coordinate system. For a one-dimensional quadratic element, the shape functions using Lagrangian multipliers are given as follows:

$$\begin{aligned}
N_1 &= \frac{x - x_2}{x_1 - x_2} \frac{x - x_3}{x_1 - x_3} \\
N_2 &= \frac{x - x_1}{x_2 - x_1} \frac{x - x_3}{x_2 - x_3} \\
N_3 &= \frac{x - x_1}{x_3 - x_1} \frac{x - x_2}{x_3 - x_2}.
\end{aligned} \tag{3.33}$$

If we substitute local coordinates, $x_1 = 0, x_2 = l/2$ and $x_3 = l$, in the above equation, we can immediately verify that the resulting equations are identical to the one derived from Equation (3.23).

Similarly, cubic elements, or any other one-dimensional higher order element shape functions, can easily be derived using the Lagrangian interpolation formula.

3.2.3   Two-dimensional Linear Triangular Element

When one-dimensional approximations are insufficient, multi-dimensional solution procedures need to be employed. In this section we introduce for the first time a two-dimensional

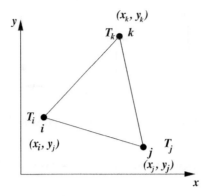

Figure 3.7   A linear triangular element.

element. The simplest geometric shape that can be employed to approximate irregular surfaces is the triangle and this is one of the popular elements currently used in finite element calculations. This is partly due to the fact that automatic mesh generation is direct and easy when a domain is filled with triangles or tetrahedron elements (see Chapter 14) (Fry and George 2008; Thompson *et al.* 1999; Zienkiewicz *et al.* 2013a).

The temperature distribution in a two-dimensional linear triangular element, also known as a simplex element, is represented by

$$T(x, y) = \alpha_1 + \alpha_2 x + \alpha_3 y, \tag{3.34}$$

where the polynomial is linear in x and y and contains three coefficients. Since a linear triangle has three nodes (Figure 3.7), the values of α_1, α_2 and α_3 are determined from

$$T_i = \alpha_1 + \alpha_2 x_i + \alpha_3 y_i$$
$$T_j = \alpha_1 + \alpha_2 x_j + \alpha_3 y_j$$
$$T_k = \alpha_1 + \alpha_2 x_k + \alpha_3 y_k, \tag{3.35}$$

which results in the following:

$$\alpha_1 = \frac{1}{2A}[(x_j y_k - x_k y_j)T_i + (x_k y_i - x_i y_k)T_j + (x_i y_j - x_j y_i)T_k]$$

$$\alpha_2 = \frac{1}{2A}[(y_j - y_k)T_i + (y_k - y_i)T_j + (y_i - y_j)T_k]$$

$$\alpha_3 = \frac{1}{2A}[(x_k - x_j)T_i + (x_i - x_k)T_j + (x_j - x_i)T_k], \tag{3.36}$$

where A is the area of the triangle given by

$$2A = det \begin{bmatrix} 1 & x_i & y_i \\ 1 & x_j & y_j \\ 1 & x_k & y_k \end{bmatrix} = (x_i y_j - x_j y_i) + (x_k y_i - x_i y_k) + (x_j y_k - x_k y_j). \tag{3.37}$$

Substituting the values of α_1, α_2 and α_3 into Equation (3.35) and collating the coefficients of T_i, T_j and T_k, we get

$$T = N_i T_i + N_j T_j + N_k T_k = [N_i \quad N_j \quad N_k] \begin{Bmatrix} T_i \\ T_j \\ T_k \end{Bmatrix}, \tag{3.38}$$

where

$$N_i = \frac{1}{2A}(a_i + b_i x + c_i y)$$

$$N_j = \frac{1}{2A}(a_j + b_j x + c_j y)$$

$$N_k = \frac{1}{2A}(a_k + b_k x + c_k y) \tag{3.39}$$

and

$$a_i = x_j y_k - x_k y_j; \quad b_i = y_j - y_k; \quad c_i = x_k - x_j$$
$$a_j = x_k y_i - x_i y_k; \quad b_j = y_k - y_i; \quad c_j = x_i - x_k$$
$$a_k = x_i y_j - x_j y_i; \quad b_k = y_i - y_j; \quad c_k = x_j - x_i. \tag{3.40}$$

If we evaluate N_i at node i, where the coordinates are (x_i, y_i), then we obtain

$$(N_i)_i = \frac{1}{2A}[(x_j y_k - x_k y_j) + (y_j - y_k)x_i + (x_k - x_j)y_i] = \frac{2A}{2A} = 1. \tag{3.41}$$

Similarly, it can readily be verified that $(N_j)_i = (N_k)_i = 0$.

Thus, we see that the shape functions have a value of unity at the designated vertex and zero at all other vertices. It is possible to show that

$$N_i + N_j + N_k = 1 \tag{3.42}$$

everywhere in the element, including the boundaries and zero elsewhere. The gradients of temperature, T, is given by

$$\frac{\partial T}{\partial x} = \frac{\partial N_i}{\partial x}T_i + \frac{\partial N_j}{\partial x}T_j + \frac{\partial N_k}{\partial x}T_k = \frac{b_i}{2A}T_i + \frac{b_j}{2A}T_j + \frac{b_k}{2A}T_k$$

$$\frac{\partial T}{\partial y} = \frac{\partial N_i}{\partial y}T_i + \frac{\partial N_j}{\partial y}T_j + \frac{\partial N_k}{\partial y}T_k = \frac{c_i}{2A}T_i + \frac{c_j}{2A}T_j + \frac{c_k}{2A}T_k \tag{3.43}$$

or

$$\{\mathbf{g}\} = \begin{Bmatrix} \dfrac{\partial T}{\partial x} \\ \dfrac{\partial T}{\partial y} \end{Bmatrix} = \frac{1}{2A}\begin{bmatrix} b_i & b_j & b_k \\ c_i & c_j & c_k \end{bmatrix}\begin{Bmatrix} T_i \\ T_j \\ T_k \end{Bmatrix} = [\mathbf{B}]\{\mathbf{T}\}. \tag{3.44}$$

It should be noted that both $\partial T/\partial x$ and $\partial T/\partial y$ are constants within an element as b_i, b_j, b_k and c_i, c_j, c_k are constants for a given triangle. Hence, the heat fluxes q_x and q_y are also constants within a linear triangular element. Since the temperature varies linearly within an element, it

Table 3.2   Data for Example 3.2.2

Node	x (cm)	y (cm)	T °C
i	0.0	0.0	50.0
j	4.0	0.0	70.0
k	0.0	2.5	100.0

is possible to draw the isotherms (constant temperature lines) within a linear triangle and this is illustrated in the following example.

Example 3.2.2   *As an illustration of the method of calculation, let us calculate the temperature, T at the point (2.0, 1.0) and heat fluxes q_x and q_y within an element for the data given in Table 3.2. Calculate the temperature T, and the heat flux components q_x and q_y if the thermal conductivity of the material is 2 W/cm K. Draw the isothermal line for 60°C in the triangle.*

The temperature at any location within the triangle is given by Equation (3.38). The shape functions at the point (2.0, 1.0) are calculated using Equation (3.39) by substituting x and y coordinates given in Table 3.2. The result is

$$N_i = \frac{1}{10}$$

$$N_j = \frac{5}{10}$$

$$N_k = \frac{4}{10}. \tag{3.45}$$

The substitution of the nodal temperatures and the above shape function values into Equation (3.38) results in the temperature of the point (2.0, 1.0) being,

$$T = N_i T_i + N_j T_j + N_k T_k = \frac{1}{10}(50) + \frac{5}{10}(70) + \frac{4}{10}(100) = 80\,°C. \tag{3.46}$$

The components of heat flux in the x and y directions are calculated as

$$\left\{ \begin{array}{c} q_x \\ q_y \end{array} \right\} = -\frac{k}{2A} \begin{bmatrix} b_i & b_j & b_k \\ c_i & c_j & c_k \end{bmatrix} \left\{ \begin{array}{c} T_i \\ T_j \\ T_k \end{array} \right\} = -\frac{2}{10} \begin{bmatrix} 50 \\ 200 \end{bmatrix}. \tag{3.47}$$

The position of the 60°C isotherm may be obtained from Figure 3.8. From the given temperature values, it is clear that one 60°C point lies on the side ij (point P) and another lies on the side ik (point Q). It should be noted that the temperature varies linearly along these sides, that is, temperature is directly proportional to distance.

In order to determine the location of P on ij, we have the following linear relation between the distances and temperature values, viz.,

$$\frac{60.0 - 50.0}{70.0 - 50.0} = \frac{\sqrt{(x_P - x_i)^2 + (y_P - y_i)^2}}{\sqrt{(x_j - x_i)^2 + (y_j - y_i)^2}}. \tag{3.48}$$

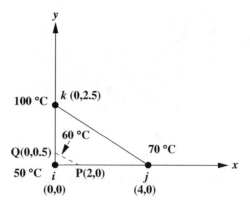

Figure 3.8   Isotherm within a linear triangular element.

From the data given, it is clear that the y coordinate on the ij side are equal to zero and thus the above equation is simplified to

$$\frac{10.0}{20.0} = \frac{(x_P - x_i)}{(x_j - x_i)},$$

(3.49)

which results in $x_P = 2.0\,cm$. The location of Q along ik can be determined in a similar fashion as

$$\frac{60.0 - 50.0}{100.0 - 50.0} = \frac{y_Q - y_i}{y_k - y_i},$$

(3.50)

which gives $y_Q = 0.5\,cm$. The x coordinate of this point is zero.

The line joining P and Q will be the $60\,°C$ isotherm (Figure 3.8). It should be noted that the same principle can be used for arbitrary triangles.

3.2.4   Area Coordinates

An area, or natural, coordinate system will now be introduced for triangular elements in order to simplify the solution process. Let us consider a point P (centroid or barycenter of the triangle) within a triangle as shown in Figure 3.9. The local coordinates L_i, L_j and L_k of this point can be established by calculating appropriate nondimensional distances or areas. For example, L_i is defined as the ratio of the perpendicular distance from point P to the side jk (OP) to the perpendicular distance of point i from the side jk (QR). Thus,

$$L_i = \frac{OP}{QR}.$$

(3.51)

Similarly, L_j and L_k are also defined. The value of L_i is also equal to the ratio of the area A_i (opposite to node i) to the total area of the triangle, that is,

$$L_i = \frac{A_i}{A} = \frac{0.5(OP)(jk)}{0.5(QR)(jk)} = \frac{OP}{QR}.$$

(3.52)

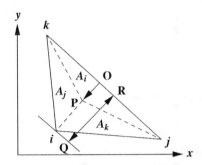

Figure 3.9   Area coordinates of a triangular element.

Thus, the local coordinates vary from 0 on the side jk to 1 at the node i (by moving point P between points O and i respectively). From Figure 3.9 it is obvious that

$$A_i + A_j + A_k = A \tag{3.53}$$

or

$$\frac{A_i}{A} + \frac{A_j}{A} + \frac{A_k}{A} = 1; \tag{3.54}$$

therefore

$$L_i + L_j + L_k = 1. \tag{3.55}$$

The relationship between the (x, y) coordinates and the natural, or area, coordinates are given by

$$x = L_i x_i + L_j x_j + L_k x_k \tag{3.56}$$

and

$$y = L_i y_i + L_j y_j + L_k y_k. \tag{3.57}$$

From Equations (3.55), (3.56) and (3.57) the following relations for the local coordinates can be derived:

$$L_i = \frac{1}{2A}(a_i + b_i x + c_i y)$$
$$L_j = \frac{1}{2A}(a_j + b_j x + c_j y)$$
$$L_k = \frac{1}{2A}(a_k + b_k x + c_k y), \tag{3.58}$$

where the constants a, b and c are defined in Equation (3.40). Comparing the Equation (3.58) with Equation (3.39) it is clear that

$$L_i = N_i$$
$$L_j = N_j$$
$$L_k = N_k. \tag{3.59}$$

Thus, the local or area coordinates in a triangle are the same as the shape functions for a linear triangular element. In general, the local coordinates and shape functions are the same for linear elements irrespective of whether they are of one, two or three dimensions.

For a two-dimensional linear triangular element with local coordinates L_i, L_j and L_k, we have a simple formula for integration over the triangle, that is,

$$\int_A L_i^a L_j^b L_k^c dA = \int_A N_i^a N_j^b N_k^c dA = \frac{a!b!c!}{(a+b+c+2)!}2A,$$ (3.60)

where A is the area of a triangle. Note that L_i, L_j and L_k happen to be the shape functions for a linear triangular element. Example 3.2.2 can also be solved using the local coordinates via Equations (3.53), (3.56) and (3.57), that is, on substituting the x and y coordinates of the three points (Table 3.2) of the triangle into Equation (3.56), we obtain

$$L_j = \frac{x}{4}$$ (3.61)

and from Equation (3.57)

$$L_k = \frac{y}{2.5}.$$ (3.62)

From Equation (3.55), we get

$$L_i = 1 - \frac{x}{4} - \frac{y}{2.5}.$$ (3.63)

At $(x, y) = (2,1)$, we have

$$L_i = 0.1 = N_i$$
$$L_j = 0.5 = N_j$$
$$L_k = 0.4 = N_k.$$ (3.64)

Note that these local coordinates are exactly the same as the shape functions calculated in Example 3.2.2.

3.2.5 Quadratic Triangular Element

We can write a quadratic approximation over a triangular element as

$$T = \alpha_1 + \alpha_2 x + \alpha_3 y + \alpha_4 x^2 + \alpha_5 y^2 + \alpha_6 xy.$$ (3.65)

Since there are six arbitrary constants, the quadratic triangle will have six nodes (Figure 3.10). The six constants α_1, α_2 ...... α_6 can be evaluated by substitution of the nodal coordinates and the corresponding nodal temperatures $T_1, T_2....T_6$. For example, we can write the following relationship for the first node:

$$T_1 = \alpha_1 + \alpha_2 x_1 + \alpha_3 y_1 + \alpha_4 x_1^2 + \alpha_5 y_1^2 + \alpha_6 x_1 y_1.$$ (3.66)

Once $\alpha_1, \alpha_2......\alpha_6$ are determined, then the substitution of these parameters into Equation (3.65) and collating the coefficients of $T_1, T_2, ...T_6$, give relations for the shape functions. The process is both tedious and unnecessary. A much superior and more general method of

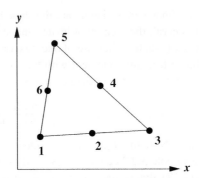

Figure 3.10   A quadratic triangular element.

establishing the shape functions exists, which is based on local coordinates. The rationale behind this is given by Silvester (1969) and can also be used to find the shape functions for a cubic triangular element.

Silvester introduced a triple-index numbering scheme $\alpha\beta\gamma$, which satisfies the following expression:

$$\alpha + \beta + \gamma = n, \tag{3.67}$$

where n is the order of the interpolation polynomial used. We can write $N_{\alpha\beta\gamma}$ to denote the interpolation function for a node as a function of the area coordinates L_i, L_j and L_k, viz.,

$$N_{\alpha\beta\gamma}(L_i, L_j, L_k) = N_\alpha(L_i)N_\beta(L_j)N_\gamma(L_k), \tag{3.68}$$

where

$$N_\alpha(L_i) = \Pi_{m=1}^{\alpha}\left[\frac{nL_i - m + 1}{m}\right] \quad \text{if} \quad \alpha \geq 1 \tag{3.69}$$

$$N_\alpha(L_i) = 1 \quad \text{if} \quad \alpha = 0. \tag{3.70}$$

Similarly, we can write relations for N_β and N_γ in terms of L_j and L_k respectively. For a quadratic triangular element shown in Figure 3.11, the shape functions are designated as

- Corner nodes: $N_1 = N_{200}; N_3 = N_{020}; N_5 = N_{002}$.

- Side nodes: $N_2 = N_{110}; N_4 = N_{011}; N_6 = N_{101}$.

For example $N_1 = N_{200}$ is calculated by substituting $\alpha = 2, \beta = 0$ and $\gamma = 0$ into Equation (3.68). Now, N_α may be calculated from Equation (3.69) as

$$N_\alpha(L_i) = \Pi_{m=1}^{2}\left[\frac{nL_i - m + 1}{m}\right] = \left[\frac{2L_i - 1 + 1}{1}\right]\left[\frac{2L_i - 2 + 1}{2}\right] = L_i(2L_i - 1) \tag{3.71}$$

and similarly

$$N_\beta(L_j) = 1 \quad \text{and} \quad N_\gamma(L_k) = 1. \tag{3.72}$$

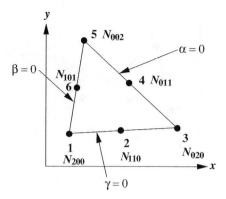

Figure 3.11 Shape function designations of a quadratic triangular element.

Hence,

$$N_{200} = N_2(L_i)N_0(L_j)N_0(L_k) = L_i(2L_i - 1) = N_1 \qquad (3.73)$$

is the shape function for the node 1. Similarly,

$$N_3 = N_{020} = L_j(2L_j - 1) \quad \text{and} \quad N_5 = N_{002} = L_k(2L_k - 1). \qquad (3.74)$$

For a middle node, with shape function N_{110}, we have ($\alpha = 1, \beta = 1, \gamma = 0$)

$$\begin{aligned}
N_{110} &= N_\alpha(L_i)N_\beta(L_j)N_\gamma(L_k) \\
&= \left[\Pi_{m-1}^1 \left(\frac{2L_i - m + 1}{m} \right) \right] \left[\Pi_{m-1}^1 \left(\frac{2L_j - m + 1}{m} \right) \right] \\
&= \left(\frac{2L_i - 1 + 1}{1} \right) \left(\frac{2L_j - 1 + 1}{1} \right).
\end{aligned} \qquad (3.75)$$

Thus,

$$N_2 = N_{110} = 4L_iL_j. \qquad (3.76)$$

Similarly,

$$\begin{aligned}
N_4 = N_{011} = 4L_jL_k \\
N_6 = N_{101} = 4L_kL_i.
\end{aligned} \qquad (3.77)$$

We can summarize the nodal shape functions for a quadratic triangle as follows:
For corner nodes,

$$N_m = L_n(2L_n - 1) \quad \text{with} \quad m = 1, 3, 5 \quad \text{and} \quad n = i, j, k \qquad (3.78)$$

and for nodes at centers,

$$\begin{aligned}
N_2 &= 4L_iL_j \\
N_4 &= 4L_jL_k \\
N_6 &= 4L_kL_i.
\end{aligned} \qquad (3.79)$$

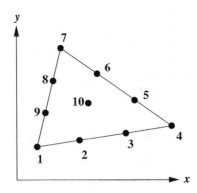

Figure 3.12   Ten-node cubic triangular element.

In a similar way, we can show that the interpolation functions for a 10-node cubic triangle are (see Figure 3.12):

For corner nodes

$$N_m = \frac{1}{2}L_n(3L_n - 1)(3L_n - 2) \quad \text{with} \quad m = 1, 4, 7 \quad \text{and} \quad n = i, j, k. \tag{3.80}$$

Side ij

$$N_2 = \frac{9}{2}L_iL_j(3L_i - 1)$$

$$N_3 = \frac{9}{2}L_iL_j(3L_j - 1). \tag{3.81}$$

Side jk

$$N_5 = \frac{9}{2}L_jL_k(3L_j - 1)$$

$$N_6 = \frac{9}{2}L_jL_k(3L_k - 1). \tag{3.82}$$

Side ki

$$N_8 = \frac{9}{2}L_kL_i(3L_k - 1)$$

$$N_9 = \frac{9}{2}L_kL_i(3L_i - 1) \tag{3.83}$$

and for the node at the center of the triangle

$$N_{10} = 27L_iL_jL_k. \tag{3.84}$$

It is possible to derive shape functions for even higher order elements using the same procedure.

3.2.6 Two-dimensional Quadrilateral Elements

The quadrilateral element has four nodes located at the verticies as shown in Figure 3.13. Eight- and nine-node quadrilaterals are also used in practice. The quadrilateral mesh resembles a finite

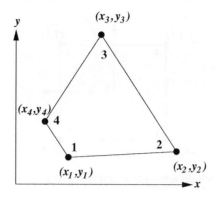

Figure 3.13   A typical quadrilateral element.

difference mesh. However, for the case of a standard finite-difference mesh, the mesh must be orthogonal, that is, all lines intersect at right angles to one another whereas in the finite element mesh, each element can be unique in shape and each side may have a different slope. In the simplest form, the quadrilateral element becomes a *rectangular element* (Figure 3.14) with the boundaries of the element parallel to a coordinate system.

The temperature within a quadrilateral is represented by

$$T = \alpha_1 + \alpha_2 x + \alpha_3 y + \alpha_4 xy \tag{3.85}$$

and thus the temperature gradients may be written as

$$\frac{\partial T}{\partial x} = \alpha_2 + \alpha_4 y$$

$$\frac{\partial T}{\partial y} = \alpha_3 + \alpha_4 x. \tag{3.86}$$

Therefore, the gradient varies within the element in a linear way. On substituting the values of T_1, T_2, T_3 and T_4 into Equation (3.85) for the nodes $(x_1, y_1)......(x_4, y_4)$ and solving,

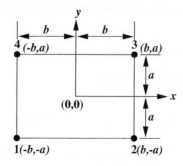

Figure 3.14   A simple rectangular element.

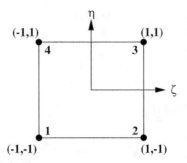

Figure 3.15   Nondimensional coordinates of a rectangular element.

we obtain the values of $\alpha_1, \alpha_2, \alpha_3$ and α_4. Substituting these relationships into Equation (3.85) and collecting the coefficients of $T_1, T_2, ...T_4$, we get

$$T = N_1 T_1 + N_2 T_2 + N_3 T_3 + N_4 T_4, \tag{3.87}$$

where for a rectangular element (Figure 3.14),

$$N_1 = \frac{1}{4ab}(b - x)(a - y)$$

$$N_2 = \frac{1}{4ab}(b + x)(a - y)$$

$$N_3 = \frac{1}{4ab}(b + x)(a + y)$$

$$N_4 = \frac{1}{4ab}(b - x)(a + y). \tag{3.88}$$

We can express these shape functions in terms of length ratios and x/b and y/a as

$$N_1 = \frac{1}{4ab}(b - x)(a - y) = \frac{1}{4}\left(1 - \frac{x}{b}\right)\left(1 - \frac{y}{a}\right) = \frac{1}{4}(1 - \zeta)(1 - \eta), \tag{3.89}$$

where

$$-1 \leq \zeta \leq 1 \quad \text{and} \quad -1 \leq \eta \leq 1 \tag{3.90}$$

are the nondimensional coordinates of an element (Figure 3.15). The shape functions can also be obtained using Lagrange interpolation functions (Equation (3.30)) as

$$N_1 = \frac{(x - b)(y - a)}{(-b - b)(-a - a)} = \frac{1}{4ab}(b - x)(a - y) = \frac{1}{4}(1 - \zeta)(1 - \eta)$$

$$N_2 = \frac{(x - (-b))(y - a)}{(b - (-b))(-a - a)} = \frac{1}{4ab}(b + x)(a - y) = \frac{1}{4}(1 + \zeta)(1 - \eta)$$

$$N_3 = \frac{(x - (-b))(y - (-a))}{(b - (-b))(-a - a)} = \frac{1}{4ab}(b + x)(a + y) = \frac{1}{4}(1 + \zeta)(1 + \eta)$$

$$N_4 = \frac{(x - b)(y - (-a))}{(-b - b)(a - (-a))} = \frac{1}{4ab}(b - x)(a + y) = \frac{1}{4}(1 - \zeta)(1 + \eta). \tag{3.91}$$

In general, the shape functions can be written as

$$N_i = (1 + \zeta\zeta_i)(1 + \eta\eta_i), \tag{3.92}$$

where (ζ_i, η_i) are the coordinates of node i. Since the shape functions are linear in the x and y directions, they are referred to as a "bilinear" configuration. The derivatives can be expressed as follows:

$$\frac{\partial T}{\partial x} = \frac{\partial N_1}{\partial x}T_1 + \frac{\partial N_2}{\partial x}T_2 + \frac{\partial N_3}{\partial x}T_3 + \frac{\partial N_4}{\partial x}T_4$$

$$= \frac{1}{4ab}\left[-(a-y)T_1 + (a-y)T_2 + (a+y)T_3 - (a+y)T_4\right]. \tag{3.93}$$

Similarly,

$$\frac{\partial T}{\partial y} = \frac{1}{4ab}\left[-(b-x)T_1 - (b+x)T_2 + (b+x)T_3 + (b-x)T_4\right]. \tag{3.94}$$

The gradient matrix can be written as

$$\{g\} = \left\{\begin{matrix} \frac{\partial T}{\partial x} \\ \frac{\partial T}{\partial y} \end{matrix}\right\} = \frac{1}{4ab}\begin{bmatrix} -(a-y) & (a-y) & (a+y) & -(a+y) \\ -(b-x) & -(b+x) & (b+x) & (b-x) \end{bmatrix}\left\{\begin{matrix} T_1 \\ T_2 \\ T_3 \\ T_4 \end{matrix}\right\}$$

$$= [\mathbf{B}]\{\mathbf{T}\}. \tag{3.95}$$

The $[\mathbf{B}]$ matrix is written as

$$[\mathbf{B}] = \frac{1}{4}\begin{bmatrix} -(1-\eta) & (1-\eta) & (1+\eta) & -(1+\eta) \\ -(1-\zeta) & -(1+\zeta) & (1+\zeta) & (1-\zeta) \end{bmatrix}. \tag{3.96}$$

Example 3.2.3 *Determine the temperature and the heat fluxes at a location (2,1) in a square plate (Figure 3.16) with the data shown in Table 3.3. Draw the isotherm for 125°C and determine the heat fluxes if $k_x = k_y = 2$ W/m°C.*

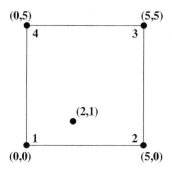

Figure 3.16   Square plate.

Table 3.3   Data for Example 3.2.3

Node no.	x (cm)	y (cm)	Temperature °C
1	0.0	0.0	100.0
2	5.0	0.0	150.0
3	5.0	5.0	200.0
4	0.0	5.0	50.0

Note that the origin is at node 1. In order to use the shape functions already derived, we can give the coordinates of the nodes with the origin at the center of the square plate.

Note that $2a = 2b = 5.0$.

The temperature at any point within the element can be expressed as

$$T = N_1 T_1 + N_2 T_2 + N_3 T_3 + N_4 T_4. \tag{3.97}$$

The location of the point (2,1), using the nondimensional coordinates and new origin at the center, is (−0.5,−1.5).

The shape functions at this point are calculated by substituting the new coordinates of point (2,1), that is,

$$N_1 = \frac{1}{4ab}(b - x)(a - y) = \frac{12}{25}$$

$$N_2 = \frac{1}{4ab}(b + x)(a - y) = \frac{8}{25}$$

$$N_3 = \frac{1}{4ab}(b + x)(a + y) = \frac{2}{25}$$

$$N_4 = \frac{1}{4ab}(b - x)(a + y) = \frac{3}{25}. \tag{3.98}$$

Note that $N_1 + N_2 + N_3 + N_4 = 1$.

Therefore, the temperature at the point (−0.5, −1.5) is

$$T_{(-0.5,-1.5)} = \frac{12}{25}(100) + \frac{8}{25}(150) + \frac{2}{25}(200) + \frac{3}{25}(50) = 118\,^{\circ}C \tag{3.99}$$

Table 3.4   Nondimensional coordinates for Example 3.2.3

1	−2.5	−2.5
2	2.5	−2.5
3	2.5	2.5
4	−2.5	2.5

The heat fluxes can be calculated from Equation (3.95) as follows:

$$\left\{ \begin{array}{c} q_x \\ q_y \end{array} \right\} = - \left\{ \begin{array}{c} k_x \frac{\partial T}{\partial x} \\ k_y \frac{\partial T}{\partial y} \end{array} \right\}$$

$$= -\frac{2}{25} \begin{bmatrix} -4.0 & 4.0 & 1.0 & -1.0 \\ -3.0 & -2.0 & 2.0 & 3.0 \end{bmatrix} \left\{ \begin{array}{c} 100.0 \\ 150.0 \\ 200.0 \\ 50.0 \end{array} \right\} \qquad (3.100)$$

$$= \left\{ \begin{array}{c} -28.0 \\ 4.0 \end{array} \right\} W/cm^2.$$

The isotherm of 125°C will not normally be a straight-line due to the bilinear nature of the elements. Thus, we need more than two points to represent an isotherm. It is certain that one point on side 1–2 and one on 3–4 will contain a point with a temperature of 125°C. We know the y coordinates of both the sides 1–2 and 3–4. Thus, the x coordinate of the point on side 1–2 which has a temperature of 125°C is calculated by substituting y = 0.0 into the temperature distribution of Equation (3.97), that is,

$$125.0 = \frac{1}{25}[(2.5 - x)(2.5 - 0.0)100.0 + (2.5 + x)(2.5 - 0.0)150$$
$$+(2.5 + x)(2.5 + 0.0)200.0 + (2.5 - x)(2.5 + 0.0)50.0. \qquad (3.101)$$

which gives x = 2.5 and similarly, if we substitute a value of y = 5.0 for the side 3–4 the results is x = 2.5. These coordinates can be written in a nondimensional form as (0.0, −2.5) and (0.0, 2.5). From the two points found, it is clear that the 125°C isotherm crosses all horizontal lines between the bottom and top sides. Therefore, to determine another point, we can assume a y value of 2.5 (0.0, in nondimensional form) and on substituting into Equation (3.97) results in an x coordinate of 2.5 (0.0, in nondimensional form). Connecting all three points will generate the 125°C isotherm.

3.2.7   Isoparametric Elements

Many practical problems have curved boundaries, and it is often necessary to use a large number of straight-sided elements along the curved boundaries in order to achieve a reasonable geometric representation. The number of elements needed can be reduced considerably if curved elements are used with a consequential reduction in the total number of variables in the system. In the case of three-dimensional problems the total number of variables is inherently large and a reduction in the total number of variables is very important, especially when there is a limitation on the computer memory/cost involved. While there are many methods of creating curved elements, the method most extensively used in practice involves isoparametric mapping from regular elements (Figure 3.17). Since the shape functions of the regular parent element are known in terms of a local coordinate system, those of the generated curvilinear element can also be determined. The mapping is simple and straightforward.

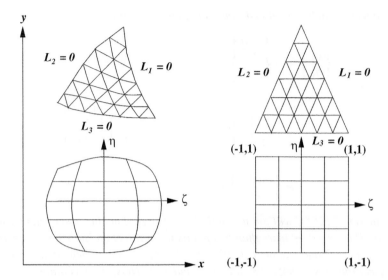

Figure 3.17   Isoparametric mapping of triangular and quadrilateral elements.

There are two sets of relations that must be defined when using the finite element method. One set determines the shape of the element and the other set defines the order of the interpolation function for the field variable. It is not necessary to use the same shape functions for the coordinate transformation and the interpolation equation. Thus, two different sets of global nodes can exist. Both sets of global nodes are identical in the case of isoparametric elements.

3.2.7.1   One-dimensional Elements

The natural coordinate system for the one-dimensional element is the length ratio defined such that $-1 \leq \zeta \leq 1$, where ζ is the natural coordinate. The origin of the coordinate is at the mid-point of the line segment. For a one-dimensional linear element (substituting $x = \zeta$, $x_1 = -1$ and $x_2 = 1$ into Equation (3.31)), we obtain

$$N_i = \frac{\zeta - 1}{-1 - 1} = \frac{1}{2}(1 - \zeta)$$

$$N_j = \frac{\zeta - (-1)}{1 - (-1)} = \frac{1}{2}(1 + \zeta), \tag{3.102}$$

where i and j are the two nodes of a one-dimensional element. For a one dimensional quadratic element, we have (Equation (3.33))

$$N_i = \frac{(\zeta - 0)(\zeta - 1)}{(-1 - 0)(-1 - 1)} = -\frac{\zeta}{2}(1 - \zeta)$$

$$N_j = \frac{(\zeta - (-1))(\zeta - 1)}{(0 - (-1))(0 - 1)} = (1 - \zeta^2)$$

$$N_k = \frac{(\zeta - (-1))\,(\zeta - 0)}{(1 - (-1))\,(1 - 0)} = \frac{\zeta}{2}(1 + \zeta), \tag{3.103}$$

where i, j and k represent the three nodes of the quadratic element. In order to calculate the stiffness matrix, we need the derivative of the shape functions with respect to the global coordinate, that is, with regard to x in this case. Therefore, a coordinate transformation of the type shown in Figure 3.17 should be determined. In either case the functions $g(\zeta)$ and $g(x)$ are assumed to be one-to-one mappings.

The coordinate transformation can be written using the same functions as given in Equation (3.103), but substituting the coordinate value for the nodal parameter. Thus, the coordinate transformation becomes

$$x = N_i x_i + N_j x_j + N_k x_k, \tag{3.104}$$

where N_i, N_j and N_k are given by Equation (3.103). The derivative of N_i is

$$\frac{dN_i}{d\zeta} = \frac{dN_i}{dx} \frac{dx}{d\zeta} = \frac{dN_i}{dx} [J], \tag{3.105}$$

which gives

$$\frac{dN_i}{dx} = [J]^{-1} \frac{dN_i}{d\zeta}. \tag{3.106}$$

The quantity $(dx/d\zeta)$ is called the Jacobian matrix of the coordinate transformation and is denoted by $[J]$. For a one-dimensional coordinate transformation $[J]$ is calculated using

$$[J] = \frac{dx}{d\zeta} = \frac{dN_i}{d\zeta} x_i + \frac{dN_j}{d\zeta} x_j + \frac{dN_k}{d\zeta} x_k. \tag{3.107}$$

Example 3.2.4 *Derive the shape function derivatives for a one-dimensional quadratic element that has nodal coordinates $x_i = 2$, $x_j = 4$ and $x_k = 6$.*
The Jacobian is written as

$$
\begin{aligned}
[J] &= \frac{dx}{d\zeta} \\
&= \frac{dN_i}{d\zeta} x_i + \frac{dN_j}{d\zeta} x_j + \frac{dN_k}{d\zeta} x_k \\
&= \left(-\frac{1}{2} + \zeta\right) 2 + (-2\zeta)4 + \left(\frac{1}{2} + \zeta\right) 6 = 2.0;
\end{aligned}
\tag{3.108}
$$

thus,

$$[J]^{-1} = \frac{1}{2}. \tag{3.109}$$

The shape function derivatives are written as follows:

$$
\begin{Bmatrix} \dfrac{dN_i}{dx} \\[2mm] \dfrac{dN_j}{dx} \\[2mm] \dfrac{dN_k}{dx} \end{Bmatrix} = [J]^{-1} \begin{Bmatrix} \dfrac{dN_i}{d\zeta} \\[2mm] \dfrac{dN_j}{d\zeta} \\[2mm] \dfrac{dN_k}{d\zeta} \end{Bmatrix} = \frac{1}{2} \begin{Bmatrix} -\dfrac{1}{2} + \zeta \\[2mm] -2\zeta \\[2mm] \dfrac{1}{2} + \zeta \end{Bmatrix}.
\tag{3.110}
$$

3.2.7.2   Two-dimensional Elements

For two-dimensional cases, we may express x and y as functions of ζ and η, that is,

$$x = x(\zeta, \eta) \quad \text{and} \quad y = y(\zeta, \eta). \tag{3.111}$$

Since we deal with Cartesian derivatives for the calculation of the stiffness matrix, we transform the derivatives of the shape functions using the chain rule as

$$\frac{\partial N_i}{\partial \zeta}(x, y) = \frac{\partial N_i}{\partial x} \frac{\partial x}{\partial \zeta} + \frac{\partial N_i}{\partial y} \frac{\partial y}{\partial \zeta}$$

$$\frac{\partial N_i}{\partial \eta}(x, y) = \frac{\partial N_i}{\partial x} \frac{\partial x}{\partial \eta} + \frac{\partial N_i}{\partial y} \frac{\partial y}{\partial \eta}, \tag{3.112}$$

which can be written as

$$\left\{ \begin{array}{c} \frac{\partial N_i}{\partial \zeta} \\ \frac{\partial N_i}{\partial \eta} \end{array} \right\} = \left[\begin{array}{cc} \frac{\partial x}{\partial \zeta} & \frac{\partial y}{\partial \zeta} \\ \frac{\partial x}{\partial \eta} & \frac{\partial y}{\partial \eta} \end{array} \right] \left\{ \begin{array}{c} \frac{\partial N_i}{\partial x} \\ \frac{\partial N_i}{\partial y} \end{array} \right\} = [\mathbf{J}] \left\{ \begin{array}{c} \frac{\partial N_i}{\partial x} \\ \frac{\partial N_i}{\partial y} \end{array} \right\}, \tag{3.113}$$

where

$$[\mathbf{J}] = \left[\begin{array}{cc} \frac{\partial x}{\partial \zeta} & \frac{\partial y}{\partial \zeta} \\ \frac{\partial x}{\partial \eta} & \frac{\partial y}{\partial \eta} \end{array} \right]. \tag{3.114}$$

Therefore, we can write

$$\left\{ \begin{array}{c} \frac{\partial N_i}{\partial x} \\ \frac{\partial N_i}{\partial y} \end{array} \right\} = [\mathbf{J}]^{-1} \left\{ \begin{array}{c} \frac{\partial N_i}{\partial \zeta} \\ \frac{\partial N_i}{\partial \eta} \end{array} \right\}. \tag{3.115}$$

Note that the inverse of the Jacobian matrix $[\mathbf{J}]^{-1}$ is calculated as

$$[\mathbf{J}]^{-1} = \frac{1}{det[\mathbf{J}]} \left[\begin{array}{cc} \frac{\partial y}{\partial \eta} & -\frac{\partial y}{\partial \zeta} \\ -\frac{\partial x}{\partial \eta} & \frac{\partial x}{\partial \zeta} \end{array} \right]. \tag{3.116}$$

The derivatives have to be numerically evaluated at each integration point, as a closed form solution does not exist.

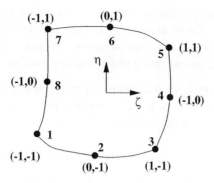

Figure 3.18   Eight-node isoparametric element.

For an eight-node isoparametric element (Figure 3.18), the values of the temperature T at any point are given by

$$T = \sum_{i=1}^{8} N_i T_i. \tag{3.117}$$

The coordinate values of x and y at any point within an element are given by the following expressions:

$$x(\zeta, \eta) = \sum_{i=1}^{8} N_i(\zeta, \eta) x_i$$

$$y(\zeta, \eta) = \sum_{i=1}^{8} N_i(\zeta, \eta) y_i, \tag{3.118}$$

where (x_i, y_i) are the coordinates of the node i and the quadratic shape functions are given by

$$N_1 = -\frac{1}{4}(1 - \zeta)(1 - \eta)(1 + \zeta + \eta)$$

$$N_2 = \frac{1}{2}(1 - \zeta^2)(1 - \eta)$$

$$N_3 = \frac{1}{4}(1 + \zeta)(1 - \eta)(\zeta - \eta - 1)$$

$$N_4 = \frac{1}{2}(1 + \zeta)(1 - \eta^2)$$

$$N_5 = \frac{1}{4}(1 + \zeta)(1 + \eta)(\zeta + \eta - 1)$$

$$N_6 = \frac{1}{2}(1 - \zeta^2)(1 + \eta)$$

$$N_7 = \frac{1}{4}(1 - \zeta)(1 + \eta)(-\zeta + \eta - 1)$$

$$N_8 = \frac{1}{2}(1 - \zeta)(1 - \eta^2). \tag{3.119}$$

The ζ and η variables are curvilinear coordinates and as such their direction will vary with position. The nodes of the element are input in an anti-clockwise sequence starting from any corner node. The directions of ζ and η are indicated on Figure 3.18 that is, positive ζ in the direction from nodes 1 to 3 and positive η in the direction from nodes 3 to 5.

Example 3.2.5   *Evaluate the partial derivatives of the shape functions at $\zeta = 1/2$, $\eta = 1/2$ of a quadrilateral element given in Example 3.2.3, assuming that the temperature is approximated by (a) bilinear and (b) quadratic interpolating polynomials.*

(a) Bilinear

The shape function derivatives in local coordinates are (Equation (3.91):

$$\frac{\partial N_1}{\partial \zeta} = -\frac{1-\eta}{4}; \quad \frac{\partial N_1}{\partial \eta} = -\frac{1-\zeta}{4}$$

$$\frac{\partial N_2}{\partial \zeta} = \frac{1-\eta}{4}; \quad \frac{\partial N_2}{\partial \eta} = -\frac{1+\zeta}{4}$$

$$\frac{\partial N_3}{\partial \zeta} = \frac{1+\eta}{4}; \quad \frac{\partial N_3}{\partial \eta} = \frac{1+\zeta}{4}$$

$$\frac{\partial N_4}{\partial \zeta} = -\frac{1+\eta}{4}; \quad \frac{\partial N_4}{\partial \eta} = \frac{1-\zeta}{4}. \tag{3.120}$$

The Jacobian matrix and its inverse are calculated from Equations (3.113) and (3.116), that is,

$$[\mathbf{J}] = \begin{bmatrix} \sum_{i=1}^{4} \frac{\partial N_i}{\partial \zeta} x_i & \sum_{i=1}^{4} \frac{\partial N_i}{\partial \zeta} y_i \\ \sum_{i=1}^{4} \frac{\partial N_i}{\partial \eta} x_i & \sum_{i=1}^{4} \frac{\partial N_i}{\partial \eta} y_i \end{bmatrix} = \frac{1}{8} \begin{bmatrix} 20 & 0 \\ 0 & 20 \end{bmatrix}. \tag{3.121}$$

The determinant of the Jacobian matrix is

$$det[\mathbf{J}] = \frac{400}{8}. \tag{3.122}$$

Employing Equation (3.116)

$$[\mathbf{J}]^{-1} = \frac{8}{400} \begin{bmatrix} 20 & 0 \\ 0 & 20 \end{bmatrix}. \tag{3.123}$$

Substituting $\zeta = 1/2$ and $\eta = 1/2$ into Equation (3.120)

$$\frac{\partial N_1}{\partial \zeta} = -\frac{1}{8} \quad and \quad \frac{\partial N_1}{\partial \eta} = -\frac{1}{8}. \tag{3.124}$$

Substituting into Equation (3.115)

$$\left\{ \begin{array}{c} \frac{\partial N_1}{\partial x} \\ \frac{\partial N_1}{\partial y} \end{array} \right\} = -\frac{1}{20} \left\{ \begin{array}{c} 1 \\ 1 \end{array} \right\}. \tag{3.125}$$

In a similar fashion all other nodal derivatives can be calculated.

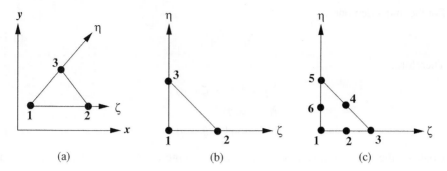

Figure 3.19 Isoparametric transformation of a single triangular element: (a) global; (b) local – linear; and (c) local – quadratic.

(b) Quadratic variation
The shape function at node 1 is

$$N_1 = -\frac{1}{4}(1 - \zeta)(1 - \eta)(\zeta + \eta + 1). \tag{3.126}$$

The derivatives with respect to transformed coordinates are

$$\frac{\partial N_1}{\partial \zeta} = \frac{3}{16} \quad and \quad \frac{\partial N_1}{\partial \eta} = \frac{3}{16}. \tag{3.127}$$

The derivatives with respect to global coordinates are

$$\left\{ \begin{array}{c} \frac{\partial N_1}{\partial x} \\ \frac{\partial N_1}{\partial y} \end{array} \right\} = \frac{1}{1120} \left\{ \begin{array}{c} 3 \\ 6 \end{array} \right\}. \tag{3.128}$$

Other derivatives can be established in a similar manner.

It is a simple matter to transform the area coordinate system for triangular elements (L_i, $i = 1, 2, 3$) to the $\zeta - \eta$ coordinates. The shape functions for the three-node linear triangle can be expressed in the ζ and η coordinate system as shown in Figure 3.19, that is,

$$\begin{aligned} N_1 &= L_1 = 1 - \zeta - \eta \\ N_2 &= L_2 = \zeta; 0 \leq \zeta \leq 1 \\ N_3 &= L_3 = \eta; 0 \leq \eta \leq 1. \end{aligned} \tag{3.129}$$

For a quadratic triangle with six nodes, the shape functions at the corner codes are

$$N_i = L_i(2L_i - 1) \quad i = 1,3,5. \tag{3.130}$$

Substituting Equation (3.129) into Equation (3.130),

$$\begin{aligned} N_1 &= [2(1 - \zeta - \eta) - 1](1 - \zeta - \eta) \\ N_3 &= \zeta(2\zeta - 1) \\ N_5 &= \eta(2\eta - 1). \end{aligned} \tag{3.131}$$

For the mid-side nodes,

$$N_i = 4L_iL_j. \tag{3.132}$$

Therefore,

$$N_2 = 4\zeta(1 - \zeta - \eta)$$
$$N_4 = 4\zeta\eta$$
$$N_6 = 4\eta(1 - \zeta - \eta). \tag{3.133}$$

Consider the linear triangular element shown in Figure 3.19(a) and express coordinates as

$$x(L_1, L_2) = N_1(L_1, L_2)x_1 + N_2(L_1, L_2)x_2 + N_3(L_1, L_2)x_3$$
$$y(L_1, L_2) = N_1(L_1, L_2)y_1 + N_2(L_1, L_2)y_2 + N_3(L_1, L_2)y_3. \tag{3.134}$$

Where x_1, x_2, x_3, y_1, y_2 and y_3 are the global coordinates of the three-noded triangular element, which are used for representing the geometry. Replacing the shape functions by the area coordinate gives

$$x(L_1, L_2) = x_1L_1 + x_2L_2 + x_3(1 - L_1 - L_2)$$
$$y(L_1, L_2) = y_1L_1 + y_2L_2 + y_3(1 - L_1 - L_2). \tag{3.135}$$

The components of the Jacobian matrix are

$$[\mathbf{J}] = \begin{bmatrix} \frac{\partial x}{\partial L_1} & \frac{\partial y}{\partial L_1} \\ \frac{\partial x}{\partial L_2} & \frac{\partial y}{\partial L_2} \end{bmatrix} = \begin{bmatrix} (x_1 - x_3) & (y_1 - y_3) \\ (x_2 - x_3) & (y_2 - y_3) \end{bmatrix}. \tag{3.136}$$

The determinant of the Jacobian matrix is

$$det[\mathbf{J}] = (x_1 - x_3)(y_2 - y_3) - (x_2 - x_3)(y_1 - y_3) = 2A, \tag{3.137}$$

where A is the area of the element. The inverse of the Jacobian matrix is

$$[\mathbf{J}]^{-1} = \frac{1}{det[\mathbf{J}]}\begin{bmatrix} (y_2 - y_3) & -(y_1 - y_3) \\ -(x_2 - x_3) & (x_1 - x_3) \end{bmatrix} = \frac{1}{2A}\begin{bmatrix} (y_2 - y_3) & -(y_1 - y_3) \\ -(x_2 - x_3) & (x_1 - x_3) \end{bmatrix}. \tag{3.138}$$

Finally the derivatives in global coordinates are written as

$$\begin{Bmatrix} \frac{\partial N_1}{\partial x} \\ \frac{\partial N_1}{\partial y} \end{Bmatrix} = [\mathbf{J}]^{-1}\begin{Bmatrix} \frac{\partial N_1}{\partial L_1} \\ \frac{\partial N_1}{\partial L_2} \end{Bmatrix}. \tag{3.139}$$

Example 3.2.6   *Calculate $\partial N_4/\partial x$ and $\partial N_4/\partial y$ at a point $(1, 4)$ for a quadratic triangle element shown in Figure 3.20 using local coordinates.*
    The coordinates are expressed as

$$x = x_1L_1 + x_2L_2 + x_3L_3$$
$$y = y_1L_1 + y_2L_2 + y_3L_3. \tag{3.140}$$

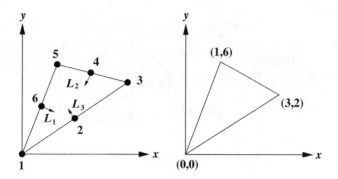

Figure 3.20   Triangular elements.

After substituting the coordinates of the three points, we have

$$x = 3L_2 + L_3$$
$$y = 2L_2 + 6L_3. \tag{3.141}$$

The determinant of the Jacobian matrix is (Equation (3.137)):

$$det[\mathbf{J}] = (-1)(-4) - (2)(-6) = 16. \tag{3.142}$$

The inverse of the Jacobian is therefore (Equation (3.138)):

$$[\mathbf{J}]^{-1} = \frac{1}{16}\begin{bmatrix} -4 & 6 \\ -2 & -1 \end{bmatrix}. \tag{3.143}$$

The shape function N_4 is given by $4\,L_2L_3 = 4L_2(1 - L_1 - L_2)$.

$$\begin{Bmatrix} \dfrac{\partial N_4}{\partial x} \\ \dfrac{\partial N_4}{\partial y} \end{Bmatrix} = [\mathbf{J}]^{-1} \begin{Bmatrix} \dfrac{\partial N_4}{\partial L_1} \\ \dfrac{\partial N_4}{\partial L_2} \end{Bmatrix} = \begin{Bmatrix} -0.5L_2 + 1.5L_3 \\ 0.75L_2 - 0.25L_3 \end{Bmatrix}. \tag{3.144}$$

To determine the local coordinates corresponding to $(x, y) = (1, 4)$, we have the following three equations (Eq. 3.141):

$$3L_2 + L_3 = 1$$
$$2L_2 + 6L_3 = 4$$
$$L_1 + L_2 + L_3 = 1, \tag{3.145}$$

which give

$$L_1 = \frac{1}{4}$$
$$L_2 = \frac{1}{8}$$
$$L_3 = \frac{5}{8}. \tag{3.146}$$

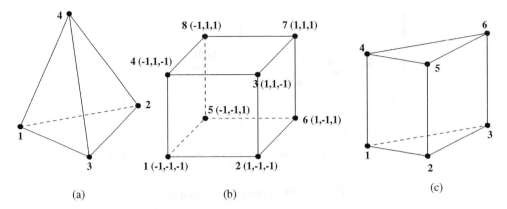

Figure 3.21   Three-dimensional elements.

Substituting into Equation (3.144) gives

$$\left\{ \begin{array}{c} \dfrac{\partial N_4}{\partial x} \\[2mm] \dfrac{\partial N_4}{\partial y} \end{array} \right\} = \left\{ \begin{array}{c} \dfrac{7}{8} \\[2mm] \dfrac{-1}{16} \end{array} \right\}. \tag{3.147}$$

Similarly, other derivatives can also be calculated.

3.2.8   Three-dimensional Elements

The amount of data required to establish the computational domain and boundary conditions become significantly greater in three dimensions than for two-dimensional problems. It is therefore obvious that the amount of computational work/cost increases by a considerable extent. Therefore, appropriate three-dimensional elements need to be used. The tetrahedron and brick-shaped hexahedron elements are developed (Figure 3.21) in this section, which are extensions of the linear triangle and quadrilateral elements in two dimensions.

The linear temperature representation for a tetrahedron element (three-dimensional linear element) is given by

$$T = \alpha_1 + \alpha_2 x + \alpha_3 y + \alpha_4 z. \tag{3.148}$$

As discussed previously for 2D elements, the constants of Equation (3.148) can be determined and may be written in the following form:

$$T = N_1 T_1 + N_2 T_2 + N_3 T_3 + N_4 T_4, \tag{3.149}$$

where

$$N_i = \frac{1}{6V}(a_i + b_i x + c_i y + d_i z) \quad \text{with} \quad i = 1, 2, 3, 4. \tag{3.150}$$

The volume of the tetrahedron is expressed as

$$6V = det \begin{bmatrix} 1 & x_1 & y_1 & z_1 \\ 1 & x_2 & y_2 & z_2 \\ 1 & x_3 & y_3 & z_3 \\ 1 & x_4 & y_4 & z_4 \end{bmatrix}.$$ (3.151)

Also, note that

$$\frac{\partial N_1}{\partial x} = \frac{b_1}{6V}$$

$$\frac{\partial N_1}{\partial y} = \frac{c_1}{6V}$$

$$\frac{\partial N_1}{\partial z} = \frac{d_1}{6V}.$$ (3.152)

Therefore, the gradient matrix of the shape functions can be written as

$$[\mathbf{B}] = \frac{1}{6V} \begin{bmatrix} b_1 & b_2 & b_3 & b_4 \\ c_1 & c_2 & c_3 & c_4 \\ d_1 & d_2 & d_3 & d_4 \end{bmatrix},$$ (3.153)

where

$$a_1 = det \begin{bmatrix} x_2 & y_2 & z_2 \\ x_3 & y_3 & z_3 \\ x_4 & y_4 & z_4 \end{bmatrix}$$ (3.154)

$$b_1 = -det \begin{bmatrix} 1 & y_2 & z_2 \\ 1 & y_3 & z_3 \\ 1 & y_4 & z_4 \end{bmatrix}$$ (3.155)

$$c_1 = -det \begin{bmatrix} x_2 & 1 & z_2 \\ x_3 & 1 & z_3 \\ x_4 & 1 & z_4 \end{bmatrix}$$ (3.156)

$$d_1 = -det \begin{bmatrix} x_2 & y_2 & 1 \\ x_3 & y_3 & 1 \\ x_4 & y_4 & 1 \end{bmatrix}.$$ (3.157)

Similarly, other terms in Equation (3.153) can also be determined using cyclic permutation. The a terms are seldom used in the calculations. We therefore summarize all the terms, except the a terms, as follows:

b terms

$$b_1 = -(y_2 - y_4)(z_3 - z_4) + (y_3 - y_4)(z_2 - z_4)$$
$$b_2 = -(y_3 - y_4)(z_1 - z_4) + (y_1 - y_4)(z_3 - z_4)$$
$$b_3 = -(y_1 - y_4)(z_2 - z_4) + (y_2 - y_4)(z_1 - z_4)$$
$$b_4 = -(b_1 + b_2 + b_3).$$ (3.158)

c terms

$$c_1 = -(x_3 - x_4)(z_2 - z_4) + (x_2 - x_4)(z_3 - z_4)$$
$$c_2 = -(x_1 - x_4)(z_3 - z_4) + (x_3 - x_4)(z_1 - z_4)$$
$$c_3 = -(x_2 - x_4)(z_1 - z_4) + (x_1 - x_4)(z_2 - z_4)$$
$$c_4 = -(c_1 + c_2 + c_3). \tag{3.159}$$

d terms

$$d_1 = -(x_2 - x_4)(y_3 - y_4) + (x_3 - x_4)(y_2 - y_4)$$
$$d_2 = -(x_3 - x_4)(y_1 - y_4) + (x_1 - x_4)(y_3 - y_4)$$
$$d_3 = -(x_1 - x_4)(y_2 - y_4) + (x_2 - x_4)(y_1 - y_4)$$
$$d_4 = -(d_1 + d_2 + d_3). \tag{3.160}$$

A volume coordinate system for the tetrahedron can be established in a similar manner as were the area coordinates for a triangle. In the tetrahedron, four distance ratios are used, each normal to a face L_1, L_2, L_3 and L_4.

Note that $L_1 + L_2 + L_3 + L_4 = 1$.

The linear shape functions are related to the volume coordinates as follows:

$$N_1 = L_1; N_2 = L_2; N_3 = L_3 \quad \text{and} \quad N_4 = L_4. \tag{3.161}$$

The volume integrals can easily be evaluated from the relationship,

$$\int_V L_1^a L_2^b L_3^c L_4^d dV = \frac{a!b!c!d!}{(a+b+c+d+3)!} 6V. \tag{3.162}$$

For a quadratic tetrahedron,

$$T = \alpha_1 + \alpha_2 x + \alpha_3 y + \alpha_4 z + \alpha_5 x^2 + \alpha_6 y^2 + \alpha_7 z^2 + \alpha_8 xy + \alpha_9 yz + \alpha_{10} zx. \tag{3.163}$$

Therefore, ten nodes will exist in quadratic tetrahedron as shown in Figure 3.22. The element may also have curved surfaces on the boundaries. As before, the temperature distribution

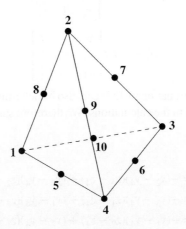

Figure 3.22   Quadratic tetrahedral element.

can be rewritten in terms of the shape functions as

$$T = N_1 T_1 + N_2 T_2 + N_3 T_3 + N_4 T_4 + N_5 T_5$$
$$+ N_6 T_6 + N_7 T_7 + N_8 T_8 + N_9 T_9 + N_{10} T_{10}. \tag{3.164}$$

The shape functions can be expressed in terms of local coordinates as

$$N_1 = L_1(2L_1 - 1)$$
$$N_2 = L_2(2L_2 - 1)$$
$$N_3 = L_3(2L_3 - 1)$$
$$N_4 = L_4(2L_4 - 1)$$
$$N_5 = 4L_4 L_1$$
$$N_6 = 4L_3 L_4$$
$$N_7 = 4L_3 L_2$$
$$N_8 = 4L_1 L_2$$
$$N_9 = 4L_2 L_4$$
$$N_{10} = 4L_1 L_3. \tag{3.165}$$

The brick, or hexahedron element as shown in Figure 3.21(b), is a simple element, which is easy to visualize when the domain is discretized. The bilinear interpolation function is

$$T = \alpha_1 + \alpha_2 x + \alpha_3 y + \alpha_4 z + \alpha_5 xy + \alpha_6 yz + \alpha_7 zx + \alpha_8 xyz, \tag{3.166}$$

which can be written as

$$T = \sum_{i=1}^{8} N_i T_i, \tag{3.167}$$

where

$$N_i = \frac{1}{8}(1 + \zeta \zeta_i)(1 + \eta \eta_i)(1 + \rho \rho_i), \tag{3.168}$$

where ζ_i, η_i and ρ_i are the local coordinates.

For a quadratic 20-noded hexahedron, which can represent arbitrary solids with curved surfaces as shown in Figure 3.23, the shape functions can be written as follows:

Corner nodes

$$N_i = \frac{1}{8}(1 + \zeta \zeta_i)(1 + \eta \eta_i)(1 + \rho \rho_i)(\zeta \zeta_i + \eta \eta_i + \rho \rho_i - 1) \quad \text{with} \quad i = 1, 2, ..8. \tag{3.169}$$

Mid-side nodes

$$N_i = \frac{1}{4}(1 - \zeta^2)(1 + \eta \eta_i)(1 + \rho \rho_i) \quad \text{with} \quad i = 9, 13, 15, 11$$

$$N_i = \frac{1}{4}(1 - \eta^2)(1 + \zeta \zeta_i)(1 + \rho \rho_i) \quad \text{with} \quad i = 10, 14, 16, 12$$

$$N_i = \frac{1}{4}(1 - \rho^2)(1 + \zeta \zeta_i)(1 + \eta \eta_i) \quad \text{with} \quad i = 18, 19, 20, 17. \tag{3.170}$$

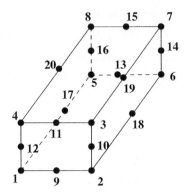

Figure 3.23   Twenty-node hexahedral element.

The shape functions for a linear pentahedran element (which is used in cylindrical geometries) can be generated from the product of triangular and one-dimensional interpolation
functions (refer to Figure 3.21(c)).

$$N_1 = \frac{1}{2}L_1(1 - w)$$

$$N_2 = \frac{1}{2}L_2(1 - w)$$

$$N_3 = \frac{1}{2}L_3(1 - w)$$

$$N_4 = \frac{1}{2}L_1(1 + w)$$

$$N_5 = \frac{1}{2}L_2(1 + w)$$

$$N_6 = \frac{1}{2}L_3(1 + w), \tag{3.171}$$

where $w = -1$ at the bottom surface and 1 at the top surface. In conclusion, isoparametric
elements are very useful as they can be used for modelling irregular solids and the element
can be mapped to a unit cube.

3.3   Formulation (Element Characteristics)

After briefly describing the various elements used in the context of finite element analysis, we
shall now focus our attention on determining the element characteristics that is, the relation
between the nodal unknowns and the corresponding loads or forces in the form of the following
matrix equation, viz.,

$$[\mathbf{K}]\{\mathbf{T}\} = \{\mathbf{f}\}, \tag{3.172}$$

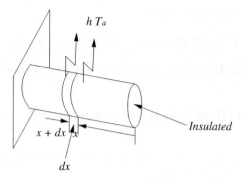

Figure 3.24   A fin problem.

where $[\mathbf{K}]$ is the thermal stiffness matrix, $\{\mathbf{T}\}$ is the vector of unknown temperatures and $\{\mathbf{f}\}$ is the thermal load, or forcing vector.

Several methods are available for the determination of the approximate solution to a given problem. We shall consider three methods in the first instance. They are:

- Ritz method (Heat balance integral),

- Rayleigh Ritz (Variational) method and

- Weighted residual methods.

In order to illustrate the above methods, we shall consider a one-dimensional fin problem as shown in Figure 3.24.

Heat balance on the differential volume of length dx as shown in Figure 3.24 gives;

$$-kA\frac{dT}{dx}|_x = hPdx(T - T_a) - kA\frac{dT}{dx}|_{x+dx}$$

$$= hPdx(T - T_a) - kA\frac{dT}{dx}|_x - kA\frac{d^2T}{dx^2}dx, \tag{3.173}$$

where k is the thermal conductivity, A is the cross-sectional area, h is the heat transfer coefficient, P is the perimeter and the suffix a represents atmospheric condition. Simplifying, the governing differential equation becomes

$$kA\frac{d^2T}{dx^2} - hP(T - T_a) = 0 \tag{3.174}$$

and the boundary conditions are:

$$\text{at } x = 0, dT/dx = 0 \text{ (tip) and at } x = L, T = T_b \text{ (base)}. \tag{3.175}$$

If we substitute $(T - T_a) = \theta, \zeta = x/L, hP/kA = m^2$ and $m^2L^2 = \mu^2$ into Equation (3.174), we obtain

$$\frac{d^2\theta}{d\zeta^2} - \mu^2\theta = 0, \tag{3.176}$$

with the following new boundary conditions

$$\text{At} \quad \zeta = 0, d\theta/d\zeta = 0 \quad \text{and at} \quad \zeta = 1, \theta = \theta_b, \tag{3.177}$$

We employ Equations (3.176) and (3.177) in the following sections.

3.3.1 Ritz Method (Heat Balance Integral Method – Goodman's Method)

An approximate solution of Equation (3.176) along with the appropriate boundary conditions, may be found using the following function:

$$T \approx \overline{T} = \overline{T}(x, a_1, a_2, ...a_n) = \sum_{i=1}^{n} a_i N_i(x), \tag{3.178}$$

which has one or more unknown parameters $a_1, a_2...a_n$ and functions $N_i(x)$ that exactly satisfy the boundary conditions given by Equation (3.177). The functions $N_i(x)$ are referred to as trial functions, which must be continuous and differentiable up to the highest order present in the integral form of the governing equation

The approximations may be carried out using one, two or n terms as follows:

$$\overline{T} = a_1 N_1(x)$$
$$\overline{T} = a_1 N_1(x) + a_2 N_2(x) \tag{3.179}$$

or

$$\overline{T} = \sum_{i=1}^{n} a_i N_i(x). \tag{3.180}$$

When approximation $\overline{T}$ is substituted into the governing differential equation, it is not satisfied exactly leaving a residual R. The exact solution results when the residual R is zero for all points in the domain. In approximate solution methods the residual is not in general zero everywhere in the domain even though it may be zero at some preferred points.

Let us select a profile which satisfies the boundary conditions (Equation (3.177)) in the global sense. By inspection, we find that

$$\frac{\theta(\zeta)}{\theta_b} = 1 - (1 - \zeta^2)B \tag{3.181}$$

satisfies the boundary conditions where B is an unknown parameter to be determined.

In the Ritz method, we insert the approximate profile in the governing differential equation, Equation (3.176), and then the integral of the residual R over the domain is equated to zero to determine the constant B, that is,

$$\int_0^1 \left(\frac{\partial^2 \theta(\zeta)}{\partial \zeta^2} - \mu^2 \theta \right) d\zeta = 0. \tag{3.182}$$

Differentiating Equation (3.181) gives

$$\frac{d^2\theta(\zeta)}{d\zeta^2} = 2B\theta_b. \tag{3.183}$$

Substituting Equation (3.183) into Equation (3.182) we have

$$\int_0^1 [2B - \mu^2(1 - \{1 - \zeta^2\}B)]\theta_b d\zeta = \left[2\theta_b B\zeta - \mu^2\theta_b\left(\zeta - B\zeta + \frac{B\zeta^3}{3}\right)\right]_0^1$$

$$= 2B\theta_b - \left(1 - B + \frac{B}{3}\right)\mu^2\theta_b$$

$$= 0, \tag{3.184}$$

which gives

$$B = \frac{\frac{\mu^2}{2}}{1 + \frac{\mu^2}{3}}. \tag{3.185}$$

Substituting Equation (3.185) into (3.181) gives following solution:

$$\frac{\theta(\zeta)}{\theta_b} = 1 - (1 - \zeta^2)\frac{\frac{\mu^2}{2}}{1 + \frac{\mu^2}{3}}. \tag{3.186}$$

For the case of a stainless steel fin ($k = 16.66\,$W/m °C of circular section with a diameter of 2 cm and length of 10 cm exposed to a convection environment with $h = 25\,$W/m² °C and $\mu^2 = 3.0$ and $m^2 = 300$, the approximate solution is

$$\frac{\bar{\theta}(\zeta)}{\theta_b} = 1 - \frac{3}{4}(1 - \zeta^2), \tag{3.187}$$

where the exact solution is

$$\frac{\theta(\zeta)}{\theta_b} = \frac{cosh[m(L - x)]}{cosh(mL)}. \tag{3.188}$$

Note that the x is taken from the tip of the fin as shown in Figure 3.24. The comparison between the exact and approximate solutions is given in Figure 3.25. As may be seen the temperatures match perfectly at the base at $x = 1$ but differ close to the insulated end at $x = 0$.

3.3.2 Rayleigh–Ritz Method (Variational Method)

In the case of the variational method we make use of an important theorem from the theory of the calculus of variations which states that:

"The function $T(x)$ that extremizes the variational integral corresponding to the governing differential equation (called Euler or Euler-Lagrange equation) is the solution of the original governing differential equation and boundary conditions." This implies that the solution

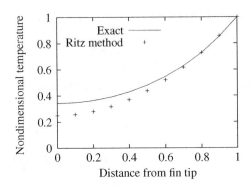

Figure 3.25   Comparison between Ritz method and exact solution.

obtained is unique, which is the case for well-posed problems. Thus, the first step is to determine the variational integral I, which corresponds to the governing differential equation and its boundary conditions. The differential equation is Equation (3.176),

$$\frac{d^2\theta}{d\zeta^2} - \mu^2\theta = 0, \tag{3.189}$$

with the following boundary conditions:

$$\frac{d\theta(0)}{d\zeta} = 0 \quad \text{and} \quad \theta(1) = \theta_b. \tag{3.190}$$

Using the differential equation as the Euler-Lagrange equation, we can write

$$\delta I = \int_0^1 \left(\frac{d^2\theta}{d\zeta^2} - \mu^2\theta\right)\delta\theta d\zeta = 0. \tag{3.191}$$

Integrating by parts gives

$$\left[\frac{d\theta}{d\zeta}\delta\theta\right]_0^1 - \int_0^1 \left(\frac{d\theta}{d\zeta}\right)\frac{d}{d\zeta}(\delta\theta)d\zeta - \mu^2 \int_0^1 \theta\delta\theta d\zeta = 0. \tag{3.192}$$

Using the relations

$$\frac{d}{d\zeta}(\delta\theta) = \delta\left(\frac{d\theta}{d\zeta}\right)$$

$$\frac{d\theta}{d\zeta}\delta\left(\frac{d\theta}{d\zeta}\right) = \frac{1}{2}\delta\left(\frac{d\theta}{d\zeta}\right)^2$$

$$\text{and} \quad \theta\delta\theta = \frac{1}{2}\delta\theta^2, \tag{3.193}$$

Equation (3.192) is simplified to the following;

$$\left[\frac{d\theta}{d\zeta}\delta\theta\right]_0^1 - \frac{1}{2}\delta\int_0^1 \left[\left(\frac{d\theta}{d\zeta}\right)^2 + \mu^2\theta^2\right] d\zeta = 0. \tag{3.194}$$

When we apply the boundary conditions (Equation (3.190)), the first term of the above equation becomes zero. Thus, the variational formulation for the given problem is

$$\delta \int_0^1 \frac{1}{2} \left[\left(\frac{d\theta}{d\zeta} \right)^2 + \mu^2 \theta^2 \right] dx = 0 \tag{3.195}$$

and the corresponding variational integral is given by

$$I = \int_0^1 \frac{1}{2} \left[\left(\frac{d\theta}{d\zeta} \right)^2 + \mu^2 \theta^2 \right] d\zeta. \tag{3.196}$$

Now, the profile which minimizes the integral Equation (3.196) is the solution to the differential Equation (3.189) with its boundary conditions given by Equation (3.190).

Let us assume the same profile as before (Equation (3.181)) and substitute into Equation (3.196), that is,

$$I = \int_0^1 \frac{1}{2} \theta_b^2 \{ (2B\zeta)^2 + \mu^2 [1 - (1 - \zeta^2)B]^2 \} d\zeta. \tag{3.197}$$

After integration and substitution of limits, we have

$$I = \frac{1}{2} \theta_b^2 \left\{ B^2 \left(\frac{4}{3} + \mu^2 - \frac{2}{3} \mu^2 + \frac{1}{5} \mu^2 \right) + \mu^2 + B \left(-2\mu^2 + \frac{2}{3} \mu^2 \right) \right\}. \tag{3.198}$$

For I to be minimum, $\frac{\partial I}{\partial B} = 0$, that is,

$$\frac{\partial I}{\partial B} = \frac{1}{2} \theta_b^2 \left\{ 2B \left(\frac{4}{3} + \frac{8}{15} \mu^2 \right) + \left(-\frac{4}{3} \mu^2 \right) \right\} = 0, \tag{3.199}$$

which gives

$$B = \frac{\frac{\mu^2}{2}}{1 + \frac{2}{5} \mu^2}. \tag{3.200}$$

Substituting into Equation (3.181) gives the solution as

$$\frac{\theta(\zeta)}{\theta} = 1 - (1 - \zeta^2) \frac{\frac{\mu^2}{2}}{1 + \frac{2}{5} \mu^2}. \tag{3.201}$$

For the fin problem of the previous subsection with $\mu^2 = 3$ and $m^2 = 300$, the comparison between the variational method and the exact solution is shown in Figure 3.26. As seen the agreement between the solutions is better than the agreement between the exact and Ritz solutions.

It can be observed from the variational Integral Equation (3.196) that it contains only a first-order derivative even though the original differential Equation (3.189) contains a second-order derivative.

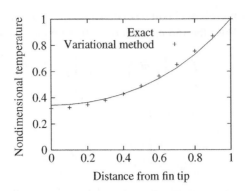

Figure 3.26   Comparison between variational method and exact solution.

3.3.3   The Method of Weighted Residuals

For those differential equations for which we cannot write a variational formulation, there is a need to find an alternative method of formulation. The method of weighted residual provides a very powerful approximate solution procedure that is applicable to a wide variety of problems and thus makes it unnecessary to search for variational formulations in order to apply the finite element method for these problems.

Let the governing equations be represented by

$$L(T) = 0 \quad \text{in} \quad \Omega. \tag{3.202}$$

Let

$$T \approx \overline{T} = \sum_{i=1}^{n} a_i N_i(x). \tag{3.203}$$

Substitution of the above equation into Equation (3.202) results in,

$$L(\overline{T}) \neq 0$$
$$= R \text{ (residual)}. \tag{3.204}$$

The method of weighted residual requires that the parameters $a_1, a_2...a_n$ be determined by satisfying

$$\int_{\Omega} w_i(x) R dx = 0 \quad \text{with} \quad i = 1, 2....n, \tag{3.205}$$

where the functions $w_i(x)$ are the n arbitrary weighting functions. There are an infinite number of choices for $w_i(x)$ but four particular functions are most often used. Depending on the choice of the weighting functions, different names are given

- *Collocation:* $w_i = \delta(x - x_i)$, here δ is the Dirac delta function:

$$\int_{\Omega} R\delta(x - x_i) dx = R_{x=x_i} = 0. \tag{3.206}$$

- *Subdomain:* $w_i = 1$ (Note the subdomain Ω_i in the integration):

$$\int_{\Omega_i} R dx = 0 \text{ with } i = 1, 2, ...n. \tag{3.207}$$

- *Galerkin:* $w_i(x) = N_i(x)$, that is, the same trial functions as used in $T(x)$:

$$\int_{\Omega} R N_i(x) dx = 0 \text{ with } i = 1, 2, ..n. \tag{3.208}$$

- *Least Squares:* $w_i = \partial R / \partial a_i$:

$$\int_{\Omega} R \frac{\partial R}{\partial a_i} dx = 0 \text{ with } i = 1, 2, ...n. \tag{3.209}$$

For illustration purposes the fin problem is re-solved with each of the above methods.

3.3.3.1 Collocation Method

The weight is $w_i = \delta(x - x_i)$. Let $\zeta_i = 1/2$ as there is only one unknown in the fin problem. Rewriting the equation in collocation form in the nondimensional coordinates gives us the following:

$$\int_0^1 \left[\frac{d^2\bar{\theta}}{d\zeta^2} - \mu^2 \bar{\theta} \right] \delta(\zeta - \zeta_i) d\zeta = 0. \tag{3.210}$$

From the above equation we can write

$$\left[\frac{d^2\bar{\theta}}{d\zeta^2} \quad \mu^2\bar{\theta} \right]_{\zeta_i = \frac{1}{2}} = 0. \tag{3.211}$$

Substituting Equation (3.181) into (3.211), with $\zeta = 1/2$, we have

$$2B - \mu^2 \left[1 - \frac{3}{4}B \right] = 0, \tag{3.212}$$

which gives

$$B = \frac{\left(\frac{\mu^2}{2} \right)}{1 + \frac{3}{8}\mu^2}. \tag{3.213}$$

Substituting into Equation (3.181), the solution is obtained as

$$\frac{\bar{\theta}(\zeta)}{\theta_b} = 1 - (1 - \zeta^2) \frac{\left(\frac{\mu^2}{2} \right)}{1 + \frac{3}{8}\mu^2}. \tag{3.214}$$

For a problem with $\mu^2 = 3$, then

$$\frac{\bar{\theta}(\zeta)}{\theta_b} = 1 - \frac{12}{17}(1 - \zeta^2). \tag{3.215}$$

3.3.3.2 Subdomain Method

The weighting function $w_i = 1$ which results in the subdomain formulation being

$$\int_0^1 (1) \left[\frac{d^2\overline{\theta}}{d\zeta^2} - \mu^2\overline{\theta} \right] d\zeta = 0. \tag{3.216}$$

Substituting Equation (3.181) and integrating, we get

$$B = \frac{\frac{\mu^2}{2}}{1 + \frac{\mu^2}{3}}. \tag{3.217}$$

The solution becomes

$$\frac{\overline{\theta}(\zeta)}{\theta_b} = 1 - (1 - \zeta^2) \frac{\left(\frac{\mu^2}{2} \right)}{1 + \frac{\mu^2}{3}}. \tag{3.218}$$

For the particular case of $\mu^2 = 3$

$$\frac{\overline{\theta}(\zeta)}{\theta_b} = 1 - \frac{3}{4}(1 - \zeta^2). \tag{3.219}$$

The result from the subdomain method coincides with the heat balance integral solution (Ritz method) as in the present case integration is carried out over the entire domain in view of only one constant being involved.

3.3.3.3 Galerkin Method

This is one of the most important methods used in finite element analysis. The weight function is $N_i(x) = (1 - \zeta^2)$. The Galerkin formulation of the fin equation is

$$\int_0^1 N_i(x) \left[\frac{d^2\overline{\theta}}{d\zeta^2} - \mu^2\overline{\theta} \right] d\zeta = 0. \tag{3.220}$$

Substituting Equation (3.181) and integrating, we obtain

$$2B - \frac{2B}{3} + \mu^2 \left(\frac{8}{15}B \right) - \frac{2\mu^2}{3} = 0 \tag{3.221}$$

and

$$B = \frac{\frac{\mu^2}{2}}{1 + \frac{2}{5}\mu^2}. \tag{3.222}$$

Thus, the solution is

$$\frac{\overline{\theta}(\zeta)}{\theta_b} = 1 - (1 - \zeta^2) \frac{\left(\frac{\mu^2}{2} \right)}{1 + \frac{2}{5}\mu^2}. \tag{3.223}$$

It can be observed that the solution using Galerkin's method is exactly the same as that obtained by the variational method. It can also be shown that the variational and Galerkin methods give the same results provided the problem has a classical variational statement. In fact, when the finite element formulation is carried out on a quasi-harmonic equation using both the variational and Galerkin methods, the same results are obtained since a classical variational principle does exist for a quasi-harmonic equation.

3.3.3.4 Least-squares Method

In this case the minimization of the error is carried out in a least-squares sense, that is,

$$\frac{1}{2}\frac{\partial}{\partial B}\int_{\Omega} R^2 dx = 0, \tag{3.224}$$

which can also be written as

$$\int_{\Omega} \frac{\partial R}{\partial B} R dx = 0, \tag{3.225}$$

where the weighting function here is

$$w_i(x) = \frac{\partial R}{\partial B} \tag{3.226}$$

From Equation (3.224), error E in the transformed coordinate is given by

$$E = \int_0^1 R^2 d\zeta$$

$$= \int_0^1 \left[\frac{d^2\overline{\theta}}{d\zeta^2} - \mu^2\overline{\theta}\right]^2 d\zeta. \tag{3.227}$$

Substituting Equation (3.181) into Equation (3.227) and integrating we have

$$E = \left[4B^2 - 4B\mu^2\left(1 - \frac{2}{3}B\right) + \mu^4 - 2B\mu^4\left(\frac{2}{3}\right) + B^2\left(\frac{8}{15}\right)\mu^4\right]\theta_b. \tag{3.228}$$

The error is minimized by satisfying $\partial E/\partial B = 0$, that is,

$$\frac{\partial E}{\partial B} = 8B - \frac{4\mu^4}{3} + \frac{16B\mu^4}{15} - 4\mu^2 + \frac{16B\mu^2}{3} = 0, \tag{3.229}$$

which gives

$$B = \frac{\frac{\mu^2}{2}\left(1 + \frac{\mu^2}{3}\right)}{1 + 2\mu^2\left(\frac{1}{3} + \frac{\mu^2}{15}\right)}. \tag{3.230}$$

Therefore, the solution is given by

$$\frac{\overline{\theta}(\zeta)}{\theta_b} = 1 - (1 - \zeta^2)\frac{\frac{\mu^2}{2}\left(1 + \frac{\mu^2}{3}\right)}{1 + 2\mu^2\left(\frac{1}{3} + \frac{\mu^2}{15}\right)}. \tag{3.231}$$

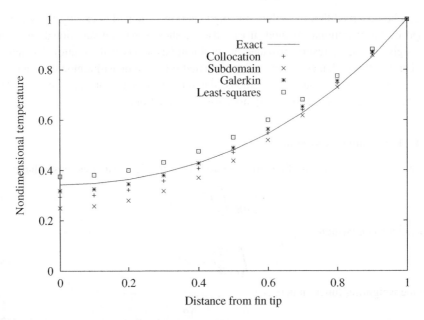

Figure 3.27   Comparison between various weighted residual methods and exact solution.

For the particular problem where $\mu^2 = 3$, then

$$\frac{\overline{\theta}(\zeta)}{\theta_b} = 1 - \frac{15}{21}(1 - \zeta^2). \tag{3.232}$$

Figure 3.27 shows the comparison between all the different weighted residual methods. As can be seen, the Galerkin method is the most accurate method.

3.3.4   Galerkin Finite Element Method

We shall solve the fin problem shown in Figure 3.24 by using the Galerkin finite element method and discretizing the domain into five linear elements with a total of six nodal points as shown in Figure 3.28. Unlike the methods discussed in the previous sections, we need no *a priori* assumption of temperature profile here.

For a linear element

$$\overline{\theta} = N_i \theta_i + N_j \theta_j \tag{3.233}$$

and the derivative in nondimensional coordinate may be written as

$$\frac{d\theta}{d\zeta} = \frac{dN_i}{d\zeta}\theta_i + \frac{dN_j}{d\zeta}\theta_j = -\frac{1}{\zeta_e}\theta_i + \frac{1}{\zeta_e}\theta_j, \tag{3.234}$$

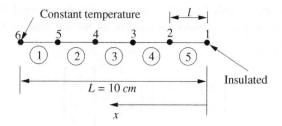

Figure 3.28   Heat dissipation from a fin (Figure 3.24). Spatial discretization. Nodes: 6; Elements: 5.

where $\zeta_e = l/L$ is the nondimensional element length. The Galerkin method requires that

$$\int_\zeta N_k \left(\frac{d^2\bar\theta}{d\zeta^2} - \mu^2\bar\theta \right) d\zeta = 0, \tag{3.235}$$

where the subscript k represents all the nodes in the domain. Integration of the above equation by parts for one element, with the weight being the first node of the element, results in the following:

$$\left[N_i \frac{d\theta}{d\zeta} \right]_0^{\zeta_e} - \int_0^{\zeta_e} \frac{dN_i}{d\zeta} \left[\frac{dN_i}{d\zeta} \quad \frac{dN_j}{d\zeta} \right] d\zeta \{\theta\} - \int_o^{\zeta_e} N_i \mu^2 (N_i\theta_i + N_j\theta_j) d\zeta. \tag{3.236}$$

In one dimension, the magnitude of $\tilde n$ is unity but the sign changes appropriately. Note the following (see Appendix C):

$$\int_0^{\zeta_e} N_i^2 d\zeta = \frac{2!0!\zeta_e}{(2+0+1)!} = \frac{\zeta_e}{3}$$

$$\int_0^{\zeta_e} N_i N_j d\zeta = \frac{1!1!\zeta_e}{(1+1+1)!} = \frac{\zeta_e}{6}. \tag{3.237}$$

For the first element, with N_i being the weight, Equation (3.236) simplifies to

$$\frac{1}{\zeta_e} \begin{bmatrix} 1 & -1 \end{bmatrix} \begin{Bmatrix} \theta_i \\ \theta_j \end{Bmatrix} + \frac{\mu^2\zeta_e}{6} \begin{bmatrix} 2 & 1 \end{bmatrix} \begin{Bmatrix} \theta_i \\ \theta_j \end{Bmatrix} - \begin{Bmatrix} \frac{d\theta}{d\zeta} \\ 0 \end{Bmatrix}. \tag{3.238}$$

Note that the gradient terms of Equation (3.236) become zero at node j as $N_i = 0$ at j. Now weighting the equation using N_j, we have

$$\frac{1}{\zeta_e} \begin{bmatrix} -1 & 1 \end{bmatrix} \begin{Bmatrix} \theta_i \\ \theta_j \end{Bmatrix} + \frac{\mu^2\zeta_e}{6} \begin{bmatrix} 1 & 2 \end{bmatrix} \begin{Bmatrix} \theta_i \\ \theta_j \end{Bmatrix} - \begin{Bmatrix} 0 \\ \frac{-d\theta}{d\zeta} \end{Bmatrix}. \tag{3.239}$$

In this case the gradient term disappears for node i as N_j is zero at node i. The element characteristics are given by combining Equation (3.238) and (3.239) as

$$\left\{ \frac{1}{\zeta_e} \begin{bmatrix} 1 & -1 \\ -1 & 1 \end{bmatrix} + \frac{\mu^2\zeta_e}{6} \begin{bmatrix} 2 & 1 \\ 1 & 2 \end{bmatrix} \right\} \begin{Bmatrix} \theta_i \\ \theta_j \end{Bmatrix} - \begin{Bmatrix} \frac{d\theta}{d\zeta} \\ -\frac{d\theta}{d\zeta} \end{Bmatrix}. \tag{3.240}$$

For the given problem with $\zeta_e = 0.2$, which is a nondimensional element length, l/L (Figure 3.28), and $\mu^2 = 3$ the element characteristics for the first element are derived as follows:

$$\begin{bmatrix} 5.2 & -4.9 \\ -4.9 & 5.2 \end{bmatrix} \begin{Bmatrix} \theta_i \\ \theta_j \end{Bmatrix} - \begin{Bmatrix} \dfrac{d\theta}{d\zeta} \\ -\dfrac{d\theta}{d\zeta} \end{Bmatrix}. \tag{3.241}$$

In a similar fashion we can write the element characteristics equation for all the other four elements. On assembling over all the five elements and making the resulting system equal to zero, we obtain (see Appendix D)

$$\begin{bmatrix} 5.2 & -4.9 & 0.0 & 0.0 & 0.0 & 0.0 \\ -4.9 & 10.4 & -4.9 & 0.0 & 0.0 & 0.0 \\ 0.0 & -4.9 & 10.4 & -4.9 & 0.0 & 0.0 \\ 0.0 & 0.0 & -4.9 & 10.4 & -4.9 & 0.0 \\ 0.0 & 0.0 & 0.0 & -4.9 & 10.4 & -4.9 \\ 0.0 & 0.0 & 0.0 & 0.0 & -4.9 & 5.2 \end{bmatrix} \begin{Bmatrix} \theta_1 \\ \theta_2 \\ \theta_3 \\ \theta_4 \\ \theta_5 \\ \theta_6 \end{Bmatrix} = \begin{Bmatrix} 0.0 \\ 0.0 \\ 0.0 \\ 0.0 \\ 0.0 \\ -\dfrac{d\theta}{d\zeta} \end{Bmatrix}, \tag{3.242}$$

where $\theta_1, \theta_2, \ldots \theta_6$ are the temperature values at all the six nodes. The assembly procedure has already been discussed in the previous chapter. Further details on the assembly procedure are given in Appendix D. Note that $\partial\theta/\partial\zeta$ at node 1 is zero due to the zero flux boundary condition (insulated) but we also have the boundary condition at $\zeta = 1$, as $\theta = 1$. The resulting nodal simultaneous equations can be written as

$$5.2\theta_1 - 4.9\theta_2 = 0.0$$
$$-4.9\theta_1 + 10.4\theta_2 - 4.9\theta_3 = 0.0$$
$$-4.9\theta_2 + 10.4\theta_3 - 4.9\theta_4 = 0.0$$
$$-4.9\theta_3 + 10.4\theta_4 - 4.9\theta_5 = 0.0$$
$$-4.9\theta_4 + 10.4\theta_5 - 4.9\theta_6 = 0.0$$
$$\theta_6 = 1.0. \tag{3.243}$$

Note that the last equation arises due to the constant temperature boundary condition at node 6. On solving the system of equations using Gaussian elimination (see Appendix A), we finally obtain all the θ values. Table 3.5 shows the comparison between the exact result and all the other computations from each of the different methods.

It can be observed from Table 3.5 that the methods used in conjunction with the assumed profile satisfying the boundary conditions for the entire domain are less accurate compared to the finite element method solution even with only five linear elements. It can also be observed that the nodal values in the finite element method solution are very close to those of the exact solution.

Table 3.5   Comparison of solutions obtained from different methods

Location (ζ)	Exact	FEM 5 linear elements	Collocation	Subdomain	Variational or Galerkin	Least squares
0.0	0.343	0.340	0.294	0.250	0.318	0.375
0.1	0.348	–	0.301	0.258	0.325	0.381
0.2	0.364	0.361	0.322	0.280	0.345	0.400
0.3	0.390	–	0.358	0.316	0.380	0.431
0.4	0.429	0.426	0.407	0.370	0.427	0.475
0.5	0.480	–	0.471	0.438	0.490	0.531
0.6	0.546	0.543	0.548	0.520	0.563	0.600
0.7	0.628	–	0.640	0.618	0.652	0.681
0.8	0.729	0.727	0.746	0.730	0.755	0.755
0.9	0.851	–	0.866	0.858	0.870	0.881
1.00	1.00	1.00	1.00	1.00	1.00	1.00

The flux $-\frac{d\theta}{d\zeta}$ at node 6 may now be calculated using the 6^{15} equation in Equation (3.242) as 1.64.

3.4   Formulation for the Heat Conduction Equation

In many practical situations, finding the temperature in a solid body is of vital importance in terms of the maximum allowable temperature, for example as in semiconductor devices. The consequences of increase in temperature include structural displacements and development of thermal stresses as in steam and gas turbines. In this section, we shall give the derivation of the finite element equations, both by the variational method as well as the Galerkin method, for the three-dimensional heat conduction equation of stationary systems under steady-state conditions.

The governing differential equation, rewritten here as given in Chapter 2 is

$$\frac{\partial}{\partial x}\left(k_x \frac{\partial T}{\partial x}\right) + \frac{\partial}{\partial y}\left(k_y \frac{\partial T}{\partial y}\right) + \frac{\partial}{\partial z}\left(k_z \frac{\partial T}{\partial z}\right) + G = 0, \tag{3.244}$$

with the following boundary conditions:

$$T = T_b \quad \text{on surface } S_1$$

$$k_x \frac{\partial T}{\partial x}\tilde{l} + k_y \frac{\partial T}{\partial y}\tilde{m} + k_z \frac{\partial T}{\partial z}\tilde{n} + q = 0 \quad \text{on surface } S_2$$

$$k_x \frac{\partial T}{\partial x}\tilde{l} + k_y \frac{\partial T}{\partial y}\tilde{m} + k_z \frac{\partial T}{\partial z}\tilde{n} + h(T - T_a) = 0 \quad \text{on surface } S_3, \tag{3.245}$$

where $\tilde{l}, \tilde{m}$ and $\tilde{n}$ are surface normals, h is the heat transfer coefficient, k is the thermal conductivity and q is the heat flux.

3.4.1 Variational Approach

The variational integral, I, corresponding to the above differential equation with its boundary conditions is given by (see Section 3.3.2)

$$I(T) = \frac{1}{2} \int_\Omega \left[k_x \left(\frac{\partial T}{\partial x} \right)^2 + k_y \left(\frac{\partial T}{\partial y} \right)^2 + k_z \left(\frac{\partial T}{\partial z} \right)^2 - 2GT \right] d\Omega$$

$$+ \int_{S_2} qT ds + \int_{S_3} \frac{1}{2} h(T - T_a)^2 ds, \tag{3.246}$$

The given domain Ω is divided into n number of finite elements with each element having r nodes. The temperature is expressed in each element by

$$T^e = \sum_{i=1}^r N_i T_i = [\mathbf{N}]\{\mathbf{T}\}, \tag{3.247}$$

where $[\mathbf{N}] = [N_i, N_j......N_r] =$ shape function matrix and

$$\{\mathbf{T}\} = \begin{Bmatrix} T_i \\ T_j \\ . \\ . \\ T_r \end{Bmatrix} \tag{3.248}$$

is the vector of nodal temperatures.

The finite element solution to the problem involves selecting the nodal values of T so as to make the function $I(T)$ stationary. In order to make $I(T)$ stationary, with respect to the nodal values of T, we require that

$$\delta I(T) = \sum_{i=1}^n \frac{\partial I}{\partial T_i} = 0, \tag{3.249}$$

where n is the total number of discrete values of T assigned to the solution domain. Since T_i are arbitrary, Equation (3.249) holds only if

$$\frac{\partial I}{\partial T_i} = 0 \quad \text{for} \quad i = 1, 2, ...n. \tag{3.250}$$

The functional $I(T)$ can be written as a sum of individual functions, defined for the assembly of elements, only if the shape functions giving piecewise representation of T obey certain continuity and compatibility conditions. These conditions will be discussed later in the text.

$$I(T) = \sum_{e=1}^M I^e(T^e), \tag{3.251}$$

where M is the total number of elements. Thus, instead of working with a functional defined over the whole solution region, our attention is now focused on a functional defined for the individual elements. Hence,

$$\delta I = \sum_{e=1}^{M} \delta I^e = 0, \tag{3.252}$$

where the variation I^e is taken only with respect to the r nodal values associated with the element e. that is,

$$\left\{ \frac{\partial I^e}{\partial T} \right\} = \frac{\partial I^e}{\partial T_j} = 0 \quad \text{with} \quad j = 1, 2, \dots r. \tag{3.253}$$

Equation (3.253) comprises a set of r equations that characterize the behavior of the element e. The fact that we can represent the functional for the assembly of elements as a sum of the functional for all individual elements provides the key to formulating individual element equations from a variational principle. The complete set of assembled finite element equations for the problem is obtained by adding all the derivatives of I, as given by Equation (3.253), for all the elements. We can write the complete set of equations as

$$\frac{\partial I}{\partial T_i} = \sum_{e=1}^{M} \frac{\partial I^e}{\partial T_i} = 0 \quad \text{with} \quad i = 1, 2, \dots n. \tag{3.254}$$

The problem is complete when the n set of equations are solved simultaneously for the n nodal values of T. We now give the details for formulating the individual finite element equations from a variational principle.

$$I^e = \frac{1}{2} \int_{\Omega} \left[k_x \left(\frac{\partial T^e}{\partial x} \right)^2 + k_y \left(\frac{\partial T^e}{\partial y} \right)^2 + k_z \left(\frac{\partial T^e}{\partial z} \right)^2 - 2GT^e \right] d\Omega$$
$$+ \int_{S_{2e}} qT^e ds + \int_{S_{3e}} \frac{1}{2} h(T^e - T_a)^2 ds \tag{3.255}$$

with

$$T^e = [N]\{T\} = [N_1, N_2, \dots N_r] \begin{Bmatrix} T_1 \\ T_2 \\ \dots \\ T_r \end{Bmatrix} = N_1 T_1 + N_2 T_2 + \dots N_r T_r \tag{3.256}$$

and

$$\frac{\partial T^e}{\partial T_1} = N_1$$
$$\frac{\partial T^e}{\partial T_2} = N_2$$
$$\qquad \cdot \quad \cdot$$
$$\qquad \cdot \quad \cdot$$
$$\frac{\partial T^e}{\partial T_r} = N_r \tag{3.257}$$

or

$$\frac{\partial T^e}{\partial \{\mathbf{T}\}} = \left\{ \begin{array}{c} N_1 \\ N_2 \\ . \\ . \\ N_r \end{array} \right\} = \{\mathbf{N}\} = [\mathbf{N}]^T. \tag{3.258}$$

The gradient matrix is written as

$$\{\mathbf{g}\}_e = \left\{ \begin{array}{c} \frac{\partial T^e}{\partial x} \\ \frac{\partial T^e}{\partial y} \\ \frac{\partial T^e}{\partial z} \end{array} \right\} = \begin{bmatrix} \frac{\partial N_1}{\partial x} & \frac{\partial N_2}{\partial x} & ..... & \frac{\partial N_r}{\partial x} \\ \frac{\partial N_1}{\partial y} & \frac{\partial N_2}{\partial y} & ..... & \frac{\partial N_r}{\partial y} \\ \frac{\partial N_1}{\partial z} & \frac{\partial N_2}{\partial z} & ..... & \frac{\partial N_r}{\partial z} \end{bmatrix} \left\{ \begin{array}{c} T_1 \\ T_2 \\ . \\ . \\ T_r \end{array} \right\} = [\mathbf{B}]_e \{\mathbf{T}\}_e. \tag{3.259}$$

Consider

$$\{\mathbf{g}\}_e^T [\mathbf{D}]_e \{\mathbf{g}\}_e = \left\{ \begin{array}{ccc} \frac{\partial T^e}{\partial x} & \frac{\partial T^e}{\partial y} & \frac{\partial T^e}{\partial z} \end{array} \right\} \begin{bmatrix} k_x & 0 & 0 \\ 0 & k_y & 0 \\ 0 & 0 & k_z \end{bmatrix} \left\{ \begin{array}{c} \frac{\partial T^e}{\partial x} \\ \frac{\partial T^e}{\partial y} \\ \frac{\partial T^e}{\partial z} \end{array} \right\}$$

$$= k_x \left(\frac{\partial T^e}{\partial x} \right)^2 + k_y \left(\frac{\partial T^e}{\partial y} \right)^2 + k_z \left(\frac{\partial T^e}{\partial z} \right)^2, \tag{3.260}$$

substituting into Equation (3.255), we have

$$I^e = \frac{1}{2} \int_{\Omega} \left[\{\mathbf{g}\}_e^T [\mathbf{D}]_e \{\mathbf{g}\}_e - 2GT^e \right] d\Omega + \int_{S_{2e}} qT^e ds + \int_{S_{3e}} \frac{1}{2} h(T^e - T_a)^2 ds. \tag{3.261}$$

Extending Equation (3.259) for the entire domain, i.e., $\{\mathbf{g}\}^T[\mathbf{D}]\{\mathbf{g}\} = \{\mathbf{T}\}^T[\mathbf{B}]^T[\mathbf{D}][\mathbf{B}]\{\mathbf{T}\}$ and minimizing the integral, we have

$$\frac{\partial I}{\partial \{\mathbf{T}\}} = \int_{\Omega} [\mathbf{B}]^T[\mathbf{D}][\mathbf{B}]\{\mathbf{T}\} d\Omega - \int_{\Omega} G[\mathbf{N}]^T d\Omega$$

$$+ \int_{S_{2e}} q[\mathbf{N}]^T ds + \int_{S_{3e}} h[\mathbf{N}]^T[\mathbf{N}]\{\mathbf{T}\} ds$$

$$- \int_{S_{3e}} h[\mathbf{N}]^T T_a ds = 0. \tag{3.262}$$

The above equation can be written in a compact form as

$$[\mathbf{K}]\{\mathbf{T}\} = \{\mathbf{f}\}, \tag{3.263}$$

where

$$[\mathbf{K}] = \int_\Omega [\mathbf{B}]^T [\mathbf{D}][\mathbf{B}] d\Omega + \int_{S_3} h[\mathbf{N}]^T [\mathbf{N}] ds$$

$$\{\mathbf{f}\} = \int_\Omega G[\mathbf{N}]^T d\Omega - \int_{S_2} q[\mathbf{N}]^T ds + \int_{S_3} hT_a[\mathbf{N}]^T ds. \qquad (3.264)$$

Equations (3.263) form the backbone of the calculation method for a finite element analysis of heat conduction problems. It can be easily noted that when there is no heat generation within an element ($G = 0$), the corresponding term disappears. Similarly, for an insulated boundary (i.e., $q = 0$ or $h = 0$) the corresponding term again disappears. Thus, for an insulated boundary we do not have to specify any contribution but leave it unattended. In this respect this is a great deal more convenient as compared to the finite difference method, where nodal equations have to be written for insulated boundaries.

3.4.2 The Galerkin Method

The method requires that the following expansion be satisfied,

$$\int_\Omega w_k \left(L(\overline{T}) + G \right) d\Omega = 0, \qquad (3.265)$$

where L is the diffusion operator given as

$$L = \left\{ \frac{\partial}{\partial x} \left(k_x \frac{\partial}{\partial x} \right) + \frac{\partial}{\partial y} \left(k_y \frac{\partial}{\partial y} \right) + \frac{\partial}{\partial z} \left(k_z \frac{\partial}{\partial z} \right) \right\}, \qquad (3.266)$$

$\overline{T}$ is the approximate temperature. The weight function w_k is also spatially discretized using the same trail functions used for the the the temperature. After simplification,

$$\int_\Omega N_k \left\{ \frac{\partial}{\partial x} \left(k_x \frac{\partial \overline{T}}{\partial x} \right) + \frac{\partial}{\partial y} \left(k_y \frac{\partial \overline{T}}{\partial y} \right) + \frac{\partial}{\partial z} \left(k_z \frac{\partial \overline{T}}{\partial z} \right) + G \right\} d\Omega = 0. \qquad (3.267)$$

Integration by parts is often essential when dealing with second-order derivatives. Using Green's lemma (see Appendix A), we can rewrite the second derivatives in two parts as

$$\int_\Omega N_k \frac{\partial}{\partial x} \left(k_x \frac{\partial \overline{T}}{\partial x} \right) d\Omega = \int_S N_k \left(k_x \frac{\partial \overline{T}}{\partial x} \right) \tilde{l} \, dS - \int_\Omega \frac{\partial N_k}{\partial x} k_x \frac{\partial N_m}{\partial x} \{T\} d\Omega, \qquad (3.268)$$

where subscripts k and m represent the nodes. With the boundary conditions (3.245), we can rewrite Equation (3.267) as

$$\int_\Omega \left(k_x \frac{\partial N_k}{\partial x} \frac{\partial N_m}{\partial x} + k_y \frac{\partial N_k}{\partial y} \frac{\partial N_m}{\partial y} + k_z \frac{\partial N_k}{\partial z} \frac{\partial N_m}{\partial z} \right) \{\overline{T}\} d\Omega$$

$$- \int_\Omega GN_k d\Omega + \int_{S_2} N_k q dS + \int_{S_3} hN_k N_m \{\overline{T}\} dS - \int_{S_3} hT_a N_k dS = 0. \qquad (3.269)$$

Now collecting the coefficient of the nodal variables $\{\overline{T}\}$, we get

$$[\mathbf{K}]\{\mathbf{T}\} = \{\mathbf{f}\} \qquad (3.270)$$

or

$$[K_{km}]\{T_m\} = \{f_k\}, \tag{3.271}$$

where

$$K_{km} = [\mathbf{K}] = \int_\Omega \left(k_x \frac{\partial N_k}{\partial x} \frac{\partial N_m}{\partial x} + k_y \frac{\partial N_k}{\partial y} \frac{\partial N_m}{\partial y} + k_z \frac{\partial N_k}{\partial z} \frac{\partial N_m}{\partial z} \right) d\Omega + \int_{S_3} h N_k N_m dS$$

$$f_k = \{\mathbf{f}\} = \int_\Omega G N_k d\Omega - \int_{S_2} q N_k dS + \int_{S_3} h T_a N_k dS. \tag{3.272}$$

It may be observed that Equations (3.263) and (3.270) are identical, which substantiates the fact that both the variational and Galerkin methods give the same result because there exists a classical variational integral for the heat conduction equation.

3.5   Requirements for Interpolation Functions

The procedure for formulating the individual element equations from a variational principle and the assemblage of these equations relies on the assumption that the interpolation functions satisfy the following requirements. This arises from the need to ensure that Equation (3.251) holds and that our approximate solution converges to the correct solution when we use an increasing number of elements, that is, when we refine the mesh.

(a) *Compatibility:* At element interfaces the field variable T and any of its partial derivatives up to one order less than the highest-order derivative appearing in $I(T)$ must be continuous

(b) *Completeness:* All uniform states of T and its partial derivatives up to the highest order appearing in $I(T)$ should have representation in T, when, in the limit the element size decreases to zero.

If the field variables are continuous at the element interfaces, then we have C° continuity. If, in addition, the first derivatives are continuous, we have C^1 continuity and if the second derivatives are continuous then we have C^2 continuity etc. If the functions appearing in the integrals of the element equations contain derivatives up to the $(r + 1)$th order, then to have a rigorous assurance of convergence as the element size decreases, we must satisfy the following requirements.

For compatibility: At the element interfaces we must have C^r continuity.

For completeness: Within an element we must have C^{r+1} continuity.

These requirements will hold regardless of whether the element equations (integral expressions) were derived using the variational method, the Galerkin method, the energy balance methods, or any other method yet to be devised. These requirements govern the selection of proper interpolation functions depending on the order of the differential equation. Thus, for a conduction heat transfer problem, the highest derivative in I is of the first order. Thus, the shape function selected should provide for the continuity of temperature at the interface between two elements and also ensures the continuity of temperature and heat flux within each element.

In addition to the requirements of continuity of the field variable and convergence to the correct solution as the element size reduces, we require that the field variable representation (polynomials used) within an element remain unchanged under a linear transformation from one Cartesian coordinate system to another. Polynomials, which exhibit this invariance property are said to possess "Geometric Isotropy". Clearly, we cannot expect a realistic approximation if our field variable representation changes with respect to a movement in origin, or in the orientation of the coordinate system. Hence, the need to ensure geometric isotropy in our polynomial interpolation functions is apparent. Fortunately, we have two simple guidelines that allow us to construct polynomial series with geometric isotropy. These are:

(i) Polynomials of order n that are complete, that is, those that contain all terms have geometric isotropy. The triangle family satisfies this condition whether it be a linear, quadratic or cubic form.

(ii) Polynomials of order n that are incomplete, yet contain the appropriate terms to preserve "symmetry" have geometric isotropy We neglect only these terms that occur in symmetric pairs, that is, $(x^3, y^3), (x^2y, xy^2)$ etc.

Example: For an eight-node element the following polynomial, P, satisfies geometric isotropy, that is,

$$P(x, y) = \alpha_1 + \alpha_2 x + \alpha_3 y + \alpha_4 x^2 + \alpha_5 xy + \alpha_6 y^2 \tag{3.273}$$

with either

$$\alpha_7 x^3 + \alpha_8 y^3 \tag{3.274}$$

or

$$\alpha_7 x^2 y + \alpha_8 y^2 x \tag{3.275}$$

added to it.

Example 3.5.1   *Before concluding this chapter it is important to consider a numerical problem for illustrating the theory presented. For this purpose we consider again a fin problem as shown in Figure 3.29. The linear variation for the temperature within each finite element is assumed. We shall derive the element equations from the most general formulation given*

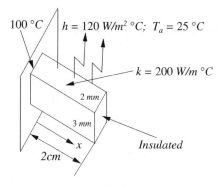

100 °C $h = 120\ W/m^2\ °C;\ T_a = 25\ °C$

$k = 200\ W/m\ °C$

2 mm

3 mm

Insulated

2cm x

Figure 3.29   Heat transfer from a rectangular fin.

Table 3.6   Element and node numbers of linear
one-dimensional elements

Element no.	Node i	Node j
1	1	2
2	2	3
e	i	j
n	n	$n+1$

in Section 3.4 and determine the temperature distribution, heat dissipation capacity and the efficiency of the fin, assuming that the tip is insulated.

Since we are using linear elements, the element will only have two nodes. First we divide the given length of the fin into number of divisions - say n elements. Therefore we will have (n + 1) nodes to represent the fin (see Table 3.6).

The variation of temperature in the elements is linear. Hence,

$$T = N_i T_i + N_j T_j \tag{3.276}$$

and the first derivative is given by

$$\frac{dT}{dx} = \frac{dN_i}{dx} T_i + \frac{dN_j}{dx} T_j$$

$$= -\frac{1}{l} T_i + \frac{1}{l} T_j; \tag{3.277}$$

that is, the gradient matrix is

$$g = \frac{dT}{dx} = \begin{bmatrix} -\dfrac{1}{l} & \dfrac{1}{l} \end{bmatrix} \begin{Bmatrix} T_i \\ T_j \end{Bmatrix} = [\mathbf{B}]\{\mathbf{T}\}; \tag{3.278}$$

where, l is the element length and

$$[\mathbf{B}] = \frac{1}{l}[-1 \quad 1]. \tag{3.279}$$

With the above relationships we can write the relevant element matrices as follows (Equation (3.264)):

$$[\mathbf{K}]_e = \int_l [\mathbf{B}]^T [\mathbf{D}][\mathbf{B}] d\Omega + \int_S h[\mathbf{N}]^T [\mathbf{N}] dS$$

$$= \int_l \frac{1}{l} \begin{bmatrix} -1 \\ 1 \end{bmatrix} [k_x] \frac{1}{l} [-1 \quad 1] A dx + \int_S h \begin{bmatrix} N_i \\ N_j \end{bmatrix} [N_i \quad N_j] P dx. \tag{3.280}$$

where A is the cross-sectional area of the fin and P is the perimeter of the fin from which convection takes place. Note that $[D] = k_x$ for one-dimensional problems, $d\Omega = Adx$ and $dS = Pdx$.

Rearranging Equation (3.280), we have

$$[\mathbf{K}]_e = \int_l \frac{Ak_x}{l^2} \begin{bmatrix} 1 & -1 \\ -1 & 1 \end{bmatrix} dx + \int_l hP \begin{bmatrix} N_i^2 & N_i N_j \\ N_i N_j & N_j^2 \end{bmatrix} dx. \tag{3.281}$$

Here $N_i = L_i$ and $N_j = L_j$, which is generally true for all linear elements. Hence, we can make use of the formula (Appendix C):

$$\int_l L_i^a L_j^b \, dl = \frac{a!b!l}{(a+b+1)!}.$$ (3.282)

For example,

$$\int_l N_i^2 \, dl = \int_l L_i^2 \, dl = \frac{2!0!l}{(2+0+1)!} = \frac{l}{3}$$ (3.283)

and other terms can be similarly integrated.

If A, k_x, P and h are all assumed to be constant throughout the element, we obtain the following $[\mathbf{K}]$ matrix:

$$[\mathbf{K}]_e = \frac{Ak_x}{l} \begin{bmatrix} 1 & -1 \\ -1 & 1 \end{bmatrix} + \frac{hPl}{6} \begin{bmatrix} 2 & 1 \\ 1 & 2 \end{bmatrix}.$$ (3.284)

Let us next consider the thermal loading. From Equation (3.264) we can write:

$$\{\mathbf{f}\}_e = \frac{GAl}{2} \begin{Bmatrix} 1 \\ 1 \end{Bmatrix} - \frac{qPl}{2} \begin{Bmatrix} 1 \\ 1 \end{Bmatrix} + \frac{hT_a Pl}{2} \begin{Bmatrix} 1 \\ 1 \end{Bmatrix}.$$ (3.285)

Note that the heat flux (q) boundary is assumed along the surface. This is not necessarily always the case. In the problem considered, no heat flux boundary condition exists. Thus, the heat flux term (the second term) in Equation (3.285) is zero. Similarly no heat generation is assumed and thus the first term in Equation (3.285) is also zero.

The solution of the given problem may be found by substitution of the numerical values.
(a) First let us consider a one-element solution for the case where $l = 2\,cm$, as shown in Figure 3.30. The element stiffness matrix is

$$\begin{aligned}
[\mathbf{K}]_e &= \frac{Ak_x}{l} \begin{bmatrix} 1 & -1 \\ -1 & 1 \end{bmatrix} + \frac{hPl}{6} \begin{bmatrix} 2 & 1 \\ 1 & 2 \end{bmatrix} \\
&= \begin{bmatrix} 0.06 & -0.06 \\ -0.06 & 0.06 \end{bmatrix} + \begin{bmatrix} 0.008 & 0.004 \\ 0.004 & 0.008 \end{bmatrix} \\
&= \begin{bmatrix} 0.068 & -0.056 \\ -0.056 & 0.068 \end{bmatrix}
\end{aligned}$$ (3.286)

and the loading term is given by

$$\begin{aligned}
\{\mathbf{f}\} &= \frac{hPlT_a}{2} \begin{Bmatrix} 1 \\ 1 \end{Bmatrix} \\
&= \begin{Bmatrix} 0.30 \\ 0.30 \end{Bmatrix}.
\end{aligned}$$ (3.287)

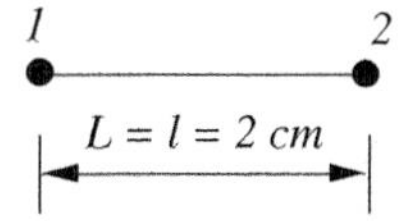

Figure 3.30   Heat transfer from a rectangular fin. One linear element.

Since only one element is employed, no assemblage of element contribution is necessary. Thus, the simultaneous equation system may be written as

$$
\begin{bmatrix} 0.068 & -0.056 \\ -0.056 & 0.068 \end{bmatrix} \begin{Bmatrix} T_1 \\ T_2 \end{Bmatrix} = \begin{Bmatrix} 0.30 + Q \\ 0.30 \end{Bmatrix}, \tag{3.288}
$$

where Q is the total heat flow at node 1. We now incorporate the known base temperature of $100\,^{\circ}C$ *at node 1. It is done in such a way that the symmetry of the* **[K]** *matrix is retained. This essential if the symmetric matrix solution procedure is employed in the solution of simultaneous equations. The following steps give a typical implementation procedure for the temperature boundary condition.*

(i) *The diagonal element of the first row is assigned a value of 1 and the remaining elements on that row are zero.*

(ii) *Replace the first row value of the loading vector* **f** *by the known value of* T_1, *that is, 100.*

(iii) *In order to retain the symmetry, the first term of the second row in the* **[K]** *matrix is transferred to the right-hand side and replace with a zero value as given below:*

$$
\begin{bmatrix} 1.0 & 0.0 \\ 0.0 & 0.068 \end{bmatrix} \begin{Bmatrix} T_1 \\ T_2 \end{Bmatrix} = \begin{Bmatrix} 100.0 \\ 0.30 + 0.056(100.0) \end{Bmatrix}. \tag{3.289}
$$

The simultaneous equation to be solved is

$$
0.068 T_2 = 0.3 + 0.056(100). \tag{3.290}
$$

Therefore the solution is $T_1 = 100\,^{\circ}C$ *and* $T_2 = 86.765\,^{\circ}C$.
Heat dissipated is calculated using the nodal equation for the the node number 1 as given in Equation (3.288), that is,

$$
Q = 0.068 T_1 - 0.056 T_2 - 0.3 = 1.64W. \tag{3.291}
$$

The heat dissipated may also be determined by using the following convection condition, that is,

$$
Q = \sum_{e=1}^{M} hPl \left(\frac{T_i + T_j}{2} - T_a \right) = 1.64W, \tag{3.292}
$$

where M is the total number of elements (subscripts i and j indicate the two nodes of an element). The maximum theoretically possible heat transfer is (when $T_i = T_j$)

$$
Q_{max} = \sum_{e=1}^{M} hPl \left(T_i - T_a \right) = 1.8W. \tag{3.293}
$$

The efficiency is defined as

$$\eta_f = \frac{Q}{Q_{max}} = \frac{1.64}{1.80} = 91.11\%.$$ (3.294)

The exact solution for this problem is

$$Q_{exact} = kAm \ tan(kml) = 1.593W,$$ (3.295)

where $m = \sqrt{hP/kA} = 31.62$. *Therefore, the exact fin efficiency is*

$$(\eta_f)_{exact} = \frac{Q_{exact}}{Q_{max}} = 88.48\%.$$ (3.296)

(b) Let us consider a two-element solution of the same problem (3 nodes).
The length of the fin is divided equally into two elements, that is, $l = 1.0$ cm.
The stiffness matrix calculation is similar to the one for the single element case, that is,

$$[K_1] = [K_2] = \begin{bmatrix} 0.124 & -0.118 \\ -0.118 & 0.124 \end{bmatrix}$$ (3.297)

and the loading vectors are

$$\{f_1\} = \{f_2\} = \begin{Bmatrix} 0.15 \\ 0.15 \end{Bmatrix}.$$ (3.298)

On assembly we obtain

$$\begin{bmatrix} 0.124 & -0.118 & 0.0 \\ -0.118 & 0.124+0.124 & -0.118 \\ 0.0 & -0.118 & 0.124 \end{bmatrix} \begin{Bmatrix} T_1 \\ T_2 \\ T_3 \end{Bmatrix} = \begin{Bmatrix} 0.15 \\ 0.15+0.15 \\ 0.15 \end{Bmatrix}.$$ (3.299)

Now we have to incorporate the known values of base temperature, that is, $T_1 = 100\,°C$.

$$\begin{bmatrix} 1.0 & 0.0 & 0.0 \\ 0.0 & 0.248 & -0.118 \\ 0.0 & -0.118 & 0.124 \end{bmatrix} \begin{Bmatrix} T_1 \\ T_2 \\ T_3 \end{Bmatrix} = \begin{Bmatrix} 100.0 \\ 0.30+0.118(100) \\ 0.15 \end{Bmatrix}.$$ (3.300)

Therefore, the two equations to be solved are
$0.248T_2 - 0.118T_3 = 12.1$
and
$-0.118T_2 + 0.124T_3 = 0.15$.
Solving these equations we get $T_2 = 90.209\,°C$, $T_3 = 87.057\,°C$.
Results, which have been generated using different number of elements, are tabulated in Tables 3.7 and 3.8.
As can be seen, the two-element solution is very good and is further improved with the use of four elements. As a first idealization even the one element solution is reasonably good considering the small effort involved.

Table 3.7   Summary of results – temperatures

x mm	Exact	1 element	2 elements	4 elements
0.0	100.00	100.00	100.00	100.00
5.0	94.28	–	–	94.26
10.0	90.28	–	90.209	90.25
15.0	87.93	–	–	87.908
20.0	87.15	86.77	87.07	87.128

Table 3.8   Summary of results – heat dissipated and efficiency

Case	Q (W)	η_f
1 element	1.64	91.11
2 elements	1.60	89.11
4 elements	1.60	88.65
Exact	1.59	88.48

3.6   Summary

In this chapter, we have discussed the basic principles of the finite element method as applied to heat conduction problems. Different types of elements have been discussed and various examples presented. In the authors opinion, this is the most important chapter for beginners. Readers already familiar with the topic of finite element may find it trivial to follow but it would be beneficial for the novice to work out the exercises provided in the following section.

3.7   Exercises

Exercise 3.7.1   *A one-dimensional linear element is used to approximate the temperature variation in a fin. The solution gives the temperature at two nodes of an element as 100°C and 80°C. The distances from the origin to node i is 6 cm and node j is 10 cm. Determine the temperature at a point 9 cm from the origin. Also calculate the temperature gradient in the elements. Show that the sum of the shape functions at the location 9 cm from origin is unity.*

Exercise 3.7.2   *A one-dimensional quadratic element is used to approximate the temperature distribution in a long fin. The solution gives the temperature at three nodes as 100, 90, and 80°C at distances of 10 cm, 15 cm and 20 cm respectively from the origin. Calculate the temperature and temperature gradient at a location of 12 cm from the origin.*

Exercise 3.7.3 *During the implementation of the finite element method, the evaluation of the integrals that contain shape functions and their derivatives is required. Evaluate the following integrals for a linear one-dimensional element*

$$\int_l N_i dl; \quad \int_l N_i^2 dl; \quad \int_l \frac{dN_i}{dx}\frac{dN_j}{dx} dl; \quad \int_l N_i^3 dl; \quad \int_l N_i N_j dl. \tag{3.301}$$

Exercise 3.7.4 *Derive the shape functions for a one-dimensional linear element in which both the temperature and the heat fluxes should be continuously varying in the element. (Note that degrees of freedom for a one-dimensional linear element are T_i, q_i, T_j, q_j.)*

Exercise 3.7.5 *The solution for temperature distribution in a linear triangle gives the nodal temperature as $T_i = 200\,°C$, $T_j = 180\,°C$ and $T_k = 160\,°C$. The coordinates of i, j and k are ($x_i = 2\,cm$, $y_i = 2\,cm$), ($x_j = 6\,cm$, $y_j = 4\,cm$) and ($x_k = 4\,cm$, $y_k = 6\,cm$). Calculate the temperature at a location given by $x = 3\,cm$ and $y = 4\,cm$. Calculate coordinates of the isotherm corresponding to $170\,°C$. Calculate the heat flux in the x and y directions at ($x = 3$, $y = 4\,cm$) if the thermal conductivity is $0.5\,W/m\,°C$. Also show that the sum of the shape functions at ($x = 3\,cm$, $y = 4\,cm$) is unity.*

Exercise 3.7.6 *For a one-dimensional quadratic element evaluate the integrals. (Note: convert N_i, N_j and N_k to local coordinates and then integrate.)*

$$\int_l N_i dl; \quad \int_l N_j dl; \quad \int_l N_k dl; \quad \int_l N_i N_j dl \tag{3.302}$$

Exercise 3.7.7 *The nodal values for a rectangular element is given as follows, $x_i = 0.25\,cm$, $y_l - 0.20\,cm$, $x_j - 0.30\,cm$, $y_m - 0.25\,cm$, $T_l - 150\,°C$, $T_j - 120\,°C$, $T_k - 100\,°C$, $T_m - 110\,°C$ Calculate (a) The temperature at the point $C(x = 0.27\,cm$, $y = 0.22\,cm)$. (b) x, y coordinates of the isotherm $130\,°C$ (c) Evaluate $\partial T/\partial x$ and $\partial T/\partial y$ at the point C.*

Exercise 3.7.8 *Calculate the shape functions for a six-noded rectangle shown in Figure 3.31.*

Exercise 3.7.9 *Evaluate the partial derivatives of shape functions at $\zeta = 1/4$ and $\eta = 1/2$ of a quadrilateral element shown in Figure 3.32, assuming the temperature is approximated by (a) bilinear, (b) quadratic interpolating polynomials.*

Exercise 3.7.10 *Calculate the derivatives $\partial N_6/\partial x$ and $\partial N_6/\partial y$ at a point $(2,5)$ for a quadratic triangle element shown in Figure 3.33 using local coordinates.*

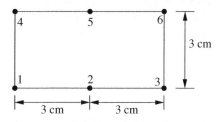

Figure 3.31   Rectangular element.

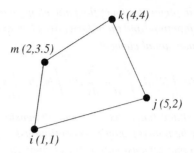

Figure 3.32   Quadrilateral element.

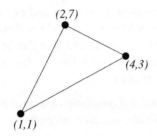

Figure 3.33   Triangular element.

Exercise 3.7.11   *In a double pipe heat exchanger, hot fluid flows inside a pipe and cold fluid flows outside in the annular space. The heat exchange between the two fluids is given by the differential equations, (refer to Exercise 2.5.12)*

$$C_1 \frac{dT_h}{dA} = -U(T_h - T_c)$$
$$C_2 \frac{DT_c}{dA} = U(T_h - T_c). \tag{3.303}$$

Develop the stiffness matrix and forcing vector using (a) subdomain method (b) Galerkin method.

Exercise 3.7.12   *Calculate (using one, two and four elements) the temperature distribution and the heat dissipation capacity of a fin of length 4 cm and cross-sectional dimensions of 6 mm × 4 mm with a heat transfer coefficient of 0.1 W/m² °C and a thermal conductivity of the material of the fin as 0.5 W/m °C. Base temperature is 90 °C, and the tip is insulated.*

References

Baker, AJ (1995) *Finite Element Computational Fluid Mechanics,* International Student Edition. McGraw-Hill, New York.

Bathe KJ (1982) *Finite Element Procedures in Engineering Analysis.* Prentice-Hall, Englewood Cliffs, New Jersey.

Chandrupatla TR and Belegundu AD (1991) *Introduction to Finite Elements in Engineering.* Prentice-Hall of India, New Delhi.

Clough RW (1960) The finite element analysis in plane stress analysis, *Proceedings of the 2nd ASCE Conference on Electronic Computation* Pittsburgh, PA, September.

Fry PJ and George PL (2008) *Mesh Generation: Application to Finite Elements.* John Wiley & Sons, Inc., New York.

Gupta KK and Meek JL (1996) A brief history of the beginning of the finite element method. *International Journal for Numerical Methods in Engineering,* **39**, 3761–3774.

Huebner K and Thornton EA (1982) *The Finite Element Method for Engineers,* 2nd Edition. John Wiley & Sons, Inc., New York.

Hughes TJR (2000) *The Finite Element Method: Linear Static and Dynamic Finite Element Analysis.* Dover Publications, New York.

Lewis RW, Morgan K, Thomas HR and Seetharamu KN (1996) *Finite Element Methods in Heat Transfer Analysis.* John Wiley & Sons, Inc., New York.

Malan AG, Lewis, RW and Nithiarasu P (2002) An improved unsteady, unstructured, artificial compressibility, finite volume scheme for viscous incompressible flows, Part I - Theory and implementation. *International Journal for Numerical Methods in Engineering,* **54**, 695–714.

Oden JT (1996) Finite elements: an introduction. In PG Ciarlet and JL Lions (eds), *Handbook of Numerical Analysis,* **2**, 3–16, Elsevier, Amsterdam.

Ozisik MN and Czisik MN (1994), *Finite Difference Methods in Heat Transfer.* CRC Press, London.

Patankar SV (1980) *Numerical Heat Transfer and Fluid Flow.* Hemisphere, Arlington, VA.

Rao SS (1989) *The Finite Element Methods in Engineering,* 2nd Edition. Pergamon Press, New York.

Segerlind LJ (1984) *Applied Finite Element Analysis,* 2nd Edition. John Wiley & Sons, Inc., New York.

Silvester P (1969) Higher-order polynomial triangular finite elements for potential problems. *International Journal of Engineering Science,* 7, 849–861.

Thompson JF, Soni BK and Weatherill NP (1999) *Handbook of Grid Generation.* CRC Press, London.

Zienkiewicz OC (1996) Origins, milestones and directions of the finite element method – a personal view. In PG Ciarlet and JL Lions (eds), *Handbook of Numerical Analysis,* **4**, 7–67, Elsevier, Amsterdam.

Zienkiewicz OC and Cheung K, (1965) Finite elements in the solution of field problems. *Engineer,* **200**, 507–510.

Zienkiewicz OC and Morgan K (1983) *Finite Elements and Approximation.* Wiley Inter Science, New York.

Zienkiewicz OC and Taylor RL (2000) *The Finite Element Method, Vol. 1, The Basis.* Butterworth & Heinemann, London.

Zienkiewicz OC, Taylor RL and Zhu JZ (2013) *The Finite Element Method. Vol. 1. The Basis,* 7th Edition. Elsevier, Amsterdam.

Zienkiewicz OC, Taylor RL and Nithiarasu P (2013) *The Finite Element Method. Vol. 3. Fluid Dynamics,* 7th Edition. Elsevier, Amsterdam.

4

Steady-State Heat Conduction in One-dimension

4.1 Introduction

A one-dimensional approximation of the heat conduction equation is feasible for many physical problems, e.g. plane walls, fins, etc. (Bejan 1993; Holman 1989; Incropera and Dewitt 1990; Ozisik 1968). In these problems, any major temperature variation is in one direction only and the variation in all other directions may be ignored. Other examples of one-dimensional heat transfer occur in cylindrical and spherical solids where the temperature variation takes place only in the radial direction. In this chapter, such one-dimensional problems are considered for steady-state conditions, in which the temperature does not depend on time. Time-dependent and multi-dimensional problems will be discussed in later chapters.

4.2 Plane Walls

4.2.1 Homogeneous Wall

The differential equations which govern the heat conduction through plane walls have already been discussed in Chapter 1. The steady-state heat conduction equation for a plane wall, shown in Figure 4.1, is

$$kA\frac{d^2T}{dx^2} = 0, \tag{4.1}$$

where k is the thermal conductivity and A is the cross-sectional area perpendicular to the direction of heat flow. The problem is complete with the following description of the boundary

Fundamentals of the Finite Element Method for Heat and Mass Transfer, Second Edition.
P. Nithiarasu, R. W. Lewis, and K. N. Seetharamu.
© 2016 John Wiley & Sons, Ltd. Published 2016 by John Wiley & Sons, Ltd.

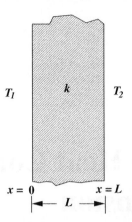

Figure 4.1   Heat conduction through a homogeneous wall.

conditions, that is,

$$\text{at} \quad x = 0, T = T_1; \quad \text{and at} \quad x = L, T = T_2.$$

The exact solution to Equation (4.1) is

$$kAT = C_1 x + C_2. \tag{4.2}$$

On applying the appropriate boundary conditions to Equation (4.2), we obtain

$$C_2 = kAT_1 \tag{4.3}$$

and

$$C_1 = -\frac{kA(T_1 - T_2)}{L}. \tag{4.4}$$

Therefore, substituting constants C_1 and C_2 into Equation (4.2) results in

$$T = -\frac{(T_1 - T_2)}{L} x + T_1. \tag{4.5}$$

The above equation indicates that the temperature distribution within the wall is linear. The heat flow, Q, can be written as (from Equation (4.5))

$$Q = -kA\frac{dT}{dx} = -\frac{kA}{L}(T_2 - T_1). \tag{4.6}$$

It is easy to verify that heat, Q, flows in the positive x direction if $T_1 > T_2$ and vice versa.

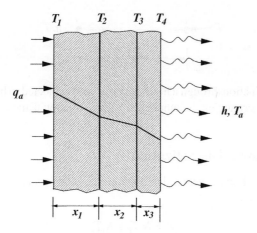

Figure 4.2   Heat conduction in a composite wall.

4.2.2   Composite Wall

Even if more than one material is used to construct the plane wall, as shown in Figure 4.2, at steady-state the heat flow will be constant (conservation of energy), that is,

$$Q = -\frac{k_1 A}{x_1}(T_2 - T_1) = -\frac{k_2 A}{x_2}(T_3 - T_2) = -\frac{k_3 A}{x_3}(T_4 - T_3). \tag{4.7}$$

Noting that

$$T_1 - T_2 = \frac{Q}{\frac{k_1 A}{x_1}}; T_2 - T_3 = \frac{Q}{\frac{k_2 A}{x_2}}; \quad \text{and} \quad T_3 - T_4 = \frac{Q}{\frac{k_3 A}{x_3}}, \tag{4.8}$$

the total heat flow may be calculated as

$$Q = \frac{(T_1 - T_4)}{\left[\frac{x_1}{k_1 A} + \frac{x_2}{k_2 A} + \frac{x_3}{k_3 A}\right]} \tag{4.9}$$

The numerator in the above equation is often referred to as the *thermal potential difference* and the denominator is known to as the *thermal resistance*. In general all x/kA terms are called thermal resistances (see Figure 4.2).

If there is a convective resistance, say on the right face as shown in Figure 4.2, then we have

$$Q = \frac{(T_4 - T_a)}{\frac{x_1}{k_1 A} + \frac{x_2}{k_2 A} + \frac{x_3}{k_3 A} + \frac{1}{hA}}, \tag{4.10}$$

where h is the heat transfer coefficient between atmosphere and the left wall surface and T_a is the atmospheric temperature. Let us now consider a finite element solution for Equation (4.1). As shown in Equation (4.5) the temperature distribution is linear for a homogeneous material.

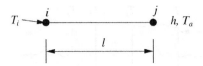

Figure 4.3   Heat conduction through a homogeneous wall subjected to heat convection on one side and constant temperature on the other side. Approximation using a single linear element.

4.2.3   Finite Element Discretization

If we consider a typical homogeneous slab, with nodes i and j on either side (see Figure 4.3), we can write

$$T = N_i T_i + N_j T_j, \tag{4.11}$$

where

$$N_i = \frac{x_j - x}{x_j - x_i} \quad \text{and} \quad N_j = \frac{x - x_i}{x_j - x_i}. \tag{4.12}$$

In local coordinates

$$N_i = 1 - \frac{x}{l} \quad \text{and} \quad N_j = \frac{x}{l} \tag{4.13}$$

and the temperature derivative is

$$\frac{dT}{dx} = -\frac{1}{l}T_i + \frac{1}{l}T_j = \left[-\frac{1}{l} \quad \frac{1}{l} \right] \left\{ \begin{array}{c} T_i \\ T_j \end{array} \right\} = [\mathbf{B}]_e \{\mathbf{T}\}_e, \tag{4.14}$$

where l is the length of the element.

The global stiffness matrix (Chapter 3) is given as

$$[\mathbf{K}] = \int_{\Omega} [\mathbf{B}]^T [\mathbf{D}][\mathbf{B}] d\Omega + \int_{A_r} h[\mathbf{N}]^T [\mathbf{N}] dA_r = \int_{l} [\mathbf{B}]^T [\mathbf{D}][\mathbf{B}] A dx + \int_{A_r} h[\mathbf{N}]^T [\mathbf{N}] dA_r, \tag{4.15}$$

where Ω is the volume integral; A_r here indicates convective surface area on the right or left face of the wall; and h is the convective heat transfer coefficient. After integration (see Figure 4.3),

$$[\mathbf{K}]_e = \frac{Ak_x}{l} \left[\begin{array}{cc} 1 & -1 \\ -1 & 1 \end{array} \right] + hA_r \left[\begin{array}{cc} 0 & 0 \\ 0 & 1 \end{array} \right]. \tag{4.16}$$

Note that the convective heat transfer boundary condition is assumed to act on the right face where $N_i = 0$ and $N_j = 1$. This is nonzero only when the last element at the right face is considered. This is the reason why we have hA_r added to the last nodal equation in Equation (4.16). In the composite wall problem considered here (Figure 4.2), the cross-sectional area A and convective surface area A_r are equal.

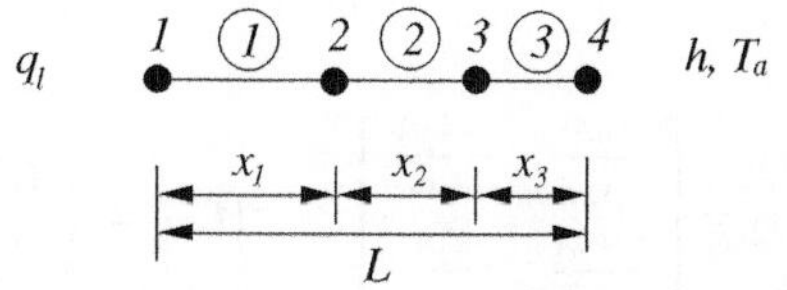

Figure 4.4   Heat conduction through a composite wall subjected to heat convection on one side and constant heat flux on the other side. Approximation using three linear elements.

The global forcing vector can be written as

$$\{\mathbf{f}\} = \int_{\Omega} G[\mathbf{N}]^T d\Omega - \int_{A_s} q_s[\mathbf{N}]^T dA_s + \int_{A_s} hT_a[\mathbf{N}]^T dA_s + \begin{Bmatrix} q_l A_l \\ \dots \\ \dots \\ \dots \\ -q_r A_r \end{Bmatrix}, \qquad (4.17)$$

where G is the internal heat generation per unit volume, q_s is the heat flux leaving the lateral surface (if it exists), q_l is the heat flux entering the wall at the left side and q_r is the heat flux leaving the right side, T_a is the atmospheric temperature and subscripts l and r respectively indicate the left and right faces of the wall. If $G = 0$, then there is no heat generation inside the slab. The zero heat flux boundary condition is denoted by $q = 0$. Since the lateral surface effect is not important in plane walls, the global load vector may be reduced to

$$\{\mathbf{f}\} = \int_{\Omega} G[\mathbf{N}]^T d\Omega + \begin{Bmatrix} q_l A_l \\ \dots \\ \dots \\ \dots \\ -q_r A_r \end{Bmatrix}. \qquad (4.18)$$

If neither internal heat generation nor external heat flux boundary conditions exist, as shown in Figure 4.3, then the finite element equation for a homogeneous slab with only one element becomes

$$\left\{ \frac{k_x A}{l} \begin{bmatrix} 1 & -1 \\ -1 & 1 \end{bmatrix} + hA \begin{bmatrix} 0 & 0 \\ 0 & 1 \end{bmatrix} \right\} \begin{Bmatrix} T_i \\ T_j \end{Bmatrix} = \begin{Bmatrix} 0 \\ hT_a A \end{Bmatrix}. \qquad (4.19)$$

Note that $qrAv = hA(T - T_a)$

The element equations can now be separately written for each slab of the composite wall shown in Figure 4.2 but now with a heat flux boundary condition at the left side and convection boundary condition on the right side. If we assume a discrete system as shown in Figure 4.4, we obtain the following element equations:

Element 1 (Slab 1):

$$[\mathbf{K}]_1 = \begin{bmatrix} \dfrac{k_1 A}{x_1} & -\dfrac{k_1 A}{x_1} \\ -\dfrac{k_1 A}{x_1} & \dfrac{k_1 A}{x_1} \end{bmatrix}; \quad \{\mathbf{f}\}_1 = \begin{Bmatrix} q_l A \\ 0 \end{Bmatrix}. \qquad (4.20)$$

Element 2 (Slab 2):

$$[\mathbf{K}]_2 = \begin{bmatrix} \dfrac{k_2 A}{x_2} & -\dfrac{k_2 A}{x_2} \\[2mm] -\dfrac{k_2 A}{x_2} & \dfrac{k_2 A}{x_2} \end{bmatrix}; \quad \{\mathbf{f}\}_2 = \begin{Bmatrix} 0 \\ 0 \end{Bmatrix}. \tag{4.21}$$

Element 3 (Slab 3):

$$[\mathbf{K}]_3 = \begin{bmatrix} \dfrac{k_3 A}{x_3} & -\dfrac{k_3 A}{x_3} \\[2mm] -\dfrac{k_3 A}{x_3} & \dfrac{k_3 A}{x_3} + hA \end{bmatrix}; \quad \{\mathbf{f}\}_3 = \begin{Bmatrix} 0 \\ hAT_a \end{Bmatrix}. \tag{4.22}$$

Note that $q_r = h(T - T_a)$ is substituted into the load vector, Equation (4.17), for the last node at the right face of the slab 3. Since T in this relation is the unknown, this part of the equation goes to the stiffness matrix as shown above in Equation (4.22). Assembly of all the three elemental equations gives

$$\begin{bmatrix} \dfrac{k_1 A}{x_1} & -\dfrac{k_1 A}{x_1} & 0 & 0 \\[2mm] -\dfrac{k_1 A}{x_1} & \left(\dfrac{k_1 A}{x_1} + \dfrac{k_2 A}{x_2}\right) & -\dfrac{k_2 A}{x_2} & 0 \\[2mm] 0 & -\dfrac{k_2 A}{x_2} & \left(\dfrac{k_2 A}{x_2} + \dfrac{k_3 A}{x_3}\right) & \dfrac{k_3 A}{x_3} \\[2mm] 0 & 0 & -\dfrac{k_3 A}{x_3} & \dfrac{k_3 A}{x_3} + hA \end{bmatrix} \begin{Bmatrix} T_1 \\ T_2 \\ T_3 \\ T_4 \end{Bmatrix} = \begin{Bmatrix} q_l A \\ 0 \\ 0 \\ hAT_a \end{Bmatrix}. \tag{4.23}$$

A solution of the above system of simultaneous equations (Appendix A) will result in the values of T_1, T_2, T_3 and T_4. In a similar way, we can extend this solution method to any number of materials that might constitute a composite wall. Note that the heat flux imposed on the left-hand face is q_l.

4.2.4   Wall with Varying Cross-sectional Area

Let us now consider a case where the cross-sectional area varies linearly from node i to j as shown in Figure 4.5. Let A_i and A_j be the areas of cross-section at distances x_i and x_j respectively ($l = x_j - x_i$). Therefore, the area A at an intermediate distance x is given by

$$A = A_i - \frac{x}{l}(A_i - A_j). \tag{4.24}$$

Rearranging, we obtain

$$A = A_i \left(1 - \frac{x}{l}\right) + \frac{x}{l} A_j = A_i N_i + A_j N_j. \tag{4.25}$$

Thus, the linear variation of area with distance can be represented in terms of the areas at the points i and j, using the same shape functions. The element stiffness matrix for the element

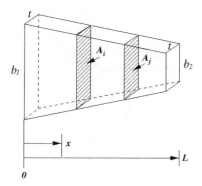

Figure 4.5   Heat conduction through a wall with linearly varying area of cross-section.

connecting i and j can be written as

$$[\mathbf{K}]_e = \int_\Omega [\mathbf{B}]_e^T [\mathbf{D}]_e [\mathbf{B}]_e d\Omega$$

$$= \int_l \frac{k}{l^2} \begin{bmatrix} 1 & -1 \\ -1 & 1 \end{bmatrix} (N_i A_i + N_j A_j) dx$$

$$= \frac{k}{l} \left(\frac{A_i + A_j}{2} \right) \begin{bmatrix} 1 & -1 \\ -1 & 1 \end{bmatrix}, \tag{4.26}$$

where l is the distance between nodes i and j (element length). In the above equation, it has been assumed that convection is absent.

Thus, when the area varies linearly, we can substitute an average area value and use the constant area formulation if there is no heat dissipation from the perimeter. This assumption will not hold good if the body is circular in cross-section, in which case the cross-sectional area varies quadratically with the axial distance. This case can be dealt with by the use of a quadratic variation within the element.

Example 4.2.1   *A composite wall, with three layers of different material as shown in Figure 4.2, has the following properties for the different layers.*

- *Layer-1: Gypsum, $k_1 = 0.05 \, W/m\,°C$, $x_1 = 1 \, cm$ and $q_l = 15 \, W/m^2$.*

- *Layer-2: Fiber-glass, $k_2 = 0.0332 \, W/m\,°C$ and $x_2 = 5 \, cm$.*

- *Layer-3: Concrete, $k_3 = 1.2 \, W/m\,°C$, $x_3 = 15 \, cm$, $h = 15 \, W/m^2\,°C$ and $T_a = 25\,°C$.*

Calculate the temperatures T_1, T_2, T_3 and T_4 assuming a heat flow area of $1 \, m^2$. On substituting the given parameter values into Equation (4.23), we obtain

$$\begin{bmatrix} 5.0 & -5.0 & 0.0 & 0.0 \\ -5.0 & 5.66 & -0.66 & 0.0 \\ 0.0 & -0.66 & 8.66 & -8 \\ 0.0 & 0.0 & -8.0 & 23.0 \end{bmatrix} \begin{Bmatrix} T_1 \\ T_2 \\ T_3 \\ T_4 \end{Bmatrix} = \begin{Bmatrix} 15 \\ 0.0 \\ 0.0 \\ 375 \end{Bmatrix}. \tag{4.27}$$

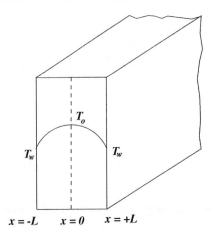

Figure 4.6 Plane wall with heat source.

The solution of the above simultaneous equations results in $T_1 = 53.6\,°C$, $T_2 = 50.60\,°C$, $T_3 = 27.875\,°C$, and $T_4 = 26\,°C$. The balance of heat may be verified by computing the total heat transfer at the right face as $Q_r = A_r h(T_4 - T_a) = 15$ W. This should be equal to the heat supplied at the left side $Q_l = A_l q_l = 15$ W. Thus, the system satisfies the equilibrium conditions. In the above relationships, A_r and A_l are respectively the surface areas at the right and left face of the composite wall.

4.2.5 Plane Wall with a Heat Source: Solution by Linear Elements

Many examples of heat transfer problems involve internal heat generation, for example, in nuclear reactors, electrical conductors, chemical and biological reactors etc. In this section, the heat conduction through a wall is considered with internal heat generation as shown in Figure 4.6. Let us assume that the one-dimensional approximation is valid and G W/cm^3 represents the quantity of heat generated per unit volume inside the wall. Therefore, under steady-state conditions, the applicable differential equation is

$$\frac{d^2 T}{dx^2} + \frac{G}{k} = 0. \tag{4.28}$$

The boundary conditions are

$$\text{at} \quad x = \pm L, T = T_w. \tag{4.29}$$

Integrating twice, we get

$$T = -\frac{G}{k}\frac{x^2}{2} + C_1 x + C_2. \tag{4.30}$$

From the symmetry of the problem, we find at $x = 0$, $dT/dx = 0$ and $C_1 = 0$ and $C_2 = T_o$. Therefore, Equation (4.30) becomes

$$T = -\frac{G}{k}\frac{x^2}{2} + T_o. \tag{4.31}$$

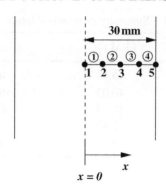

Figure 4.7   Finite element discritization.

The temperature, T_w, at both ends can be obtained by substituting $x = \pm L$, which results in

$$T_w = -\frac{G}{k}\frac{L^2}{2} + T_o \tag{4.32}$$

or

$$T_o = T_w + \frac{GL^2}{2k}. \tag{4.33}$$

From Equations (4.31) and (4.32) we can write

$$\frac{T - T_o}{T_w - T_o} = \left(\frac{x}{L}\right)^2 \tag{4.34}$$

which shows that the temperature distribution is parabolic.

In the case of a finite element formulation, we have to account for the heat generation in the forcing vector such that (see Equation (4.17))

$$\{\mathbf{f}\}_e = \int_v G[\mathbf{N}]^T dV = \int_l G\left\{\begin{array}{c} N_i \\ N_j \end{array}\right\} A dx = \frac{GAl}{2}\left\{\begin{array}{c} 1 \\ 1 \end{array}\right\}. \tag{4.35}$$

In the above equation, the heat generated is distributed equally between the two nodes i and j. In all linear elements we observe that the heat generated, or any other type of load, is equally distributed among the participating nodes. Because of the symmetry of the problem, it is sufficient in this case if we consider only one half of the domain.

Example 4.2.2   *Determine the temperature distribution in a plane wall of thickness 60 mm, which has an internal heat source of 0.3 MW/m³ and the thermal conductivity of the material is 21 W/m°C. Assume that the surface temperature of the wall is 40 °C.*

Because of symmetry, we may consider only one half of the plane wall as shown in Figure 4.7. Let us consider four elements, each of length 7.5 mm. Let the cross-sectional area for heat flow, A = 1 m².

The element stiffness matrix is

$$[\mathbf{K}]_e = \frac{kA}{l}\begin{bmatrix} 1 & -1 \\ -1 & 1 \end{bmatrix} = \begin{bmatrix} 2800 & -2800 \\ -2800 & 2800 \end{bmatrix}, \tag{4.36}$$

Table 4.1   Summary of results – temperatures

T	FEM (°C)	Exact (°C)
T_1	46.43	46.43
T_2	46.03	46.03
T_3	44.83	44.82
T_4	42.82	42.81
T_5	40.0	40.0

which is identical for every element and

$$\{\mathbf{f}\}_e = \frac{GAl}{2}\left\{\begin{array}{c} 1 \\ 1 \end{array}\right\} = \left\{\begin{array}{c} 1125 \\ 1125 \end{array}\right\}, \tag{4.37}$$

which also is identical for all elements. Assembly gives

$$\begin{bmatrix} 2800 & -2800 & 0.0 & 0.0 & 0.0 \\ -2800 & 5600 & -2800 & 0.0 & 0.0 \\ 0.0 & -2800 & 5600 & -2800 & 0.0 \\ 0.0 & 0.0 & -2800 & 5600 & -2800 \\ 0.0 & 0.0 & 0.0 & -2800 & 2800 \end{bmatrix} \left\{\begin{array}{c} T_1 \\ T_2 \\ T_3 \\ T_4 \\ T_5 \end{array}\right\} = \left\{\begin{array}{c} 1125 \\ 2250 \\ 2250 \\ 2250 \\ 1125 \end{array}\right\}. \tag{4.38}$$

Applying the boundary condition, $T_5 = 40°$, the modifications are necessary to retain the symmetry of the stiffness matrix, as discussed in Chapter 3.

$$\begin{bmatrix} 2800 & -2800 & 0.0 & 0.0 & 0.0 \\ -2800 & 5600 & -2800 & 0.0 & 0.0 \\ 0.0 & -2800 & 5600 & -2800 & 0.0 \\ 0.0 & 0.0 & -2800 & 5600 & 0.0 \\ 0.0 & 0.0 & 0.0 & 0.0 & 1 \end{bmatrix} \left\{\begin{array}{c} T_1 \\ T_2 \\ T_3 \\ T_4 \\ T_5 \end{array}\right\} = \left\{\begin{array}{c} 1125 \\ 2250 \\ 2250 \\ 2250 + 2800 \times 40 \\ 40 \end{array}\right\}. \tag{4.39}$$

Solving the above system of equations, we obtain the temperature distribution as shown in Table 4.1.

We observe that the finite element method results are either very close or equal to the exact solution. The method can be extended for the case of a known wall heat flux, or a convection boundary condition at the wall, as shown in Example 4.2.3.

Example 4.2.3   In Example 4.2.2, the left-hand face is insulated and the right-hand face is subjected to a convection environment at 93 °C with a surface heat transfer coefficient of 570 W/m²°C (see Figure 4.8). Determine the temperature distribution within the wall.

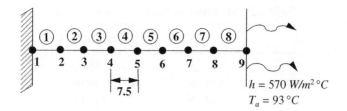

Figure 4.8 Finite element discritization for the example with convection.

Since there is no symmetry, we have to consider the entire domain. Let us subdivide the domain into eight elements (Figure 4.8), each of 7.5 mm length. Then

$$[\mathbf{K}]_1 = [\mathbf{K}]_2 = ...[\mathbf{K}]_7 = \begin{bmatrix} 2800 & -2800 \\ -2800 & 2800 \end{bmatrix} \tag{4.40}$$

$$[\mathbf{K}]_8 = \begin{bmatrix} 2800 & -2800 \\ -2800 & 2800 \end{bmatrix} + 570 \begin{bmatrix} 0 & 0 \\ 0 & 1 \end{bmatrix} = \begin{bmatrix} 2800 & -2800 \\ -2800 & 3370 \end{bmatrix}. \tag{4.41}$$

The elemental forcing vectors are the same as for Example 4.2.2, except for the last element, which is

$$\{\mathbf{f}\}_8 = \left\{ \begin{array}{c} 1125 \\ 1125 \end{array} \right\} + hAT_a \left\{ \begin{array}{c} 0 \\ 1 \end{array} \right\} = \left\{ \begin{array}{c} 1125 \\ 54135 \end{array} \right\}. \tag{4.42}$$

Assembly may be carried out as in Example 4.2.2. The solution of the assembled equation results in the temperature distribution within the wall. The FEM solution is compared with the analytical[1] results as shown in Table 4.2 and compare very favorably.

4.2.6 Plane Wall with Heat Source: Solution by Quadratic Elements

We have seen from the previous section that the analytical solution to the problem of a plane wall with a heat source gives a quadratic temperature distribution. Thus, it is appropriate to solve such a problem using quadratic elements. Let us consider the problem shown in Figure 4.6. We require three nodes for each element in order to represent a quadratic variation as discussed in Section 3.2.2, that is,

$$T = N_i T_i + N_j T_j + N_k T_k \tag{4.43}$$

---

[1] Analytical solution is obtained by solving

$$\frac{d^2 T}{dx^2} + \frac{G}{k} = 0,$$

subjected to boundary conditions. The final exact relation is

$$T = \frac{G}{2k}(L^2 - x^2) + \left(\frac{GL}{h} + T_a \right).$$

Table 4.2   Summary of results – temperatures

T	FEM °C	Analytical °C
T_1	150.28	150.29
T_2	149.88	149.89
T_3	148.68	148.68
T_4	146.67	146.67
T_5	143.86	143.86
T_6	140.24	140.24
T_7	135.82	135.83
T_8	130.60	130.60
T_9	124.59	124.59

with

$$N_i = \left[1 - \frac{3x}{l} + \frac{2x^2}{l^2}\right]$$

$$N_j = \frac{4x}{l} - \frac{4x^2}{l^2}$$

$$N_k = \frac{2x^2}{l^2} - \frac{x}{l}. \tag{4.44}$$

From Chapter 3, the stiffness matrix is defined as

$$[\mathbf{K}]_e = \int_l [\mathbf{B}]_e^T [\mathbf{D}]_e [\mathbf{B}]_e d\Omega = \frac{Ak}{6l} \begin{bmatrix} 14 & -16 & 2 \\ -16 & 32 & -16 \\ 2 & -16 & 14 \end{bmatrix}, \tag{4.45}$$

where $d\Omega = Adx$ and

$$[\mathbf{B}]_e = \left[\left(\frac{4x}{l^2} - \frac{3}{l}\right) \quad \left(\frac{4}{l} - \frac{8x}{l^2}\right) \quad \left(\frac{4x}{l^2} - \frac{1}{l}\right) \right]. \tag{4.46}$$

The loading vector is

$$\{\mathbf{f}\}_e = \int_l G[\mathbf{N}]^T d\Omega = \int_l G \begin{Bmatrix} L_i(2L_i - 1) \\ 4L_iL_j \\ L_j(2L_j - 1) \end{Bmatrix} Adx = \frac{GAl}{6} \begin{Bmatrix} 1 \\ 4 \\ 1 \end{Bmatrix}. \tag{4.47}$$

In the above equation, the shape functions N_i, N_j and N_k are expressed in terms of the local coordinate system L_i and L_j, the use of which will facilitate the integration process by using

$$\int_l L_i^a L_j^b dx = \frac{a!b!}{(a+b+1)!}l. \tag{4.48}$$

Example 4.2.4   *We shall now solve Example 4.2.2 using one quadratic element only as shown in Figure 4.9. As before, we consider only one half of the wall, where L is equal to 30 mm.*

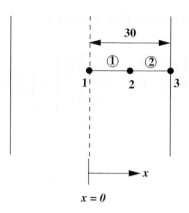

Figure 4.9   Quadratic finite element discretization.

Substituting values into Equations (4.45) and (4.47), we obtain

$$[\mathbf{K}] = \begin{bmatrix} 1633.33 & -1866.66 & 233.33 \\ -1866.66 & 3733.33 & -1866.66 \\ 233.33 & -1866.66 & 1633.33 \end{bmatrix} \tag{4.49}$$

and

$$\{\mathbf{f}\} = \left\{ \begin{array}{c} 1500 \\ 6000 \\ 1500 \end{array} \right\}. \tag{4.50}$$

Incorporating the boundary condition, that is, $T_3 = 40\,^{\circ}C$, results in the following set of equations,

$$\begin{bmatrix} 1633.33 & -1866.66 & 0.0 \\ -1866.66 & 3733.33 & 0.0 \\ 0.0 & 0.0 & 1.0 \end{bmatrix} \left\{ \begin{array}{c} T_1 \\ T_2 \\ T_3 \end{array} \right\} = \left\{ \begin{array}{c} 1500 - 233.33(40) \\ 6000 + 1866.66(40) \\ 40.0 \end{array} \right\}. \tag{4.51}$$

The solution to the above system gives, $T_1 = 46.43\,^{\circ}C$ and $T_2 = 44.82\,^{\circ}C$, which are identical to the exact solution.

4.2.7   Plane Wall with a Heat Source: Solution by Modified Quadratic Equations (Static Condensation)

In many transient and nonlinear problems, it will be necessary to obtain the temperature distribution several times. Hence, any possible reduction in the number of nodes, without sacrificing accuracy, is important. For one-dimensional quadratic elements it is possible to transfer the central node contribution to the side nodes. Thus, there will be only two nodes but the influence of the quadratic variation is inherently present. This process is referred to as static condensation and the procedure will be demonstrated by considering a typical quadratic element equation, viz.,

$$
\begin{bmatrix} K_{11} & K_{12} & K_{13} \\ K_{21} & K_{22} & K_{23} \\ K_{31} & K_{32} & K_{33} \end{bmatrix} \begin{Bmatrix} T_1 \\ T_2 \\ T_3 \end{Bmatrix} = \begin{Bmatrix} f_1 \\ f_2 \\ f_3 \end{Bmatrix}.
\tag{4.52}
$$

In order to eliminate the middle node, that is, node 2, we transfer its contribution to nodes 1 and 3. This is accomplished by expressing the temperature at node 2 in terms of the temperatures at nodes 1 and 3, that is

$$
T_2 = \frac{f_2}{K_{22}} - \left[\frac{K_{21}T_1}{K_{22}} + \frac{K_{23}T_3}{K_{22}} \right].
\tag{4.53}
$$

Now, on substituting the above relation into the first and third nodal equations, we have

$$
\left[K_{11} - \frac{K_{21}}{K_{22}}K_{12} \right] T_1 + \left[K_{13} - \frac{K_{23}}{K_{22}}K_{12} \right] T_3 = \left[f_1 - f_2 \frac{K_{12}}{K_{22}} \right]
\tag{4.54}
$$

for the first node, and

$$
\left[K_{31} - \frac{K_{21}}{K_{22}}K_{32} \right] T_1 + \left[K_{33} - \frac{K_{23}}{K_{22}}K_{32} \right] T_3 = \left[f_3 - f_2 \frac{K_{32}}{K_{22}} \right]
\tag{4.55}
$$

for the second node. Now the matrix form of the equation can be rewritten as

$$
\begin{bmatrix} \left(K_{11} - \frac{K_{21}}{K_{22}}K_{12} \right) & \left(K_{13} - \frac{K_{23}}{K_{22}}K_{12} \right) \\ \left(K_{31} - \frac{K_{21}}{K_{22}}K_{32} \right) & \left(K_{33} - \frac{K_{23}}{K_{22}}K_{32} \right) \end{bmatrix} \begin{Bmatrix} T_1 \\ T_3 \end{Bmatrix} = \begin{Bmatrix} f_1 - f_2 \frac{K_{12}}{K_{22}} \\ f_3 - f_2 \frac{K_{32}}{K_{22}} \end{Bmatrix}.
\tag{4.56}
$$

Note that the number of equations have been reduced, which leads to a small decrease in computational cost. This procedure is therefore often employed when higher order approximations are used for temperature.

Example 4.2.5   *Repeat Example 4.2.4 using the static condensation procedure.*

Substituting all relevant values into Equation (4.56) and applying the boundary condition $(T_3 = 40\,^\circ C)$, *leads to the following*

$$
\begin{bmatrix} 700.0 & 0.0 \\ 0.0 & 1 \end{bmatrix} \begin{Bmatrix} T_1 \\ T_3 \end{Bmatrix} = \begin{Bmatrix} 4499.89 + 700 \times 40 \\ 40.0 \end{Bmatrix}.
\tag{4.57}
$$

The solution to the above equation results in $T_1 = 46.43\,^\circ C$, *which is identical to the exact solution.*

4.3   Radial Heat Conduction in a Cylinder Wall

Many problems in industry, such as heat exchangers, crude oil transport etc., involve the flow of hot fluids in very long pipes which have uniform boundary conditions along the circumference, both inside and outside as shown in Figure 4.10. In such problems, the heat transfer mainly

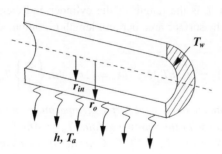

Figure 4.10   Radial heat conduction in an infinitely long cylinder.

takes place along the radial direction apart from the end effects. The governing differential equation for heat flow in cylindrical geometries is

$$\frac{1}{r}\frac{d}{dr}\left(rk\frac{dT}{dr}\right) = 0 \qquad (4.58)$$

The boundary conditions are as follows:

$$At \quad r = r_{in}, T = T_w$$
$$and \ at \quad r = r_o, -k\frac{dT}{dr} = h(T_o - T_a), \qquad (4.59)$$

where T_w is the inside wall temperature, T_o is the outside wall temperature, k is the thermal conductivity, h is the heat transfer coefficient at the outside surface and T_a is the atmospheric temperature.

Integrating Equation (4.58) we obtain

$$kT = C_1 \ln r + C_2. \qquad (4.60)$$

Subjecting the above equation to the boundary conditions of Equation (4.59) results in

$$C_1 = -hr_o(T_o - T_a) \quad and \quad C_2 = kT_w - C_1 \ln r_{in}. \qquad (4.61)$$

Substituting the constants and rearranging Equation (4.60), we obtain the exact solution as

$$\frac{(T - T_w)}{(T_o - T_a)} = \frac{hr_o}{k} \ln \frac{r_{in}}{r}. \qquad (4.62)$$

With the use of the finite element method and assuming a linear variation of temperature, the resulting stiffness matrix is given by

$$[\mathbf{K}]_e = \int_\Omega [\mathbf{B}]_e^T [\mathbf{D}]_e [\mathbf{B}]_e d\Omega + \int_{A_s} h[\mathbf{N}]_e^T [\mathbf{N}]_e dA_s$$

$$= \int_{r_i}^{r_j} \begin{bmatrix} -\frac{1}{l} \\ \frac{1}{l} \end{bmatrix} k \begin{bmatrix} -\frac{1}{l} & \frac{1}{l} \end{bmatrix} L(2\pi r)dr + \int_0^L h \begin{bmatrix} N_i \\ N_j \end{bmatrix} \begin{bmatrix} N_i & N_j \end{bmatrix} (2\pi r_o)dx$$

$$= \frac{2\pi k L}{l} \frac{(r_i + r_j)}{2} \begin{bmatrix} 1 & -1 \\ -1 & 1 \end{bmatrix} + (2\pi r_o)Lh \begin{bmatrix} 0 & 0 \\ 0 & 1 \end{bmatrix}. \qquad (4.63)$$

In the above equation, L is the length of the cylinder and element length in the radial direction is $l = (r_j - r_i)$. The surface area per unit length is computed as $A_s = 2\pi r_o$. The load vector is

$$\{f\}_e = \int_{A_s} hT_a[N]^T dA_s = (2\pi r_o)hL \begin{Bmatrix} 0 \\ 1 \end{Bmatrix} T_a. \tag{4.64}$$

Example 4.3.6 *Calculate the outer wall surface temperature and the temperature distribution in a thick wall cylinder with the following data: $T_w = 100\,°C$, $r_{in} = 40\,cm$, $r_o = 60\,cm$, $k = 10\,W/m\,°C$, $h_o = 10\,W/m^2 C$, $T_a = 30\,°C$. Consider a one element solution with an element length of $l = 60 - 40 = 20\,cm$. Assume a unit cylinder length.*

The element stiffness matrix and the loading vectors are given by

$$[K]_e = \frac{2\pi k}{l}\left(\frac{r_i + r_j}{2}\right)\begin{bmatrix} 1 & -1 \\ -1 & 1 \end{bmatrix} + (2\pi r_o)h\begin{bmatrix} 0 & 0 \\ 0 & 1 \end{bmatrix} = \pi\begin{bmatrix} 50 & -50 \\ -50 & 62 \end{bmatrix} \tag{4.65}$$

and

$$\{f\}_e = \pi\begin{Bmatrix} 0 \\ 360 \end{Bmatrix}. \tag{4.66}$$

The complete system of equations can be written as

$$\pi\begin{bmatrix} 50 & -50 \\ -50 & 62 \end{bmatrix}\begin{Bmatrix} T_i \\ T_j \end{Bmatrix} = \pi\begin{Bmatrix} 0 \\ 360 \end{Bmatrix}. \tag{4.67}$$

The solution to the above system, with $T_i = 100\,°C$ results in $T_j = T_o = 86.45\,°C$, which is greater than the analytical solution, that is $86.30\,°C$. A more accurate solution may be obtained if two elements, each 10 cm long are employed. The assembled equation for the two element system is

$$\begin{bmatrix} 90 & -90 & 0 \\ -90 & 200 & -110 \\ 0 & -110 & 122 \end{bmatrix}\begin{Bmatrix} T_1 \\ T_2 \\ T_3 \end{Bmatrix} = \begin{Bmatrix} 0 \\ 0 \\ 360 \end{Bmatrix}. \tag{4.68}$$

The solution to the above equation, with boundary condition $T_1 = 100\,°C$, gives $T_2 = 92.46\,°C$ and $T_3 = T_o = 86.35\,°C$. The accuracy of the outer wall temperature has been greatly improved by using two elements.

4.4 Solid Cylinder with Heat Source

Consider a homogeneous solid cylinder of radius r_o and length L with uniformly distributed heat source. If we assume a very long cylinder, the temperature in the cylinder will be a function of radius only. Thus,

$$k\left(\frac{d^2T}{dr^2} + \frac{1}{r}\frac{dT}{dr}\right) + \frac{G}{r} = 0. \tag{4.69}$$

The boundary conditions are:

$$\text{at} \quad r = 0, \frac{dT}{dr} = 0 \quad \text{and} \quad r = r_o, T = T_o \tag{4.70}$$

and the heat generated will be equal to the heat lost at the surface, that is,

$$G(\pi r_o^2)L = -k(2\pi r_o)L\left(\frac{dT}{dr}\right)_{r_o}. \tag{4.71}$$

Equation (4.69) can be rewritten as

$$\frac{k}{r}\frac{d}{dr}\left(r\frac{dT}{dr}\right) + G = 0. \tag{4.72}$$

The analytical solution for this problem is:

$$T - T_o = \frac{G}{4k}(r_o^2 - r^2). \tag{4.73}$$

Substituting $r = 0$ and $T = T_c$ (temperature at the center of the solid cylinder) gives

$$T_c - T_o = \frac{Gr_o^2}{4k}. \tag{4.74}$$

Thus,

$$\frac{T - T_o}{T_c - T_o} = 1 - \left(\frac{r}{r_o}\right)^2 \tag{4.75}$$

and

$$\frac{dT}{dr} = -\frac{Gr}{2k}. \tag{4.76}$$

Let us now consider a finite element solution employing linear elements. The stiffness matrix is

$$[\mathbf{K}]_e = \frac{2\pi k}{l}\left(\frac{r_i + r_j}{2}\right)\begin{bmatrix} 1 & -1 \\ -1 & 1 \end{bmatrix} \tag{4.77}$$

and the forcing vector is

$$\{\mathbf{f}\}_e = \int_r G[\mathbf{N}]^T(2\pi r)dr \tag{4.78}$$

per unit length.

In cylindrical coordinates, r may be expressed as

$$r = N_i r_i + N_j r_j. \tag{4.79}$$

Substituting the above equation into Equation (4.78) and integrating between r_i and r_j we obtain

$$\{\mathbf{f}\}_e = \frac{2\pi Gl}{6}\begin{Bmatrix} 2r_i + r_j \\ r_i + 2r_j \end{Bmatrix}, \tag{4.80}$$

where the length of an element l is equal to $(r_j - r_i)$.

Example 4.4.1   *Calculate the surface temperature in a circular solid cylinder of radius 25 mm with a volumetric heat generation of 35.3 MW/m³. The external surface of the cylinder is exposed to a liquid at a temperature of 20°C with a surface heat transfer coefficient of 4000 W/m²°C. The thermal conductivity of the material is 21 W/m°C.*

Let us divide the region into four elements, each of width 6.25 cm.

On substituting the given data into Equation (4.77), the stiffness matrix of the four elements may be calculated as follows:

$$[\mathbf{K}]_1 = 2\pi \begin{bmatrix} 10.5 & -10.5 \\ -10.5 & 10.5 \end{bmatrix} \tag{4.81}$$

$$[\mathbf{K}]_2 = 2\pi \begin{bmatrix} 31.5 & -31.5 \\ -31.5 & 31.5 \end{bmatrix} \tag{4.82}$$

$$[\mathbf{K}]_3 = 2\pi \begin{bmatrix} 52.5 & -52.5 \\ -52.5 & 52.5 \end{bmatrix} \tag{4.83}$$

and

$$[\mathbf{K}]_4 = 2\pi \begin{bmatrix} 73.5 & -73.5 \\ -73.5 & 73.5 \end{bmatrix} + 2\pi \begin{bmatrix} 0 & 0 \\ 0 & 100 \end{bmatrix} \tag{4.84}$$

Similarly the forcing vectors for all the four elements can be calculated as

$$\{\mathbf{f}\}_1 = 2\pi \begin{Bmatrix} 229.82 \\ 459.63 \end{Bmatrix} \tag{4.85}$$

$$\{\mathbf{f}\}_2 = 2\pi \begin{Bmatrix} 919.27 \\ 1149.09 \end{Bmatrix} \tag{4.86}$$

$$\{\mathbf{f}\}_3 = 2\pi \begin{Bmatrix} 1608.18 \\ 1838.54 \end{Bmatrix} \tag{4.87}$$

and

$$\{\mathbf{f}\}_4 = 2\pi \begin{Bmatrix} 2298.18 \\ 2528.00 \end{Bmatrix} + 2\pi \begin{Bmatrix} 0 \\ 2000 \end{Bmatrix} \tag{4.88}$$

Assembly gives

$$\begin{bmatrix} 10.5 & -10.5 & 0.0 & 0.0 & 0.0 \\ -10.5 & 42.0 & -31.5 & 0.0 & 0.0 \\ 0.0 & -31.5 & 84.0 & -52.5 & 0.0 \\ 0.0 & 0.0 & -52.5 & 126.0 & -73.5 \\ 0.0 & 0.0 & 0.0 & -73.5 & 173.5 \end{bmatrix} \begin{Bmatrix} T_1 \\ T_2 \\ T_3 \\ T_4 \\ T_5 \end{Bmatrix} = \begin{Bmatrix} 229.82 \\ 1378.9 \\ 2757.81 \\ 4136.72 \\ 4528.00 \end{Bmatrix}. \tag{4.89}$$

The solution obtained by solving the above system of equations is tabulated in Table 4.3

We can see that the surface temperature, T_5, is predicted very well but the deviation from the exact solution increases as we proceed towards the center. If two linear elements

Table 4.3   Summary of results – temperatures

T	FEM (°C)	Exact (°C)
$T_1 = T_c$	402.19	392.95
T_2	380.28	376.54
T_3	329.20	327.29
T_4	246.02	245.22
$T_5 = T_o$	130.32	130.31

replaces the one element near the center, then the solution for the maximum temperature is improved to 398.43 °C. It is also possible to improve the accuracy of the temperature solution by using quadratic elements. The analytical solution for the outer wall temperature T_o is computed using the energy balance per unit cylinder length at the outer surface, that is, $-k(dT/dr) = h(T_o - T_a)$ and Equation (4.76).

4.5   Conduction–convection Systems

Many physical situations involve the transfer of heat in a material by conduction and its subsequent dissipation by exchange with a fluid or the environment by convection. The heat sinks used in the electronic industry to dissipate heat from electronic components to the ambient is an example of a conduction–convection system. Other examples include the dissipation of heat in electrical windings to the coolant, the heat exchange process in heat exchangers, and the cooling of gas turbine blades where the temperature of the hot gases are greater than the melting point of the blade material. In Section 3.5 we have already demonstrated the applications of the finite element method for extended surfaces with rectangular cross-sections. Also, the problems discussed in the previous section of this chapter include the influence of convective boundary conditions. However, all the problems studied previously in this chapter assumed that the domains were of infinite length.

Figure 4.11 shows various types of fins used in practice. Let us now consider the case of a tapered fin (extended surfaces) with plane surfaces on the top and bottom. The fin also loses heat to the ambient via the tip. The thickness of the fin varies linearly from t_2 at the base to t_1 at the tip as shown in Figure 4.12. The width, b, of the fin remains constant along the whole length.

Let us consider a typical element e, with thicknesses t_i and t_j, areas A_i and A_j and perimeter P_i and P_j at locations i, and j respectively, as shown in Figure 4.13.

$$A_i = bt_i; A_j = bt_j; P_i = 2(b + t_i) \quad \text{and} \quad P_j = 2(b + t_j). \tag{4.90}$$

Since A varies linearly with x we can write

$$A = A_i - \left(\frac{A_i - A_j}{l}\right)x, \tag{4.91}$$

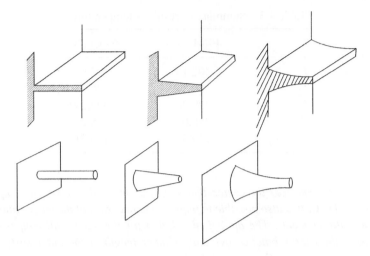

Figure 4.11   Different types of fins.

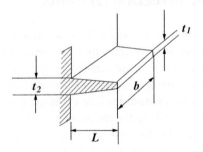

Figure 4.12   Tapered fin.

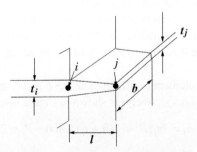

Figure 4.13   Tapered fin. Locations i and j.

where l is the length of an element. Alternatively we can write

$$A = A_i \left(1 - \frac{x}{l}\right) + A_j \frac{x}{l} = N_i A_i + N_j A_j. \tag{4.92}$$

Similarly, $P = N_i P_i + N_j P_j$. The stiffness matrix is written as

$$[\mathbf{K}]_e = \int_l \begin{bmatrix} -\frac{1}{l} \\ \frac{1}{l} \end{bmatrix} [k] \begin{bmatrix} -\frac{1}{l} & \frac{1}{l} \end{bmatrix} A dx + \int_l h \begin{bmatrix} N_i \\ N_j \end{bmatrix} \begin{bmatrix} N_i & N_j \end{bmatrix} P dx. \tag{4.93}$$

After integration and rearrangement, we have

$$[\mathbf{K}]_e = \frac{k}{l} \left(\frac{A_i + A_j}{2}\right) \begin{bmatrix} 1 & -1 \\ -1 & 1 \end{bmatrix} + \frac{hl}{12} \begin{bmatrix} 3P_i + P_j & P_i + P_j \\ P_i + P_j & P_i + 3P_j \end{bmatrix} + hA_r \begin{bmatrix} 0 & 0 \\ 0 & 10 \end{bmatrix}. \tag{4.94}$$

The load vector for this problem is

$$\{\mathbf{f}\}_e = \int_l G[\mathbf{N}]^T A dx - \int_{A_s} q[\mathbf{N}]^T dA_s + \int_{A_s} hT_a[\mathbf{N}]^T dA_s, \tag{4.95}$$

where G is the heat source per unit volume, q is the heat flux, h is the heat transfer coefficient and T_a is the atmospheric temperature. Integrating we obtain

$$\{\mathbf{f}\}_e = \frac{Gl}{6} \left\{ \begin{array}{c} 2A_i + A_j \\ A_i + 2A_j \end{array} \right\} - \frac{ql}{6} \left\{ \begin{array}{c} 2P_i + P_j \\ P_i + 2P_j \end{array} \right\} + \frac{hT_a l}{6} \left\{ \begin{array}{c} 2P_i + P_j \\ P_i + 2P_j \end{array} \right\} + hT_a A_j \left\{ \begin{array}{c} 0 \\ 1 \end{array} \right\}. \tag{4.96}$$

The last contribution is valid only for the element at the end face. For all other elements this last convective term is zero. Note that the general form of the load vector Equation (4.96) includes the lateral surface heat loss due to heat flux q and convection heat transfer. In practice only one of them is active if the lateral surface is not insulated. For the majority of the time the surface heat loss is due to convection heat transfer.

Example 4.5.1 *Let us consider an example with the fin tapering linearly from a thickness of 2 mm at the base to 1 mm at the tip (see Figure 4.14). Also, the tip and lateral surfaces lose heat to the ambient via convection, with a heat transfer coefficient, h, = 120 W/m²°C and atmospheric temperature, T_a, = 25 °C. Determine the temperature distribution if the base temperature is maintained at 100°C. The total length of the fin, L, is 20 mm, the width, b, is 3 mm and thermal conductivity is 200 W/m°C.*

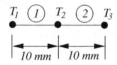

Figure 4.14   Tapered fin. Finite element discretization.

Let us divide the region into two elements of equal length 10 mm each as shown in Figure 4.14. Substituting the relevant data into Equation (4.94) we obtain the stiffness matrices for both elements as follows:

$$[\mathbf{K}]_1 = \begin{bmatrix} 0.109 & -0.103 \\ -0.103 & 0.109 \end{bmatrix} \tag{4.97}$$

and

$$[\mathbf{K}]_2 = \begin{bmatrix} 0.079 & -0.073 \\ -0.073 & 0.079 \end{bmatrix}. \tag{4.98}$$

Similarly, the forcing vectors are calculated as

$$\{\mathbf{f}\}_1 = \begin{Bmatrix} 0.145 \\ 0.140 \end{Bmatrix} \tag{4.99}$$

and

$$\{\mathbf{f}\}_2 = \begin{Bmatrix} 0.130 \\ 0.134 \end{Bmatrix}. \tag{4.100}$$

Assembly of the above equations results in

$$\begin{bmatrix} 0.109 & -0.103 & 0.0 \\ -0.103 & 0.188 & -0.073 \\ 0.0 & -0.073 & 0.079 \end{bmatrix} \begin{Bmatrix} T_1 \\ T_2 \\ T_3 \end{Bmatrix} = \begin{Bmatrix} 0.145 \\ 0.140 + 0.13 \\ 0.134 \end{Bmatrix}. \tag{4.101}$$

On applying the relevant boundary conditions and solving the above system we obtain $T_1 = 100\,°C$, $T_2 = 88.83\,°C$ and $T_3 = 83.96\,°C$.

The heat dissipation can be calculated from the following relationship:

$$Q = \Sigma_{e=1}^2 hP_e L_e \left(\frac{T_i + T_j}{2} - T_a \right). \tag{4.102}$$

Substituting the contribution from both elements results in a value of $Q = 1.42\,W$.

4.6 Summary

In this chapter, examples of one-dimensional problems have been discussed in detail. In most cases analytical solutions were available as benchmarks for the finite element solutions. There are many other application problems which can be studied in one dimension. However, the essential fundamentals of the finite element method for one dimensional heat conduction problems have been given, which may easily be extended to other forms of one-dimensional heat conduction problems.

4.7 Exercises

Exercise 4.7.1 *A composite wall with three different layers, as shown in Figure 4.2 generates $0.25\,G\,W/m^3$ of heat. Using the relevant data given in Example 4.2.1, determine the temperature distribution across the wall using both linear and quadratic variations and compare the results.*

Exercise 4.7.2 *An insulation system around a cylindrical pipe consists of two different layers. The first layer immediately on the outer surface of the pipe is made of glass wool and the second one is constructed using plaster of Paris. The pipe diameter is 10 cm and each insulating layers is 1 cm thick. The thermal conductivity of the glass wool is $0.04\,W/m\,°C$ and that of the plaster is $0.06\,W/m\,°C$. The pipe carries hot oil at a temperature of $92\,°C$ and the atmospheric temperature, outside is $15\,°C$. If the heat transfer coefficient from the outer surface of the insulation to the atmosphere is $15\,W/m^2\,°C$, calculate the temperature at the interface between the two insulating materials and on the outer surface (neglect pipe wall thickness).*

Exercise 4.7.3 *A solid cylinder of 10 cm diameter generates $0.3\,G\,W/m^3$ of heat due to nuclear reaction. If the outside temperature is $40\,°C$ and the heat transfer coefficient from the solid surface to the surrounding fluid is $30\,W/m^2\,°C$ calculate the temperature distribution using quadratic elements.*

Exercise 4.7.4 *A circular fin of inner diameter 20 cm and outer diameter of 26 cm transfers heat from a small motor cycle engine. If the average engine surface temperature is $112\,°C$ determine the temperature distribution along the fin surface. The thermal conductivity of the fin material is $21\,W/m\,°C$ and the convective heat transfer coefficient between the fin and atmosphere is $120\,W/m\,°C$. Assume a suitable atmospheric temperature.*

Exercise 4.7.5 *Consider a composite wall consisting of four different materials as shown in Figure 4.15. Assuming one-dimensional heat flow, determine the heat flow through the composite slab and the interfacial temperatures. $k_A = 200\,W/m\,°C$, $k_b = 20\,W/m\,°C$ and*

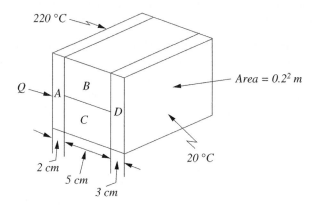

Figure 4.15   A composite wall.

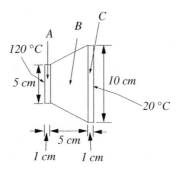

Figure 4.16   A composite wall.

$k_C = 40 \, \text{W/m}\,^\circ C$ and $k_D = 60 \, \text{W/m}\,^\circ C$. Assume areas of surfaces B and C are equal to $0.1 \, m^2$. All side surfaces are insulated.

Exercise 4.7.6   Consider a composite wall, which has one linearly varying cross-sectional area as shown in Figure 4.16. Determine the heat flow and interfacial temperatures. Thickness $= 10 \, cm$, $k_A = 200 \, \text{W/m}\,^\circ C$, $k_B = 20 \, \text{W/m}\,^\circ C$ and $k_C = 40 \, \text{W/m}\,^\circ C$. Assume top and bottom surfaces are insulated.

Exercise 4.7.7   A plane wall $(k = 20 \, \text{W/m}\,^\circ C)$ of thickness 40 cm has its outer surfaces maintained at $30\,^\circ C$. If there is uniform internal heat generation of $0.2 \, MW/m^3$ in the plane wall, determine the temperature distribution in the plane wall. Solve this problem using (a) four linear elements; (b) one quadratic element; (c) one modified quadratic element with only two nodes. Compare the results with analytical solutions.

Exercise 4.7.8   A plane wall $(k = 10 \, \text{W/m}\,^\circ C)$ of thickness 50 cm has its exterior surface subjected to convection environment of $30\,^\circ C$ with a surface heat transfer coefficient of $600 \, W/m^2\,^\circ C$. Determine the temperature distribution in the plane wall using (a) four linear elements; (b) one quadratic element; (c) one modified quadratic element with only two nodes. Compare the results with analytical solution. If the heat transfer coefficient increases to $10,000 \, W/m^2\,^\circ C$, what happens to the temperature of the exterior surface?

Exercise 4.7.9   Calculate the outer wall surface temperature and the temperature distribution in a thick walled hollow cylinder when the inner wall temperature is $120\,^\circ C$ and the outer wall is exposed to a convection environment of $25\,^\circ C$ with a surface heat transfer coefficient of $20 \, W/m^2\,^\circ C$. The inner and outer radii of the hollow cylinder are 30 cm and 60 cm respectively. The thermal conductivity of the material of the hollow cylinder is $20 \, W/m\,^\circ C$. Use one linear element and two linear elements for solution. Compare the results with the analytical solution.

Exercise 4.7.10   Calculate the surface temperature in a circular solid cylinder $(k = 20 \, \text{W/m}^2 C)$ of radius 30 mm with a volumetric heat generation of $25 \, MW/m^3$. The external surface of the cylinder is exposed to a liquid at $25\,^\circ C$ with a heat transfer coefficient of $5000 \, W/m^2\,^\circ C$. Use (a) four linear elements; (b) two quadratic elements. Compare the solution with analytical solution.

Exercise 4.7.11   *Consider a tapered fin of length 5 cm dissipating heat to an ambient at 30 °C. The heat transfer coefficient on the surface at the tip is 100 W/m² °C. The fin tapers from a thickness of 5 mm to a thickness of 2 mm at the tip. The thermal conductivity of the material of the fin is 100 W/m °C. The width of the fin is constant along the length and equal to 2 mm. Determine the heat dissipation from the fin for a base temperature of 100 °C. Use (a) two linear elements; (b) one quadratic element. Also calculate the fin efficiency.*

References

Bejan A (1993) *Heat Transfer*. John Wiley & Sons, New York.

Holman JP (1989) *Heat Transfer*. McGraw-Hill, New York.

Incropera, FP and Dewitt DP (1990) *Fundamentals of Heat and Mass Transfer*. John Wiley & Sons, Inc., New York.

Ozisik MN (1968) *Boundary Value Problems of Heat Conduction*. International Text Book Company, Scranton, PA.

5

Steady-state Heat Conduction in Multi-dimensions

5.1 Introduction

As seen in the previous chapters, a one-dimensional approximation is easy to implement and is also economical. However, the majority of heat transfer problems are multi-dimensional in nature (Bejan 1993; Holman 1989; Incropera and Dewitt 1990; Ozisik 1968). For such problems the accuracy of the solution can be improved using either a two- or three-dimensional approximation. For instance, conduction heat transfer in an infinitely long hallow rectangular tube, which is exposed to uniform but different boundary conditions inside and outside the tube (Figure 5.1(a)), and heat conduction in a thin plate, which has negligible heat transfer in the direction of the thickness, may be approximated as a two-dimensional problem.

In certain situations it is often difficult to simplify the problem to two dimensions without sacrificing accuracy. Most complex industrial heat transfer problems are three-dimensional in nature due to the complicated geometries involved. Heat transfer in aircraft structures and heat shields used in space vehicles are examples of such problems. It is, however, important to note that even simple geometries but which have complex boundary conditions become three-dimensional in nature. For example, the same hollow rectangular tube mentioned previously, but in this case having nonuniform conditions along the length, is a three-dimensional problem. Also, if the hollow rectangular tube is finite, again it may be necessary to treat it as a three-dimensional problem even if the boundary conditions are uniform along the length (Figure 5.1). One typical and simple example of three-dimensional heat conduction is that of a solid cube subjected to different boundary conditions on all six faces as shown in Figure 5.1(b).

Another approximation commonly employed in heat conduction studies is the *axisymmetric formulation*. This type of problem is often considered as A two- and-a-half-dimensional

Fundamentals of the Finite Element Method for Heat and Mass Transfer, Second Edition.
P. Nithiarasu, R. W. Lewis, and K. N. Seetharamu.
© 2016 John Wiley & Sons, Ltd. Published 2016 by John Wiley & Sons, Ltd.

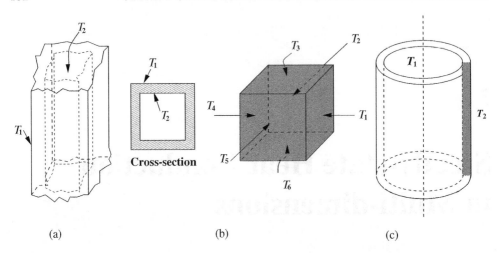

(a)   (b)   (c)

Figure 5.1   Examples of heat conduction in two-dimensional, three-dimensional and axisymmetric geometries. (a) Two dimensional plane geometry; (b) Three dimensional domain; (c) Axisymmetric configuration.

case as it has the features of both a two- and three-dimensional approximation. If a geometry is generated by revolving a surface through 360° with reference to its axis, then it is referred to as being axisymmetric. For instance, revolution of a rectangular surface through 360°, with respect to a vertical axis, produces a vertical cylinder as shown in Figure 5.1(c). Therefore, the heat conduction equations need to be written in three-dimensional cylindrical coordinates for such a system. However, if no significant variation in temperature is expected in the circumferential direction (θ direction), which is often the case, the problem can be reduced to two dimensions and a solution based on the shaded rectangular plane in Figure 5.1(c) is sufficient.

Unlike one-dimensional problems, two- and three-dimensional situations are usually geometrically complex and expensive to solve. The complexity of the problem is increased in multi-dimensions by the occurrence of irregular geometry shapes and the appropriate implementation of boundary conditions on their boundaries. In the case of complicated geometries, it is often necessary to use unstructured meshes (unstructured meshes are generated by employing arbitrarily generated points in a domain, see Chapter 14) to divide the domain into finite elements. Fortunately, due to present day computing capabilities, even complex three-dimensional problems can be solved on a standard personal computer (PC). In the following sections, we demonstrate the solution of multi-dimensional steady-state problems with relevant examples.

5.2   Two-dimensional Plane Problems

5.2.1   Triangular Elements

The simplest finite element discretization which can be employed in two dimensions is by using linear triangular elements. In Chapter 3, we discussed in detail the use of triangular elements. These elements are employed here to solve two-dimensional conduction heat transfer problems.

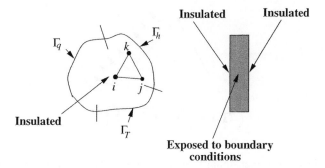

Figure 5.2   Typical two-dimensional plane geometry and triangular element. Front view (left) and side view (right).

In order to demonstrate the use of linear triangular elements, let us consider a general problem as shown in Figure 5.2. As shown in the figure, the geometry is irregular and both the flat faces of the plate are insulated. The surface in the thickness direction is exposed to various boundary conditions. This is an ideal two-dimensional heat conduction problem with no temperature variation allowed in the thickness direction. The final matrix form of the global finite element equations, as given in Chapter 3, is

$$[\mathbf{K}]\{\mathbf{T}\} = \{\mathbf{f}\}, \tag{5.1}$$

where, the stiffness matrix is

$$[\mathbf{K}] = \int_{\Omega} [\mathbf{B}]^T [\mathbf{D}][\mathbf{B}] d\Omega + \int_{\Gamma_h} h[\mathbf{N}]^T [\mathbf{N}] d\Gamma \tag{5.2}$$

and the load vector is

$$\{\mathbf{f}\} = \int_{\Omega} G[\mathbf{N}]^T d\Omega - \int_{\Gamma_q} q[\mathbf{N}]^T d\Gamma + \int_{\Gamma_h} hT_{\infty}[\mathbf{N}]^T d\Gamma. \tag{5.3}$$

In the above equation, heat is assumed to leave the domain due to imposed heat flux and convective boundary conditions.

For a linear triangular element, the temperature distribution can be written as

$$T_e = N_i T_i + N_j T_j + N_k T_k. \tag{5.4}$$

The element gradient matrix is given as

$$\{\mathbf{g}\}_e = \left\{ \begin{array}{c} \dfrac{\partial T}{\partial x} \\ \dfrac{\partial T}{\partial y} \end{array} \right\}_e = \begin{bmatrix} \dfrac{\partial N_i}{\partial x} & \dfrac{\partial N_j}{\partial x} & \dfrac{\partial N_k}{\partial x} \\ \dfrac{\partial N_i}{\partial y} & \dfrac{\partial N_j}{\partial y} & \dfrac{\partial N_k}{\partial y} \end{bmatrix} \left\{ \begin{array}{c} T_i \\ T_j \\ T_k \end{array} \right\} = [\mathbf{B}]_e \{\mathbf{T}\}_e. \tag{5.5}$$

where

$$[\mathbf{B}]_e = \begin{bmatrix} \dfrac{\partial N_i}{\partial x} & \dfrac{\partial N_j}{\partial x} & \dfrac{\partial N_k}{\partial x} \\ \dfrac{\partial N_i}{\partial y} & \dfrac{\partial N_j}{\partial y} & \dfrac{\partial N_k}{\partial y} \end{bmatrix} = \frac{1}{2A_e} \begin{bmatrix} b_i & b_j & b_k \\ c_i & c_j & c_k \end{bmatrix}. \tag{5.6}$$

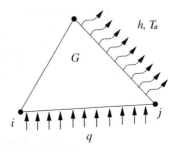

Figure 5.3 Typical two-dimensional triangular element with heat generation and heat flux and convection boundaries.

Note that G in Equation (5.3) is a uniform heat source. Assuming an anisotropic material, we have

$$[\mathbf{D}]_e = \begin{bmatrix} k_x & 0 \\ 0 & k_y \end{bmatrix}. \tag{5.7}$$

Note that the off-diagonal terms are neglected from the above equation for the sake of simplicity. Substituting $[\mathbf{D}]_e$ and $[\mathbf{B}]_e$ into Equation (5.2), we get, for a boundary element as shown in Figure 5.3,

$$[\mathbf{K}]_e = \frac{t}{4A_e}\left\{ k_x \begin{bmatrix} b_i^2 & b_ib_j & b_ib_k \\ b_ib_j & b_j^2 & b_jb_k \\ b_ib_k & b_jb_k & b_k^2 \end{bmatrix} + k_y \begin{bmatrix} c_i^2 & c_ic_j & c_ic_k \\ c_ic_j & c_j^2 & c_jc_k \\ c_ic_k & c_jc_k & c_k^2 \end{bmatrix} \right\} + \frac{htl_{jk}}{6} \begin{bmatrix} 0 & 0 & 0 \\ 0 & 2 & 1 \\ 0 & 1 & 2 \end{bmatrix}. \tag{5.8}$$

The subscript e in the above equation denotes a single element. It should be noted that $d\Omega$ in the above equation is equal to $tdxdy$ and $d\Gamma$ is equal to tdx, where t is the thickness of the plate and l is the length of an element side (or edge) on the domain boundary. In a similar fashion, the forcing vector can be written as

$$\{\mathbf{f}\}_e = \frac{GA_et}{3}\begin{Bmatrix} 1 \\ 1 \\ 1 \end{Bmatrix} + \frac{qtl_{ij}}{2}\begin{Bmatrix} 1 \\ 1 \\ 0 \end{Bmatrix} + \frac{hT_atl_{jk}}{2}\begin{Bmatrix} 0 \\ 1 \\ 1 \end{Bmatrix}. \tag{5.9}$$

Note that the heat flux term is positive in the above equation. This is due to the fact that the heat flux is going into the element, that is, $q = q_y \times n_y$. Although q_y is positive, the outward pointing normal, n_y, on along the edge $i - j$ in Figure 5.3 is -1. The integration formulae used in the derivation of stiffness matrix and loading vectors are simple, as indicated in Chapter 3. For convenience, we have listed the integration formulae in Appendix C.

As seen in the previous equations, the effect of uniform heat generation contributes to all three nodes of an element irrespective of its position. However, the convection and flux boundary conditions are applicable only on the boundaries of the domain. If we need to have a "point source" G^* instead of a "uniform source" G, the first term in Equation (5.9) is replaced with

$$G^* t \begin{Bmatrix} N_i \\ N_j \\ N_k \end{Bmatrix}_{(x_o, y_o)}, \qquad (5.10)$$

where x_o and y_o are the coordinates of the point source. In the above equations, all the shape function values must be evaluated at (x_o, y_o) (note that although G^* is a point source, in two dimensions, it is a line in the thickness direction and expressed in units of W/m). The contribution from the point source is then appropriately distributed to the three nodes of the element which contains the point source.

In order to demonstrate the characteristics of two dimensional steady-state heat transfer, the temperature distribution in a flat plate having constant temperature boundary conditions is considered in the following example.

Example 5.2.1   *A square plate of unit thickness and size 1 m, as shown in Figure 5.4, is subjected to isothermal boundary conditions of 100°C on all sides except the top side which is subjected to 500°C. If the thermal conductivity of the material is constant and equal to 10 W/m°C, determine the temperature distribution using linear triangular finite elements.*

*The square domain is first divided into eight equal-sized linear triangular elements as shown in Figure 5.5. Two sets of elemental [**K**] matrices exist due to the orientation of the triangles. For elements 1, 3, 5 and 7 we have the following elements of the [**K**] matrix, if they are numbered as shown in Figure 5.4.*

$$\begin{aligned} b_1 &= y_2 - y_4 = -0.50; & c_1 &= x_4 - x_2 = -0.50 \\ b_2 &= y_4 - y_1 = 0.50; & c_2 &= x_1 - x_4 = 0.00 \\ b_4 &= y_1 - y_2 = 0.00; & c_4 &= x_2 - x_1 = 0.50. \end{aligned} \qquad (5.11)$$

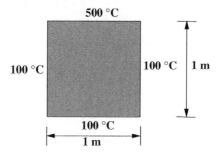

Figure 5.4   Square plate with different temperature boundary conditions.

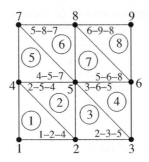

Figure 5.5   Discretization using triangular elements.

The elemental **[K]** *matrices for elements 1, 3, 5 and 7 can be written as (refer to Equation (5.8)):*

$$[\mathbf{K}]_1 = \frac{tk}{4A_e} \begin{bmatrix} b_1^2 + c_1^2 & b_1 b_2 + c_1 c_2 & b_1 b_4 + c_1 c_4 \\ b_1 b_2 + c_1 c_2 & b_2^2 + c_2^2 & b_2 b_4 + c_2 c_4 \\ b_1 b_4 + c_1 c_4 & b_2 b_4 + c_2 c_4 & b_4^2 + c_4^2 \end{bmatrix} = [\mathbf{K}]_3 = [\mathbf{K}]_5 = [\mathbf{K}]_7, \quad (5.12)$$

where the area of the elements can be written as

$$2A_e = det \begin{vmatrix} 1.0 & 0.0 & 0.0 \\ 1.0 & 0.5 & 0.0 \\ 1.0 & 0.0 & 0.5 \end{vmatrix} = 0.25 \ m^2. \quad (5.13)$$

Substituting the area into Equation (5.12) we get the final form of the elemental stiffness matrix as

$$[\mathbf{K}]_1 = [\mathbf{K}]_3 = [\mathbf{K}]_5 = [\mathbf{K}]_7 = \frac{tk}{2} \begin{bmatrix} 2.0 & -1.0 & -1.0 \\ -1.0 & 1.0 & 0.0 \\ -1.0 & 0.0 & 1.0 \end{bmatrix}. \quad (5.14)$$

Similarly, we can calculate elemental **[K]** *matrices for elements 2, 4, 6 and 8 as*

$$[\mathbf{K}]_2 = [\mathbf{K}]_4 = [\mathbf{K}]_6 = [\mathbf{K}]_8 = \frac{tk}{2} \begin{bmatrix} 1.0 & -1.0 & 0.0 \\ -1.0 & 2.0 & -1.0 \\ 0.0 & -1.0 & 1.0 \end{bmatrix}. \quad (5.15)$$

It is also easy to verify that if the connectivity of element 2 is changed to 5-4-2, the element matrix (5.15) will be identical to Equation (5.14). Now, the assembled equations are (see Appendix D):

$$\frac{tk}{2}\begin{bmatrix} 2.0 & -1.0 & 0.0 & -1.0 & 0.0 & 0.0 & 0.0 & 0.0 & 0.0 \\ -1.0 & 4.0 & -1.0 & 0.0 & -2.0 & 0.0 & 0.0 & 0.0 & 0.0 \\ 0.0 & -1.0 & 2.0 & 0.0 & 0.0 & -1.0 & 0.0 & 0.0 & 0.0 \\ -1.0 & 0.0 & 0.0 & 4.0 & -2.0 & 0.0 & -1.0 & 0.0 & 0.0 \\ 0.0 & -2.0 & 0.0 & -2.0 & 8.0 & -2.0 & 0.0 & -2.0 & 0.0 \\ 0.0 & 0.0 & -1.0 & 0.0 & -2.0 & 4.0 & 0.0 & 0.0 & -1.0 \\ 0.0 & 0.0 & 0.0 & -1.0 & 0.0 & 0.0 & 2.0 & -1.0 & 0.0 \\ 0.0 & 0.0 & 0.0 & 0.0 & -2.0 & 0.0 & -1.0 & 4.0 & -1.0 \\ 0.0 & 0.0 & 0.0 & 0.0 & 0.0 & -1.0 & 0.0 & -1.0 & 2.0 \end{bmatrix}\begin{bmatrix} T_1 \\ T_2 \\ T_3 \\ T_4 \\ T_5 \\ T_6 \\ T_7 \\ T_8 \\ T_9 \end{bmatrix} = \begin{Bmatrix} 0.0 + Q_1 \\ 0.0 + Q_2 \\ 0.0 + Q_3 \\ 0.0 + Q_4 \\ 0.0 \\ 0.0 + Q_6 \\ 0.0 + Q_7 \\ 0.0 + Q_8 \\ 0.0 + Q_9 \end{Bmatrix},$$

(5.16)

where Q_1 to Q_4 and Q_6 to Q_9 are the heat flow at the corresponding nodes. The only unknown temperature in the above equation is T_5, which can be calculated from the equation corresponding to the fifth node, that is, from

$$8T_5 = 2T_2 + 2T_4 + 2T_6 + 2T_8.$$

(5.17)

Substituting $T_2 = T_4 = T_6 = 100\,°C$ and $T_8 = 500\,°C$, we get $T_5 = 200\,°C$. The heat flow at different nodes may be computed using the nodal equations of the system 5.16. The calculated heat flow values are $Q_1 = 0$ W, $Q_2 = -1000$ W, $Q_3 = 0$ W, $Q_4 = -1000$ W, $Q_6 = -1000$ W, $Q_7 = -2000$ W, $Q_8 = 7000$ W and $Q_9 = -2000$ W. The negative sign indicates the heat leaving the domain. As seen the heat entering the domain at node 8 balances the total heat leaving the domain.

The analytical solution to this problem is given by Holman (1989).

$$T(x, y) = (T_{max} - T_{min})\frac{2}{\pi}\sum_{n=1}^{\infty}\frac{(-1)^{n+1} + 1}{n}\sin\left(\frac{n\pi x}{w}\right)\frac{\sinh\left(\frac{n\pi y}{w}\right)}{\sinh\left(\frac{n\pi H}{w}\right)} + T_{min},$$

(5.18)

where w is the width, H is the height of the plate, T_{max} is the maximum temperature on the boundary and T_{min} is the minimum temperature on the boundary. Therefore,

$$T(0.5, 0.5) = 200.11\,°C.$$

(5.19)

As seen the finite element solution is in close agreement with the analytical solution. It is interesting to note that the finite difference solution is given by

$$T_5 = \frac{T_2 + T_4 + T_6 + T_8}{4} = 200\,°C,$$

(5.20)

which is identical to the finite element solution. Figure 5.6 shows the computer generated solution on an unstructured mesh for this problem. As shown, the temperature at the center is close to that obtained from regular mesh shown in Figure 5.5, and also to the analytical solution. However, the unstructured mesh solution is not as accurate as that of the regular mesh solution. This indicates, that the accuracy of a regular structured mesh is superior to that of unstructured meshes. If a finer structured mesh as shown in Figure 5.7 is used, the

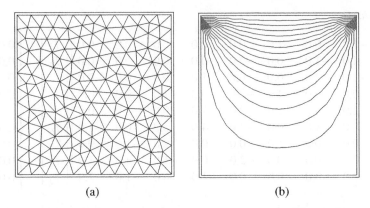

(a)	(b)

Figure 5.6   Solution for Example 5.6.1 on an unstructured mesh. The temperature obtained at the center of the plate is 200.42 °C. (a) Finite element mesh; (b) Temperature contours. Temperature varies between 100 °C and 500 °C. Interval between two contours is 25 °C.

temperature at the center is 199.99 °C. This is closer to the analytical solution than that of all other solutions presented.

Using the nodal temperature values, the temperature at any other location within an element can be determined using linear interpolation. The calculation of the temperature at any arbitrary location has been demonstrated in Chapter 3. The following two-dimensional example is given in order to further illustrate this point.

Example 5.2.2   *Calculate the temperature at point 4 (40,40) shown in Figure 5.8. The temperature values at nodes 1 2 and 3 are 100°C, 200°C and 100°C respectively. The*

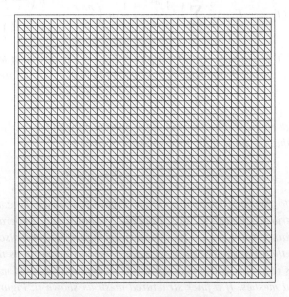

Figure 5.7   Fine structured mesh.

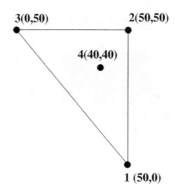

Figure 5.8   Interpolation into a triangular element.

coordinates of these points are (50,0), (50,50) and (0,50) respectively. All dimensions are in cm. Also, calculate the heat flux in both the x and y directions. Assume a thermal conductivity value of 10 W/cm°C.

The following expression can be used to describe the linear variation of temperature within the element

$$T = N_1T_1 + N_2T_2 + N_3T_3. \tag{5.21}$$

In order to calculate the temperature at node 4, the shape functions N_1, N_2 and N_3 have to be calculated at node 4. Therefore, for the first node

$$N_1 = \frac{1}{2A}(a_1 + b_1x_4 + c_1y_4), \tag{5.22}$$

where

$$a_1 = x_2y_3 - x_3y_2 = 2500.00$$
$$b_1 = y_2 - y_3 = 0.0$$
$$c_1 = x_3 - x_2 = -50.00. \tag{5.23}$$

At point 4, (x = 40, y = 40), from Equation (5.22) we get

$$N_1 = \frac{1}{5}. \tag{5.24}$$

Similarly, it can be verified that $N_2 = 3/5$ and $N_3 = 1/5$. Note that $N_1 + N_2 + N_3 = 1$. On substituting these shape function values into Equation (5.21) results in a value of $T_4 = 160°C$.

The heat flux in the x and y directions are calculated as

$$q_x = -k\frac{\partial T}{\partial x} = -\frac{k}{2A_e}(b_1T_1 + b_2T_2 + b_3T_3)$$

$$= -\frac{10}{2500}(b_1T_1 + b_2T_2 + b_3T_3) = -20 \ W/cm^2. \tag{5.25}$$

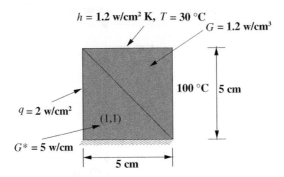

Figure 5.9   A square domain with mixed boundary conditions.

Similarly, it can be shown that $q_y = -20$ W/cm^2. It should be noted that the flux is constant over a linear triangular element.

From Examples 5.2.1 and 5.2.2, the demonstration of problems involving constant temperature boundary conditions is clear. It is therefore essential to move on to an example with more complicated boundary conditions. Thus, in the following example, a conduction problem is considered, which has mixed boundary conditions.

Example 5.2.3   *Determine the temperature distribution in a square plate of size 5 cm and unit thickness as shown in Figure 5.9. The upper triangular half has an internal heat generation of 1.2 W/cm^3 while the lower half has a point source of 5 W/cm in the thickness direction (point source on a two-dimensional plane) at the point (1,1) cm. In addition to the above heat sources, the bottom side of the plate is insulated, the right vertical side is subjected to a temperature of 100°C, the top side is subjected to a convective heat heat transfer boundary condition with a heat transfer coefficient of h = 1.2 W/cm^{2}K and $T_a = 30$°C and the left vertical side is subjected to a uniform heat flux of 2 W/cm^2 leaving the domain. Assume a thermal conductivity of 2 W/cm°C.*

To make the solution procedure simple, the plate is divided into two triangular elements as shown in Figure 5.10. The elemental equations of both elements can be set up separately using the formulation discussed (Equations (5.8) and (5.9)). For the first element, $a_1 = 25.0$, $b_1 = -5.0$, $c_1 = -5.0$, $a_2 = 0.0$, $b_2 = 5.0$, $c_2 = 0.0$, $a_3 = 0.0$, $b_3 = 0.0$, $c_3 = 5.0$.

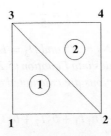

Figure 5.10   Discretization using two triangular elements.

The stiffness matrix for element 1 is

$$[\mathbf{K}]_1 = \frac{t}{4A_e} \left\{ k_x \begin{bmatrix} b_1^2 & b_1 b_2 & b_1 b_3 \\ b_1 b_2 & b_2^2 & b_2 b_3 \\ b_1 b_3 & b_2 b_3 & b_3^2 \end{bmatrix} + k_y \begin{bmatrix} c_1^2 & c_1 c_2 & c_1 c_3 \\ c_1 c_2 & c_2^2 & c_2 c_3 \\ c_1 c_3 & c_2 c_3 & c_3^2 \end{bmatrix} \right\}. \tag{5.26}$$

Substituting the values for a, b and c we obtain

$$[\mathbf{K}]_1 = \begin{bmatrix} 2.0 & -1.0 & -1.0 \\ -1.0 & 1.0 & 0.0 \\ -1.0 & 0.0 & 1.0 \end{bmatrix}. \tag{5.27}$$

The loading term for the element 1 is given by

$$\{\mathbf{f}\}_1 = -\frac{t q l_{31}}{2} \begin{Bmatrix} 1.0 \\ 0.0 \\ 1.0 \end{Bmatrix} + G^* t \begin{Bmatrix} N_1 \\ N_2 \\ N_3 \end{Bmatrix}_{(1,1)} = \begin{Bmatrix} -2.0 \\ 1.0 \\ -4.0 \end{Bmatrix}. \tag{5.28}$$

Note that the shape functions evaluated at point (1,1) are $N_1 = 3/5, N_2 = 1/5$ and $N_3 = 1/5$. In a similar way, the stiffness matrix and loading terms for the second element can be calculated. They are

$$[\mathbf{K}]_2 = \begin{bmatrix} 1.0 & -1.0 & 0.0 \\ -1.0 & 4.0 & 0.0 \\ 0.0 & 0.0 & 3.0 \end{bmatrix} \tag{5.29}$$

and

$$\{\mathbf{f}\}_2 = \begin{Bmatrix} 5.0 \\ 95.0 \\ 95.0 \end{Bmatrix}. \tag{5.30}$$

On assembling the above contributions for the two elements, we obtain the following system of simultaneous equations (see Appendix D), that is,

$$\begin{bmatrix} 2.0 & -1.0 & -1.0 & 0.0 \\ -1.0 & 2.0 & 0.0 & -1.0 \\ -1.0 & 0.0 & 4.0 & 0.0 \\ 0.0 & -1.0 & 0.0 & 4.0 \end{bmatrix} \begin{Bmatrix} T_1 \\ T_2 \\ T_3 \\ T_4 \end{Bmatrix} = \begin{Bmatrix} -2.0 \\ 6.0 + Q_2 \\ 91.0 \\ 95.0 + Q_4 \end{Bmatrix}. \tag{5.31}$$

In the above set of equations the temperature T_2 and T_4 are known and are equal to $100\,^\circ C$.

The boundary conditions can be implemented as previously explained in Chapters 2 and 3. In order to fully appreciate the application of the boundary conditions, the reader is refer to Appendix D.

Applying the boundary conditions, we get

$$\begin{bmatrix} 2.0 & -1.0 & -1.0 & 0.0 \\ 0.0 & 1.0 & 0.0 & 0.0 \\ -1.0 & 0.0 & 4.0 & 0.0 \\ 0.0 & 0.0 & 0.0 & 1.0 \end{bmatrix} \begin{Bmatrix} T_1 \\ T_2 \\ T_3 \\ T_4 \end{Bmatrix} = \begin{Bmatrix} -2.0 \\ 100.0 \\ 91.0 \\ 100.0 \end{Bmatrix}. \tag{5.32}$$

Therefore, the simultaneous equations to be solved are $2T_1 - T_3 = 98$ *and* $-T_1 + 4T_3 = 91$. *The solution to these equations results in* $T_1 = 69\,°C$ *and* $T_3 = 40\,°C$. *The heat flow at nodes 2 and 4 are calculated as* $Q_2 = 25$ W *and* $Q_4 = 205$ W. *The total heat flowing in is therefore is 230 W. The total heat of the system is calculated by adding the total heat flow into the domain with any heat generation, that is, total heat of the system is* $Q_2 + Q_4 + G×$ *volume of triangle 2-3-4* $+ G^* × t = 250$ W.

The heat flowing out of the plate may be computed by calculating the heat flow at edges 4-3 and 3-1 as

$$Q_h = -htl_{43}\left(\frac{1}{2}(T_3 + T_4) - T_a\right) = -240 \text{ W}; \quad and \quad Q_q = -qtl_{31} = -10 \text{ W} \quad (5.33)$$

respectively. As seen, the total heat flowing out of the domain is -250 W *which is identical to the total heat of the system.*

If in the above example, there is an uniform heat generation of 1.2 W/cm^3 *throughout the domain (no point heat source), then the loading term for the first element changes to*

$$\{\mathbf{f}\}_1 = -\frac{qtl_{31}}{2}\begin{Bmatrix} 1 \\ 0 \\ 1 \end{Bmatrix} + \frac{GA_et}{3}\begin{Bmatrix} 1 \\ 1 \\ 1 \end{Bmatrix} = \begin{Bmatrix} 0 \\ 5 \\ 0 \end{Bmatrix}. \quad (5.34)$$

The resulting simultaneous equations become $2T_1 - T_3 = 100$ *and* $-T_1 + 4T_3 = 95$ *and the solution becomes* $T_1 = 70.71\,°C$ *and* $T_3 = 41.42\,°C$. *We leave the heat balance calculations to the readers. The total heat flow here is slightly increased due to the uniform heat generation to approximately 255 W. This value is approximately equal to heat leaving the domain due to convection and prescribed flux along sides 4-3 and 3-1 respectively.*

5.3 Rectangular Elements

A typical rectangular element is shown in Figure 5.11 with mixed boundary conditions. The temperature distribution in a rectangular element is written as

$$T_e = N_iT_i + N_jT_j + N_kT_k + N_lT_l. \quad (5.35)$$

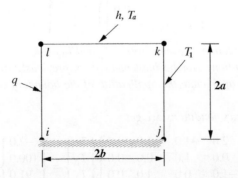

Figure 5.11   Rectangular element with different boundary conditions.

From Chapter 3 (Equation (3.91) with origin at node i), the shape functions for a rectangular element are given as (replacing x with $(x - b)$ and y with $(y - a)$ in Equation 3.88)

$$N_i = \left(1 - \frac{x}{2b}\right)\left(1 - \frac{y}{2a}\right)$$
$$N_j = \frac{x}{2b}\left(1 - \frac{y}{2a}\right)$$
$$N_k = \frac{xy}{4ab}$$
$$N_l = \frac{y}{2a}\left(1 - \frac{x}{2b}\right). \tag{5.36}$$

The element gradient matrix of the shape functions is

$$[\mathbf{B}]_e = \begin{bmatrix} \frac{\partial N_i}{\partial x} & \frac{\partial N_j}{\partial x} & \frac{\partial N_k}{\partial x} & \frac{\partial N_l}{\partial x} \\ \frac{\partial N_i}{\partial y} & \frac{\partial N_j}{\partial y} & \frac{\partial N_k}{\partial y} & \frac{\partial N_l}{\partial y} \end{bmatrix} = \frac{1}{4ab}\begin{bmatrix} -(2a - y) & (2a - y) & y & -y \\ -(2b - x) & -x & x & (2b - x) \end{bmatrix}. \tag{5.37}$$

The element stiffness matrix is given by

$$[\mathbf{K}]_e = \int_\Omega [\mathbf{B}]_e^T [\mathbf{D}]_e [\mathbf{B}]_e dV + \int_\Gamma h[\mathbf{N}]_e^T [\mathbf{N}]_e d\Gamma, \tag{5.38}$$

where

$$[\mathbf{D}]_e = \begin{bmatrix} k_x & 0 \\ 0 & k_y \end{bmatrix}_e. \tag{5.39}$$

Substituting, the $[\mathbf{B}]_e$ and $[\mathbf{D}]_e$ matrices into the above equation, results in a 4×4 matrix. We leave the algebra to the readers to work out. A typical term in the matrix is

$$\int_0^{2b}\int_0^{2a} \frac{k_x}{16a^2b^2}(2a - y)^2 t\,dx dy + \int_0^{2b}\int_0^{2a} \frac{k_y}{16a^2b^2}(2b - x)^2 t\,dx dy. \tag{5.40}$$

After integration, the matrix $[\mathbf{K}]$ becomes

$$[\mathbf{K}]_e = \frac{k_x at}{6b}\begin{bmatrix} 2.0 & -2.0 & -1.0 & 1.0 \\ -2.0 & 2.0 & 1.0 & -1.0 \\ -1.0 & 1.0 & 2.0 & -2.0 \\ 1.0 & -1.0 & -2.0 & 2.0 \end{bmatrix} + \frac{k_y at}{6b}\begin{bmatrix} 2.0 & 1.0 & -1.0 & -2.0 \\ 1.0 & 2.0 & -2.0 & -1.0 \\ -1.0 & -2.0 & 2.0 & 1.0 \\ -2.0 & -1.0 & 1.0 & 2.0 \end{bmatrix}$$

$$+ \frac{hbt}{6}\begin{bmatrix} 0.0 & 0.0 & 0.0 & 0.0 \\ 0.0 & 0.0 & 0.0 & 0.0 \\ 0.0 & 0.0 & 4.0 & 2.0 \\ 0.0 & 0.0 & 2.0 & 4.0 \end{bmatrix}. \tag{5.41}$$

The loading vector can be written as

$$\{\mathbf{f}\}_e = \int_\Omega G[\mathbf{N}]^T d\Omega = \int_0^{2b}\int_0^{2a} Gt \begin{Bmatrix} N_i \\ N_j \\ N_k \\ N_l \end{Bmatrix} dx dy = \frac{GA_e t}{4}\begin{Bmatrix} 1 \\ 1 \\ 1 \\ 1 \end{Bmatrix}. \tag{5.42}$$

The heat flux and convective heat transfer boundary integrals are evaluated as for triangular elements. In order to demonstrate the application of such elements, Example 5.2.3 will now be reconsidered using a rectangular element.

Example 5.3.1 *Determine the temperature distribution in the square plate of Example 5.2.3, using a single rectangular element. Now assume that the volumetric heat generation is extended to throughout the domain and the point source is absent.*

Substituting the relevant data into Equation (5.41) we get (see Figure 5.12)

$$
[\mathbf{K}]_e = \frac{5}{15}
\begin{bmatrix}
2.0 & -2.0 & -1.0 & 1.0 \\
-2.0 & 2.0 & 1.0 & -1.0 \\
-1.0 & 1.0 & 2.0 & -2.0 \\
1.0 & -1.0 & -2.0 & 2.0
\end{bmatrix}
+ \frac{5}{15}
\begin{bmatrix}
2.0 & 1.0 & -1.0 & -2.0 \\
1.0 & 2.0 & -2.0 & -1.0 \\
-1.0 & -2.0 & 2.0 & 1.0 \\
-2.0 & -1.0 & 1.0 & 2.0
\end{bmatrix}
$$

$$
+
\begin{bmatrix}
0.0 & 0.0 & 0.0 & 0.0 \\
0.0 & 0.0 & 0.0 & 0.0 \\
0.0 & 0.0 & 2.0 & 1.0 \\
0.0 & 0.0 & 1.0 & 2.0
\end{bmatrix}.
\qquad (5.43)
$$

Simplifying, this becomes

$$
[\mathbf{K}]_e = \frac{1}{6}
\begin{bmatrix}
8.0 & -2.0 & -4.0 & -2.0 \\
-2.0 & 8.0 & -2.0 & -4.0 \\
-4.0 & -2.0 & 20.0 & 4.0 \\
-2.0 & -4.0 & 4.0 & 20.0
\end{bmatrix}.
\qquad (5.44)
$$

The forcing vector is

$$
\{\mathbf{f}\}_e = \frac{30t}{4}
\begin{Bmatrix} 1 \\ 1 \\ 1 \\ 1 \end{Bmatrix}
- \frac{qtl_{14}}{2}
\begin{Bmatrix} 1 \\ 0 \\ 0 \\ 1 \end{Bmatrix}
+ \frac{hT_a tl_{31}}{2}
\begin{Bmatrix} 0 \\ 0 \\ 1 \\ 1 \end{Bmatrix};
\qquad (5.45)
$$

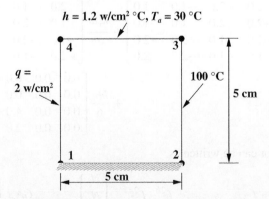

$h = 1.2 \text{ w/cm}^2 \,^{\circ}\text{C}, T_a = 30 \,^{\circ}\text{C}$

$q = 2 \text{ w/cm}^2$

$100 \,^{\circ}\text{C}$

5 cm

5 cm

Figure 5.12 Heat conduction in a square plate. Approximated using a rectangular (square) element.

again, on simplifying we obtain

$$\{\mathbf{f}\}_e = \begin{Bmatrix} 2.5 \\ 7.5 + Q_2 \\ 97.5 + Q_3 \\ 92.5 \end{Bmatrix}. \tag{5.46}$$

Therefore, the final form of the set of simultaneous equations can be written as

$$\frac{1}{6} \begin{bmatrix} 8.0 & -2.0 & -4.0 & -2.0 \\ -2.0 & 8.0 & -2.0 & -4.0 \\ -4.0 & -2.0 & 20.0 & 4.0 \\ -2.0 & -4.0 & 4.0 & 20.0 \end{bmatrix} \begin{Bmatrix} T_1 \\ T_2 \\ T_3 \\ T_4 \end{Bmatrix} = \begin{Bmatrix} 2.5 \\ 7.5 + Q_2 \\ 97.5 + Q_3 \\ 92.5 \end{Bmatrix}. \tag{5.47}$$

The temperatures at points 2 and 4 are known. On substitution into the above system, results in the following simultaneous equations:

$$\begin{bmatrix} 8 & -2 \\ -2 & 20 \end{bmatrix} \begin{Bmatrix} T_1 \\ T_4 \end{Bmatrix} = \begin{Bmatrix} 615 \\ 555 \end{Bmatrix}. \tag{5.48}$$

The solution of the above simultaneous equation gives $T_4 = 36.34\,°C$ and $T_1 = 85.96\,°C$.

The energy balance between the heat inflow and out flow is easily verified. The heat inflow is equal to $G \times volume + Q_2 + Q_3 = 240$ W. The heat outflow is due to the heat convection along side $k - l$ and hear flux along side $l - i$. These together gives approximately -240 W.

5.4 Plate with Variable Thickness

The conduction heat transfer in a plate with variable thickness is essentially a three-dimensional problem. However, if the thickness variation is small, then it is possible to express the thickness as a linear variation in the discretized triangular element as shown in Figure 5.13. If the thickness variation is assumed to be linear, we can write

$$t = N_i t_i + N_j t_j + N_k T_k \tag{5.49}$$

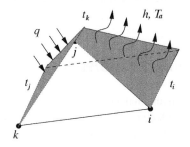

Figure 5.13 A triangular plate with linearly varying thickness.

Therefore, the element stiffness matrix can be rewritten as

$$
\begin{aligned}
[\mathbf{K}]_e &= \int_V [\mathbf{B}]_e^T [\mathbf{D}]_e [\mathbf{B}]_e dV + \int_S h[\mathbf{N}]_e^T [\mathbf{N}]_e dS \\
&= \int_A [\mathbf{B}]_e^T [\mathbf{D}]_e [\mathbf{B}]_e (N_i t_i + N_j t_j + N_k t_k)_e dA \\
&\quad + \int_{l_{ij}} h[\mathbf{N}]_e^T [\mathbf{N}]_e (N_i t_i + N_j t_j + N_k t_k)_e dx.
\end{aligned}
\tag{5.50}
$$

On substitution of the various matrices and integrating (see Appendix C), we finally obtain

$$
[\mathbf{K}]_e = \left(\frac{t_i + t_j + t_k}{12A_e} \right) \left\{ k_x \begin{bmatrix} b_i^2 & b_i b_j & b_i b_k \\ b_i b_j & b_j^2 & b_j b_k \\ b_i b_k & b_j b_k & b_k^2 \end{bmatrix} + k_y \begin{bmatrix} c_i^2 & c_i c_j & c_i c_k \\ c_i c_j & c_j^2 & c_j c_k \\ c_i c_k & c_j c_k & c_k^2 \end{bmatrix} \right\}
$$
$$
+ \frac{h l_{ij}}{12} \begin{bmatrix} 3t_i + t_j & t_i + t_j & 0.0 \\ t_i + t_j & t_i + 3t_j & 0.0 \\ 0.0 & 0.0 & 0.0 \end{bmatrix}.
\tag{5.51}
$$

The load term is calculated as

$$
\begin{aligned}
\{\mathbf{f}\}_e &= \int_A G[\mathbf{N}]_e^T (N_i t_i + N_j t_j + N_k t_k)_e dA \\
&\quad - \int_{l_{jk}} q[\mathbf{N}]_e^T (N_i t_i + N_j t_j + N_k t_k)_e dx \\
&\quad + \int_{l_{ij}} hT_a [\mathbf{N}]_e^T (N_i t_i + N_j t_j + N_k t_k)_e dx.
\end{aligned}
\tag{5.52}
$$

Again on integration we obtain,

$$
\frac{GA_e}{12} \left\{ \begin{matrix} 2t_i + t_j + t_k \\ t_i + 2t_j + t_k \\ t_i + t_j + 2t_k \end{matrix} \right\} - \frac{q l_{jk}}{6} \left\{ \begin{matrix} 0 \\ 2t_j + t_k \\ t_j + 2t_k \end{matrix} \right\} + \frac{hT_a l_{ij}}{6} \left\{ \begin{matrix} 2t_i + t_j \\ t_i + 2t_j \\ 0 \end{matrix} \right\}.
\tag{5.53}
$$

If the thickness is constant, the above relations reduce to the same set of equations as in Section 5.2.

5.5 Three-dimensional Problems

The formulation of a three-dimensional problem follows a similar approach, as explained previously for two-dimensional plane geometries, but with an additional third dimension. The finite element equation is the same as in Equation (5.1), that is,

$$
[\mathbf{K}]\{\mathbf{T}\} = \{\mathbf{f}\}.
\tag{5.54}
$$

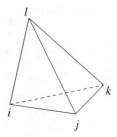

Figure 5.14   A linear tetrahedral element.

For a linear tetrahedral element as shown in Figure 5.14, the temperature distribution over an element can be written as

$$T_e = N_i T_i + N_j T_j + N_k T_k + N_l T_l. \tag{5.55}$$

The element gradient matrix is given as

$$\{g\}_e = \left\{ \begin{array}{c} \frac{\partial T}{\partial x} \\ \frac{\partial T}{\partial y} \\ \frac{\partial T}{\partial z} \end{array} \right\} = \left[\begin{array}{cccc} \frac{\partial N_i}{\partial x} & \frac{\partial N_j}{\partial x} & \frac{\partial N_k}{\partial x} & \frac{\partial N_l}{\partial x} \\ \frac{\partial N_i}{\partial y} & \frac{\partial N_j}{\partial y} & \frac{\partial N_k}{\partial y} & \frac{\partial N_l}{\partial y} \\ \frac{\partial N_i}{\partial z} & \frac{\partial N_j}{\partial z} & \frac{\partial N_k}{\partial z} & \frac{\partial N_l}{\partial z} \end{array} \right] \left\{ \begin{array}{c} T_i \\ T_j \\ T_k \\ T_l \end{array} \right\} = [\mathbf{B}]_e \{\mathbf{T}\}_e, \tag{5.56}$$

The thermal conductivity matrix becomes

$$[\mathbf{D}]_e = \left[\begin{array}{ccc} k_x & 0 & 0 \\ 0 & k_y & 0 \\ 0 & 0 & k_z \end{array} \right]_e, \tag{5.57}$$

where the off-diagonal terms are assumed to be zero for the sake of simplicity. On substituting $[\mathbf{D}]_e$ and $[\mathbf{B}]_e$ into Equation (5.2), we obtain the necessary elemental $[\mathbf{K}]_e$ equation. Similarly, the elemental equation for $\{\mathbf{f}\}_e$ can also be derived.

In Figure 5.15, an extension of Example 5.2.1 to three dimensions is given for demonstration purpose only. As seen, the geometry is extended in the third dimension by 1 m. The corresponding boundary conditions are also given. The boundary conditions remain the same but the boundary sides become boundary surfaces in 3D. Two extra surfaces, one in the front, another at the back, are also introduced when the problem is extended to three dimensions. These two extra surfaces are subjected to no heat flux conditions in order to preserve two dimensionality of the problem.

The mesh generated and the solution to this problem are shown in Figure 5.16. As seen, the solution in the plane perpendicular to the third dimension, x_3, is identical to that of the two-dimensional solution given in Figure 5.6(b). As mentioned previously, the variation of the temperature in the third dimension is suppressed by imposing no heat flux conditions on the front and back faces, perpendicular to x_3, as shown in Figure 5.15.

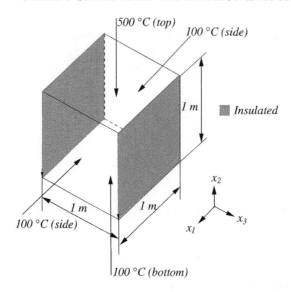

Figure 5.15   Representation of Example 5.2.1 in three dimensions.

5.6   Axisymmetric Problems

In many three-dimensional problems, there is often a geometric symmetry about a reference
axis, and such problems can be solved using two-dimensional elements, provided the boundary
conditions and all field functions are independent of the circumferential direction (θ direction).
The domain can then be represented by axisymmetric ring elements and analysed in a similar

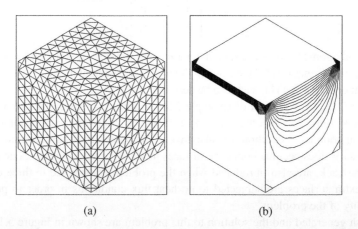

(a)   (b)

Figure 5.16   Solution for the Example 5.2.1 on a three-dimensional mesh, temperature at the
center point, (0.5,0.5,0.5), of the cube is 200.66 °C. (a) Finite element mesh; (b) Temperature
contours. Temperature varies between 100 °C and 500 °C. Interval between two contours is
25 °C.

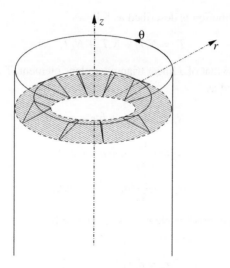

Figure 5.17   An axisymmetric problem.

fashion to that of a two-dimensional problem. Figure 5.17 shows an axisymmetric ring element in which the nodes of the finite element model lie in the $r - z$ plane.

The Galerkin formulation and the element equations are similar to those for two-dimensional plane heat transfer problems but are different due to the ring nature of the elements. The differential equation in a cylindrical coordinate system (r, z) for steady state is

$$k_r \frac{\partial^2 T}{\partial r^2} + \frac{k_r}{r} \frac{\partial T}{\partial r} + k_\theta \frac{\partial T}{\partial \theta^2} + k_z \frac{\partial^2 T}{\partial z^2} + G = 0. \qquad (5.58)$$

An axisymmetric problem is independent of the angle θ and hence Equation (5.58) reduces to

$$k_r \frac{\partial^2 T}{\partial r^2} + \frac{k_r}{r} \frac{\partial T}{\partial r} + k_z \frac{\partial^2 T}{\partial z^2} + G = 0. \qquad (5.59)$$

This can be rewritten, if the thermal conductivity in the radial direction, k_r is constant, as

$$\frac{1}{r} \left[k_r \frac{\partial}{\partial r} \left(r \frac{\partial T}{\partial r} \right) \right] + k_z \frac{\partial^2 T}{\partial z^2} + G = 0. \qquad (5.60)$$

The boundary conditions are

$$T = T_b \quad \text{on} \quad \Gamma_T$$

$$k_r \frac{\partial T}{\partial r} l + k_z \frac{\partial T}{\partial z} n + h(T - T_a) = 0 \quad \text{on} \quad \Gamma_h$$

$$k_r \frac{\partial T}{\partial r} l + k_z \frac{\partial T}{\partial z} n + q = 0 \quad \text{on} \quad \Gamma_q. \qquad (5.61)$$

The temperature distribution is described as follows:

$$T_e = N_i T_i + N_j T_j + N_k T_k,$$ (5.62)

which is similar in form to that of a linear triangular plane element. The shape functions in the above equation are defined as

$$N_i = \frac{1}{2A_e}(a_i + b_i r + c_i z)$$

$$N_j = \frac{1}{2A_e}(a_j + b_j r + c_j z)$$

$$N_k = \frac{1}{2A_e}(a_k + b_k r + c_k z).$$ (5.63)

The area, A, of an axisymmetric element is calculated from

$$2A_e = det \begin{vmatrix} 1 & r_i & z_i \\ 1 & r_j & z_j \\ 1 & r_k & z_k \end{vmatrix}.$$ (5.64)

Other constants in Equation (5.63) are defined as

$$\begin{aligned} a_i = r_j z_k - r_k z_j; & \quad b_i = z_j - z_k; \quad c_i = r_k - r_j \\ a_j = r_k z_i - r_i z_k; & \quad b_j = z_k - z_i; \quad c_j = r_i - r_k \\ a_k = r_i z_j - r_j z_i; & \quad b_k = z_i - z_j; \quad c_k = r_j - r_i. \end{aligned}$$ (5.65)

5.6.1 Galerkin Method for Linear Triangular Axisymmetric Elements

The Galerkin method for the axisymmetric equations results in the following integral form

$$\int_\Omega N_i \left[\frac{k_r}{r} \frac{\partial}{\partial r}\left(r\frac{\partial T}{\partial r}\right) + k_z \frac{\partial^2 T}{\partial z^2} + G \right] d\Omega = 0.$$ (5.66)

The spatial approximation of the temperature is given by Equation (5.62). As in the previous sections, the substitution of the spatial approximation will result in the familiar final form of the matrix equation as

$$[\mathbf{K}]\{\mathbf{T}\} = \{\mathbf{f}\},$$ (5.67)

where

$$[\mathbf{K}] = \int_\Omega [\mathbf{B}]^T [\mathbf{D}][\mathbf{B}]d\Omega + \int_\Gamma h[\mathbf{N}]^T[\mathbf{N}]d\Gamma.$$ (5.68)

The elemental $[\mathbf{B}]_e$ matrix may be defined as

$$[\mathbf{B}]_e = \begin{Bmatrix} \frac{\partial T}{\partial x} \\ \frac{\partial T}{\partial y} \end{Bmatrix} = \begin{bmatrix} \frac{\partial N_i}{\partial r} & \frac{\partial N_j}{\partial r} & \frac{\partial N_k}{\partial r} \\ \frac{\partial N_i}{\partial z} & \frac{\partial N_j}{\partial z} & \frac{\partial N_k}{\partial z} \end{bmatrix} = \frac{1}{2A_e} \begin{bmatrix} b_i & b_j & b_k \\ c_i & c_j & c_k \end{bmatrix}$$ (5.69)

and

$$[\mathbf{D}]_e = \begin{bmatrix} k_r & 0 \\ 0 & k_z \end{bmatrix}_e. \tag{5.70}$$

In Equation (5.68), the volume Ω and Γ are defined as

$$d\Omega = 2\pi r dA \text{ and } d\Gamma = 2\pi r dr \text{ respectively}, \tag{5.71}$$

where r is the radius, which varies and can be approximated by using linear shape functions over an element as

$$r_e = N_i r_i + N_j r_j + N_k r_k. \tag{5.72}$$

Substituting into Equation (5.68) and integrating, we obtain

$$[\mathbf{K}]_e = \frac{2\pi \bar{r} k_r}{4A_e} \begin{bmatrix} b_i^2 & b_i b_j & b_i b_k \\ b_i b_j & b_j^2 & b_j b_k \\ b_i b_k & b_j b_k & b_k^2 \end{bmatrix} + \frac{2\pi \bar{r} k_z}{4A_e} \begin{bmatrix} c_i^2 & c_i c_j & c_i c_k \\ c_i c_j & c_j^2 & c_j c_k \\ c_i c_k & c_j c_k & c_k^2 \end{bmatrix}$$

$$+ \frac{2\pi h l_{ij}}{12} \begin{bmatrix} 3r_i + r_j & r_i + r_j & 0.0 \\ r_i + r_j & r_i + 3r_j & 0.0 \\ 0.0 & 0.0 & 0.0 \end{bmatrix}, \tag{5.73}$$

where

$$\bar{r} = \frac{r_i + r_j + r_k}{3}. \tag{5.74}$$

Similarly,

$$\{\mathbf{f}\}_e = \int_\Omega G[\mathbf{N}]^T r_e d\Omega - \int_{\Gamma_q} q[\mathbf{N}]^T r_e d\Gamma + \int_{\Gamma_h} hT_a[\mathbf{N}]^T r_e d\Gamma$$

$$= \frac{2\pi G A_e}{12} \begin{bmatrix} 2 & 1 & 1 \\ 1 & 2 & 1 \\ 1 & 1 & 2 \end{bmatrix} \begin{Bmatrix} r_i \\ r_j \\ r_k \end{Bmatrix} - \frac{2\pi q l_{jk}}{6} \begin{Bmatrix} 0 \\ 2r_j + r_k \\ r_j + 2r_k \end{Bmatrix} + \frac{2\pi hT_a l_{ij}}{6} \begin{Bmatrix} 2r_i + r_j \\ r_i + 2r_j \\ 0 \end{Bmatrix}. \tag{5.75}$$

It is possible to approximately recover the two-dimensional plane problem by substituting a very high value for the radius r. In order to clarify the axisymmetric formulation, an example problem is solved as follows.

Example 5.6.1   *Calculate the temperature at the nodes of the axisymmetric element shown in Figure 5.18 with heat generation of $G = 1.2$ W/cm³. The coordinates of the nodes are given in cm as $i(15,10)$, $j(25,10)$ and $k(20,12)$. The heat transfer coefficient on the side ij is 1.2 W/cm²K and the ambient temperature is 30°C. The heat flux on the side jk is equal to 1 W/cm². Assume the thermal conductivities $k_r = k_z = 2$ W/cm°C.*

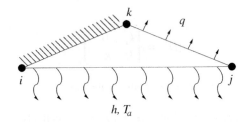

Figure 5.18   An axisymmetric triangular element.

The solution to this problem starts with the calculation of various terms in the stiffness matrix (Equation (5.73)).

$$b_i = z_j - z_k = -2.0$$
$$b_j = z_k - z_i = 2.0$$
$$b_k = z_i - z_j = 0.0$$
$$c_i = r_k - r_j = -5.0$$
$$c_j = r_i - r_k = -5.0$$
$$c_k = r_j - r_i = 10.0. \tag{5.76}$$

From Equation (5.64), the value of $2A_e$ is 20 cm². Similarly, $\bar{r}$ from Equation (5.74) is calculated as being 20 cm (a reference axis at $r = 0.0$ is assumed). Also, the coefficients used in the stiffness matrix can be calculated as

$$\frac{2\pi\bar{r}k_r}{4A_e} = \frac{2\pi\bar{r}k_z}{4A_e} = 2\pi. \tag{5.77}$$

Similarly,

$$\frac{2\pi h l_{ij}}{12} = 2\pi. \tag{5.78}$$

Note that length of the convective side l_{ij} is calculated as

$$l_{ij} = \sqrt{(x_i - x_j)^2 + (y_i - y_j)^2} = 10 \ cm. \tag{5.79}$$

Substituting into Equation (5.73) gives

$$[\mathbf{K}]_e = 2\pi \begin{bmatrix} 99 & 61 & -50 \\ 61 & 119 & -50 \\ -50 & -50 & 100 \end{bmatrix}. \tag{5.80}$$

Now, to calculate the loading vector, we need to determine the relevant coefficients, that is,

$$\frac{2\pi h T_a l_{ij}}{6} = 120\pi. \tag{5.81}$$

Similarly,

$$\frac{2\pi q l_{jk}}{6} = 1.8\pi. \tag{5.82}$$

Substituting the coefficients and other values into Equation (5.75), we obtain

$$\{\mathbf{f}\}_e = 2\pi \begin{Bmatrix} 3375.0 \\ 3922.2 \\ 21.690 \end{Bmatrix}. \tag{5.83}$$

The system of equation may be put together as

$$\begin{bmatrix} 99 & 61 & -50 \\ 61 & 119 & -50 \\ -50 & -50 & 100 \end{bmatrix} \begin{Bmatrix} T_i \\ T_j \\ T_k \end{Bmatrix} = \begin{Bmatrix} 3375.0 \\ 3922.2 \\ 21.690 \end{Bmatrix}. \tag{5.84}$$

Application of Gaussian elimination gives $T_k = 30.76\,°C$, $T_j = 29.87\,°C$ and $T_i = 31.22\,°C$. Total heat generated per unit thickness may be calculates as $G \times volume = 1.2 \times A_e = 12$ W. The total heat flowing out of the element may be calculated by adding the heat leaving due to convection along the side $i - j$ ($h \times l_{ij} \times t \times [0.5 \times (T_i + T_j) - T_a]) = -6.54$ W, and heat leaving the side along $j - k$ due to the flux leaving this side ($q \times l_{jk} \times t) = -5.4$ W. As seen the total heat leaving the element is approximately -12 W, proving the energy balance.

5.7 Summary

In this chapter, an extension of the steady-state heat conduction analysis to multi-dimensions has been given. All commonly encountered approximations, viz., two dimensional, three-dimensional and axisymmetric have been discussed. Most of the boundary conditions have also been implemented and explained via examples. We trust the reader will appreciate the difficulties associated with such multi-dimensional calculations and that the exercises given in this chapter will prove useful for further understanding of multi-dimensional steady-state heat conduction.

5.8 Exercises

Exercise 5.8.1 *A square plate size 100 cm × 100 cm is subjected to an isothermal boundary condition of 500°C on the top and to convection environment (on all the remaining three sides) of 100°C with a heat transfer coefficient of 10 W/m²K. The thermal conductivity of the material of the plate is 10 W/m²K. Assume the thickness of the plate is 1 cm. Determine the temperature distribution in the plate using (a) two triangles; (b) eight triangles. Calculate the temperature at a location (x = 30 cm, y = 30 cm) and heat fluxes in x and y directions.*

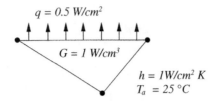

Figure 5.19 An axisymmetric element.

Exercise 5.8.2 *If in Exercise 5.8.1, there is a uniform heat generation of 2 W/cm^3 exists, and a line source of 5 W/cm at a location of $(x = 30$ cm and $y = 30$ cm), calculate the new temperature distribution using (a) two triangles; (b) eight triangles. Calculate the temperature at the location $(x = 40$ cm, $y = 40$ cm) and heat fluxes in x and y directions.*

Exercise 5.8.3 *Repeat Exercise 5.8.1 using (a) one rectangle; (b) four rectangles.*

Exercise 5.8.4 *Repeat Exercise 5.8. using (a) one rectangle; (b) four rectangles.*

Exercise 5.8.5 *In Exercise 5.8.1, if the thickness increases uniformly from 1 cm from the bottom edge to 3 cm at the top edge, rework the problem with (a) two triangles; (b) eight triangles.*

Exercise 5.8.6 *Calculate the stiffness matrix and loading vector for the axisymmetric element shown in Figure 5.19 with heat generation of $G = 1$ W/cm^3, the heat transfer coefficient on the side ij is 1.0 $W/cm^2 K$ and the ambient temperature is 25 °C. The heat flux on the side jk is equal to 0.5 W/cm^2. Assume the thermal conductivities $k_r = k_z = 1.5$ $W/m °C$ and sike k_j is insulated.*

Exercise 5.8.7 *An Internal Combustion (IC) engine cylinder is exposed to hot gases at 1000 °C on the inside wall with a heat transfer coefficient of 25 $W/m^2 C$ as shown in Figure 5.20.*

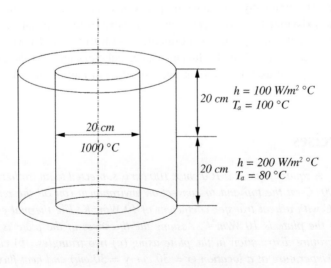

Figure 5.20 Cylinder of an IC engine.

On the external surface it is exposed to a coolant at 100°C with a heat transfer coefficient of 100 W/m²°C on the top half of the cylinder while the bottom half of the cylinder is exposed to a coolant at 80°C with a heat transfer coefficient of 200 W/m²°C. Calculate the temperature distribution in the cylinder wall with 4 axisymmetric elements. Assume a thermal conductivity of 200 W/m°C.

References

Bejan A (1993) *Heat Transfer.* John Wiley & Sons, Inc., New York.

Holman JP (1989) *Heat Transfer.* McGraw-Hill, New York.

Incropera FP and Dewitt DP (1990) *Fundamentals of Heat and Mass Transfer.* John Wiley & Sons, Inc., New York.

Ozisik MN (1968) *Boundary Value Problems of Heat Conduction.* International Text Book Company, Scranton, PA.

6

Transient Heat Conduction Analysis

6.1 Introduction

In the previous chapters, we have discussed steady-state heat conduction in which the temperature in a solid body was assumed to be invariant with respect to time. However, many practical heat transfer applications are unsteady (transient) in nature and in such problems the temperature varies with respect to time, in addition to space. For instance, in many components of industrial plants such as boilers and refrigeration and air-conditioning equipment, the heat transfer process is transient during the initial stages of operation. Other transient processes include crystal growth, casting processes, drying, heat transfer associated with the earth's atmosphere and many more. It is therefore obvious that the analysis of transient heat conduction is very important.

Analytical techniques such as variable separation, which are employed to solve transient heat conduction problems, are of limited use (Ozisik 1968), and a solution for practical heat transfer problems by these methods is difficult. Thus, it is essential to develop numerical solution procedures to solve transient heat conduction problems. In the following section, a simplified analytical method for the solution of transient problems is presented before discussing the finite element solution for such problems in Section 6.3.

6.2 Lumped Heat Capacity System

In this section, we consider the transient analysis of a body in which the temperature is assumed to be constant at any point (no spatial variation of the temperature) within and on the surface of the body at any given instant of time. It is also assumed that the temperature of the whole body

Fundamentals of the Finite Element Method for Heat and Mass Transfer, Second Edition.
P. Nithiarasu, R. W. Lewis, and K. N. Seetharamu.
© 2016 John Wiley & Sons, Ltd. Published 2016 by John Wiley & Sons, Ltd.

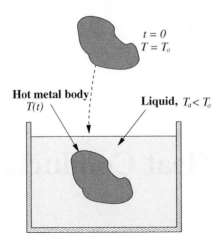

Figure 6.1   Lumped heat capacity system. A hot metal body is immersed in a liquid maintained at a constant temperature.

changes uniformly with time. Such an analysis is called a "lumped heat capacity" method and is a simple and approximate procedure in which no spatial variation in temperature is allowed but temporal variation can take place. It is therefore obvious that the lumped heat capacity analysis is limited to small-sized bodies and/or high thermal conductivity materials.

Consider a body at an initial temperature of T_o, immersed in a liquid maintained at a constant temperature of T_a as shown in Figure 6.1. At any instant in time, the conduction heat loss from the surface of the body is at the expense of the internal energy of the body. Therefore, the internal energy of the body at any time will be equal to the heat convected to the surrounding medium, that is,

$$-\rho c_p V \frac{dT(t)}{dt} = hA(T(t) - T_a), \tag{6.1}$$

where ρ is the density, c_p is the specific heat, V is the total volume of the hot metal body, A is the surface area of the body, h is the heat transfer coefficient between the body surface and surrounding medium, t is the time and $T(t)$ is the instantaneous temperature of the body. The negative sign in Equation 6.1 indicates reduction in internal energy.

Equation (6.1) is a first-order ordinary differential equation in time, which requires an initial condition to obtain a solution. As mentioned previously, the initial temperature of the body at time $t = 0$, is T_o. Rearranging Equation (6.1), we get

$$\frac{dT}{T(t) - T_a} = -\frac{hA}{\rho c_p V} dt. \tag{6.2}$$

Integrating between temperatures T_o and $T(t)$, we obtain

$$\int_{T_o}^{T(t)} \frac{dT}{T(t) - T_a} = -\int_0^t \frac{hA}{\rho c_p V} dt. \tag{6.3}$$

Note that the temperature changes from T_o to $T(t)$ as the time changes from 0 to t. Integration of the above equation results in transient temperature distribution as follows:

$$ln\left(\frac{T-T_a}{T_o-T_a}\right) = -\frac{hAt}{\rho c_p V}$$ (6.4)

or

$$\frac{T-T_a}{T_o-T_a} = e^{\left[-\frac{hA}{\rho c_p V}\right]t}.$$ (6.5)

The quantity $\rho C_p V/hA$ is referred to as the time constant of the system because it has the dimensions of time. When $t = \rho C_p V/hA$, it can be observed that the temperature difference $(T(t) - T_a)$ has a value of 36.78% of the initial temperature difference $(T_o - T_a)$.

The lumped heat capacity analysis gives results within an accuracy of 5% when

$$\frac{h(V/A)}{k_s} < 0.1,$$ (6.6)

where k_s is the thermal conductivity of the solid. It should be observed that (V/A) represents a characteristic dimension of the body. The above nondimensional parameter can thus be rewritten as hL/k_s, which is known as the Biot number. The Biot number represents a ratio between conduction resistance within the body and convection resistance at the surface of the hot body (readers should consult Chapter 1 for the meaning of conduction and convection resistances).

Due to the high variability of the convection heat transfer coefficient, a lumped system analysis is often considered as a realistic approximation even if the Biot number is as high as 0.1. However, for higher Biot numbers, this method is certainly not valid. In such situations, numerical methods such as the finite element method, are ideal in obtaining solutions with a better accuracy.

6.3   Numerical Solution

Heat conduction solutions for many geometric shapes of practical interest cannot be found using the charts available for regular geometries. Because of the time-dependent boundary or interface conditions, prevalent in many transient heat conduction problems, analytical or lumped solutions are also difficult to obtain. In such complex situations, it is essential to develop approximate time-stepping procedures to determine the transient temperature distribution.

6.3.1   Transient Governing Equations and Boundary and Initial Conditions

The transient heat conduction equation for a stationary medium is given by (Chapter 1):

$$\frac{\partial}{\partial x}\left[k_x(T)\frac{\partial T}{\partial x}\right] + \frac{\partial}{\partial y}\left[k_y(T)\frac{\partial T}{\partial y}\right] + \frac{\partial}{\partial z}\left[k_z(T)\frac{\partial T}{\partial z}\right] + G = \rho c_p\frac{\partial T}{\partial t},$$ (6.7)

where $k_x(T)$, $k_y(T)$ and $k_z(T)$ are the temperature dependent thermal conductivities in the x, y and z directions respectively. The boundary conditions for this type of problem are

$$T = T_b \quad \text{on} \quad \Gamma_b \tag{6.8}$$

$$k_x(T)\frac{\partial T}{\partial x}l + k_y(T)\frac{\partial T}{\partial y}m + k_z(T)\frac{\partial T}{\partial z}n + q = 0 \quad \text{on} \quad \Gamma_q \tag{6.9}$$

and

$$k_x(T)\frac{\partial T}{\partial x}l + k_y(T)\frac{\partial T}{\partial y}m + k_z(T)\frac{\partial T}{\partial z}n + h(T - T_a) = 0 \quad \text{on} \quad \Gamma_h, \tag{6.10}$$

where, Γ represents the boundary. In the above equation, l, m and n are direction cosines, h is the heat transfer coefficient, T_a is the atmospheric temperature and q is the boundary heat flux. The initial condition for the problem is

$$T = T_o \quad \text{at} \quad t = 0.0. \tag{6.11}$$

It is now possible to solve the above system, providing that appropriate spatial and temporal discretizations are available. Before dealing with the temporal discretization, we introduce in the following subsection, the standard Galerkin weighted residual form for the transient equations.

6.3.2   The Galerkin Method

In this subsection, the application of the Galerkin method for the transient equations subjected to appropriate boundary and initial conditions is addressed. The temperature is approximated over space as follows:

$$\hat{T} = \sum_{i=1}^{n} N_i T_i, \tag{6.12}$$

where N_i are the shape functions, n is the number of nodes in a domain, and T_i are the time-dependent nodal temperatures. The Galerkin representation of Equation (6.7) is

$$\int_{\Omega} N_i \left[\frac{\partial}{\partial x}\left(k_x\frac{\partial \hat{T}}{\partial x}\right) + \frac{\partial}{\partial y}\left(k_y\frac{\partial \hat{T}}{\partial y}\right) + \frac{\partial}{\partial z}\left(k_z\frac{\partial \hat{T}}{\partial z}\right) + G - \rho c_p \frac{\partial \hat{T}}{\partial t} \right] d\Omega = 0. \tag{6.13}$$

Employing integration by parts on the first three terms of Equation (6.13), we get

$$-\int_{\Omega} \left[\frac{\partial N_i}{\partial x}\left(k_x\frac{\partial \hat{T}}{\partial x}\right) + \frac{\partial N_i}{\partial y}\left(k_y\frac{\partial \hat{T}}{\partial y}\right) + \frac{\partial N_i}{\partial z}\left(k_z\frac{\partial \hat{T}}{\partial z}\right) - N_i G + N_i \left(\rho c_p\frac{\partial \hat{T}}{\partial t}\right) \right] d\Omega$$

$$+ \int_{\Gamma_{q+h}} N_i \left[\left(k_x\frac{\partial \hat{T}}{\partial x}l\right) + \left(k_y\frac{\partial \hat{T}}{\partial y}m\right) + \left(k_z\frac{\partial \hat{T}}{\partial z}n\right) \right] d\Gamma = 0. \tag{6.14}$$

Note that from Equation (6.9) (assuming heat leaving a domain),

$$\int_{\Gamma_q} N_i \left[\left(k_x\frac{\partial T}{\partial x}l\right) + \left(k_y\frac{\partial T}{\partial y}m\right) + \left(k_z\frac{\partial T}{\partial z}n\right) \right] d\Gamma = -\int_{\Gamma_q} N_i q d\Gamma$$

$$\int_{\Gamma_h} N_i \left[\left(k_x\frac{\partial T}{\partial x}l\right) + \left(k_y\frac{\partial T}{\partial y}m\right) + \left(k_z\frac{\partial T}{\partial z}n\right) \right] d\Gamma = -\int_{\Gamma_h} N_i h(T - T_a) d\Gamma. \tag{6.15}$$

On substituting the spatial approximation from Equation (6.12), then Equation (6.14) finally becomes,

$$
-\int_{\Omega}\left[\frac{\partial N_i}{\partial x}\left(k_x\frac{\partial N_j}{\partial x}T_j\right)+\frac{\partial N_i}{\partial y}\left(k_y\frac{\partial N_j}{\partial y}T_j\right)+\frac{\partial N_i}{\partial z}\left(k_z\frac{\partial N_j}{\partial z}T_j\right)\right]d\Omega
$$

$$
+\int_{\Omega}\left[N_iG-N_i\left(\rho c_p\frac{\partial N_j}{\partial t}T_j\right)\right]d\Omega-\int_{\Gamma_q}N_iqd\Gamma-\int_{\Gamma_h}N_ih(T-T_a)d\Gamma=0, \qquad (6.16)
$$

where i and j are representing the nodes. Equation (6.16) can be written in a more convenient form as

$$
[\mathbf{C}]\left\{\frac{d\mathbf{T}}{dt}\right\}+[\mathbf{K}]\{\mathbf{T}\}=\{\mathbf{f}\} \qquad (6.17)
$$

or

$$
C_{ij}\left\{\frac{dT}{dt}\right\}+K_{ij}T_j=f_i, \qquad (6.18)
$$

where

$$
C_{ij}=\int_{\Omega}(\rho c_p)N_iN_jd\Omega \qquad (6.19)
$$

is the capacitance matrix,

$$
K_{ij}=\int_{\Omega}\left[\frac{\partial N_i}{\partial x}\left(k_x\frac{\partial N_j}{\partial x}\right)+\frac{\partial N_i}{\partial y}\left(k_y\frac{\partial N_j}{\partial y}\right)+\frac{\partial N_i}{\partial z}\left(k_z\frac{\partial N_j}{\partial z}\right)\right]d\Omega+\int_{\Gamma_h}hN_iN_jd\Gamma \qquad (6.20)
$$

is the stiffness matrix and

$$
f_i=\int_{\Omega}N_iGd\Omega-\int_{\Gamma_q}qN_id\Gamma+\int_{\Gamma_h}N_ihT_ad\Gamma \qquad (6.21)
$$

is the load vector.

In matrix form,

$$
[\mathbf{C}]=\int_{\Omega}\rho c_p[\mathbf{N}]^T[\mathbf{N}]d\Omega \qquad (6.22)
$$

$$
[\mathbf{K}]=\int_{\Omega}[\mathbf{B}]^T[\mathbf{D}][\mathbf{B}]d\Omega+\int_{\Gamma_h}h[\mathbf{N}]^T[\mathbf{N}]d\Gamma \qquad (6.23)
$$

and

$$
\{\mathbf{f}\}=\int_{\Omega}G[\mathbf{N}]^Td\Omega-\int_{\Gamma_q}q[\mathbf{N}]^Td\Gamma+\int_{\Gamma_h}hT_a[\mathbf{N}]^Td\Gamma. \qquad (6.24)
$$

Since $k_x(T), k_y(T)$ and $k_z(T)$ are functions of temperature, Equation (6.17) is nonlinear and requires an iterative solution. If k_x, k_y and k_z are independent of temperature, then Equation (6.17) is linear in form.

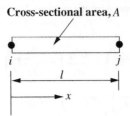

Figure 6.2   One-dimensional linear element.

6.4   One-dimensional Transient State Problem

The relation derived in Equation (6.17) is employed here locally on an element to illustrate the application to a one-dimensional transient problem using a linear element as shown in Figure 6.2.

The temperature T is represented in the element by

$$T_e = N_i T_i + N_j T_j = [\mathbf{N}]_e \{\mathbf{T}\}_e. \tag{6.25}$$

Note that i and j in the above equation represent the nodes i and j of the element shown in Figure 6.2. The shape functions in Equation (6.25) are defined as

$$N_i = 1 - \frac{x}{l} \quad \text{and} \quad N_j = \frac{x}{l}. \tag{6.26}$$

The spatial derivative of temperature is given as

$$\frac{\partial T}{\partial x} = \frac{\partial N_i}{\partial x} T_i + \frac{\partial N_j}{\partial x} T_j = -\frac{1}{l} T_i + \frac{1}{l} T_j = [\mathbf{B}]_e \{\mathbf{T}\}_e. \tag{6.27}$$

The relevant matrices, as discussed in the previous section (Equation (6.17)), are

$$[\mathbf{C}]_e = \int_\Omega (\rho c_p)[\mathbf{N}]_e^T [\mathbf{N}]_e d\Omega = \int_l (\rho c_p) A \begin{bmatrix} N_i^2 & N_i N_j \\ N_i N_j & N_j^2 \end{bmatrix} dx. \tag{6.28}$$

Note that $d\Omega$ is replaced by $A dx$ in the above equation. Here A is the uniform cross-sectional area of a one-dimensional body. The integration of Equation (6.28) results in (for details of the integration, refer to Chapter 3 and Appendix C):

$$[\mathbf{C}]_e = \frac{\rho c_p l A}{6} \begin{bmatrix} 2 & 1 \\ 1 & 2 \end{bmatrix}. \tag{6.29}$$

Similarly, the elemental $[\mathbf{K}]$ matrix and load vector $\{\mathbf{f}\}$ can be written as

$$[\mathbf{K}]_e = \frac{A k_x}{l} \begin{bmatrix} 1 & -1 \\ -1 & 1 \end{bmatrix} + \frac{hPl}{6} \begin{bmatrix} 2 & 1 \\ 1 & 2 \end{bmatrix} \tag{6.30}$$

and

$$\{\mathbf{f}\}_e = \frac{GAl}{2} \begin{Bmatrix} 1 \\ 1 \end{Bmatrix} - \frac{qPl}{2} \begin{Bmatrix} 1 \\ 1 \end{Bmatrix} + \frac{hT_a Pl}{2} \begin{Bmatrix} 1 \\ 1 \end{Bmatrix}, \tag{6.31}$$

where P is the perimeter of the one-dimensional body. Substituting Equations (6.29) to (6.31) into Equation (6.17), for a domain with only one element, gives (multiple elements require assembly):

$$\frac{\rho c_p l A}{6} \begin{bmatrix} 2 & 1 \\ 1 & 2 \end{bmatrix} \left\{ \begin{matrix} \frac{dT_i}{dt} \\ \frac{dT_j}{dt} \end{matrix} \right\} + \left(\frac{Ak_x}{l} \begin{bmatrix} 1 & -1 \\ -1 & 1 \end{bmatrix} + \frac{hPl}{6} \begin{bmatrix} 2 & 1 \\ 1 & 2 \end{bmatrix} \right) \left\{ \begin{matrix} T_i \\ T_j \end{matrix} \right\}$$

$$= \frac{GAl}{2} \left\{ \begin{matrix} 1 \\ 1 \end{matrix} \right\} - \frac{qPl}{2} \left\{ \begin{matrix} 1 \\ 1 \end{matrix} \right\} + \frac{hT_a Pl}{2} \left\{ \begin{matrix} 1 \\ 1 \end{matrix} \right\}. \tag{6.32}$$

The above equation is a general representation of a one-dimensional problem with one linear element. All the terms are included irrespective of whether or not boundary fluxes and heat generation are present. We shall do appropriate modification to Equation (6.32), when solving numerical problems.

Equation (6.32) is semi-discrete as it is discretized only in space. We now require a method of discretizing the transient terms of Equation (6.32). The following subsections give the details of how the transient terms will be discretized.

6.4.1 Time Discretization-Finite Difference Method (FDM)

As may be seen from the semi-discrete form of Equation (6.32) (or (6.17)), the differential operator involving the time-dependent term remains to be discretized. In this section, a numerical approximation of the transient term, using the Finite Difference Method (FDM), is considered.

Figure 6.3 shows a model temperature variation in the time domain between the n and $(n + 1)$ time levels. Using a Taylor series, we can write the temperature at the $(n + 1)^{th}$ level as

$$T^{n+1} = T^n + \Delta t \frac{dT^n}{dt} + \frac{\Delta t^2}{2} \frac{d^2 T^n}{dt^2} + \cdots \tag{6.33}$$

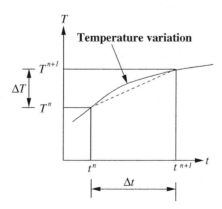

Figure 6.3   Temperature variation within a time step.

Table 6.1   Different time steeping schemes

θ	Name of the scheme	Comments
0.0	Fully explicit scheme	Forward difference method.
1.0	Fully implicit scheme	Backward difference method
0.5	Semi-implicit scheme	Crank-Nicolson method

If the second- and higher-order terms in the above equation are neglected then

$$\frac{dT^n}{dt} \approx \frac{T^{n+1} - T^n}{\Delta t} + O(\Delta t), \tag{6.34}$$

which is first-order accurate in time (forward difference). If we now introduce a parameter θ such that

$$T^{n+\theta} = \theta T^{n+1} + (1 - \theta)T^n \tag{6.35}$$

into Equation (6.17) then, along with Equation (6.34), we have

$$[\mathbf{C}] \left\{ \frac{T^{n+1} - T^n}{\Delta t} \right\} + [\mathbf{K}]\{\mathbf{T}\}^{n+\theta} = \{\mathbf{f}\}^{n+\theta} \tag{6.36}$$

or

$$[\mathbf{C}] \left\{ \frac{T^{n+1} - T^n}{\Delta t} \right\} + [\mathbf{K}] \left\{ \theta T^{n+1} + (1 - \theta)T^n \right\} = \theta \{\mathbf{f}\}^{n+1} + (1 - \theta)\{\mathbf{f}\}^n. \tag{6.37}$$

The above equation can be rearranged as follows:

$$([\mathbf{C}] + \theta \Delta t[\mathbf{K}]) \{\mathbf{T}\}^{n+1} = ([\mathbf{C}] - (1 - \theta)\Delta t[\mathbf{K}]) \{\mathbf{T}\}^n + \Delta t \left(\theta\{\mathbf{f}\}^{n+1} + (1 - \theta)\{\mathbf{f}\}^n \right). \tag{6.38}$$

Equation (6.38) gives the nodal values of temperature at the $(n + 1)$ time level. These temperature values are calculated using the n time level values. However, both the $(n + 1)$ and n time level values of the forcing vector $\{\mathbf{f}\}$ must be known. By varying the parameter θ, different transient schemes can be constructed, which are shown in Table 6.1 for varying values of θ.

In the following numerical example, we demonstrate how the Crank-Nicolson time-stepping scheme can be used to solve a one-dimensional transient problem.

Example 6.4.1   *In Example 3.5.1 let us assume that the initial temperature of the fin is equal to the atmospheric temperature, 25 °C. If the base temperature is suddenly raised to a temperature of 100 °C, and maintained at that value, determine the temperature distribution in the fin with respect to time. Assume a heat capacity (ρc_p) value of 2.42×10^7 W/m³°C.*

Let us assume that this problem is to be solved using the Crank-Nicolson method, where θ is equal to 0.5. Assume a time step, Δt, of 20 s. Equation (6.38) can be rewritten with the given value for θ and Δt as

$$([\mathbf{C}] + 0.5 \times 20[\mathbf{K}])\{\mathbf{T}\}^{n+1} = ([\mathbf{C}] - 0.5 \times 20[\mathbf{K}])\{\mathbf{T}\}^n + 20\{\mathbf{f}\}. \tag{6.39}$$

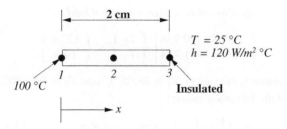

Figure 6.4   One-dimensional transient heat transfer. Two elements and three nodes.

If we consider two elements, as shown in Figure 6.4, we have from Example 3.5.1(b),

$$[\mathbf{K}]_1 = [\mathbf{K}]_2 = \begin{bmatrix} 0.124 & -0.118 \\ -0.118 & 0.124 \end{bmatrix} \tag{6.40}$$

and

$$\{\mathbf{f}\}_1 = \{\mathbf{f}\}_2 = \begin{Bmatrix} 0.15 \\ 0.15 \end{Bmatrix}. \tag{6.41}$$

The [C] matrix can be calculated as

$$[\mathbf{C}]_1 = [\mathbf{C}]_2 = \frac{\rho c_p A l}{6} \begin{bmatrix} 2 & 1 \\ 1 & 2 \end{bmatrix} = \begin{bmatrix} 0.484 & 0.242 \\ 0.242 & 0.484 \end{bmatrix}. \tag{6.42}$$

On assembling the stiffness matrix and load vector we obtain

$$[\mathbf{K}] = \begin{bmatrix} 0.124 & 0.118 & 0.00 \\ -0.118 & 0.248 & -0.118 \\ 0.00 & -0.118 & 0.124 \end{bmatrix} \tag{6.43}$$

and

$$\{\mathbf{f}\} = \begin{Bmatrix} 0.15 \\ 0.30 \\ 0.15 \end{Bmatrix}. \tag{6.44}$$

The global capacitance matrix is

$$[\mathbf{C}] = \begin{bmatrix} 0.484 & 0.242 & 0 \\ 0.242 & 0.968 & 0.242 \\ 0 & 0.242 & 0.484 \end{bmatrix}. \tag{6.45}$$

Substituting into Equation (6.39), we get at $\Delta t = 20\,s$

$$\begin{bmatrix} 1.724 & -0.938 & 0.0 \\ -0.938 & 3.448 & -0.938 \\ 0.0 & -0.938 & 1.724 \end{bmatrix} \begin{Bmatrix} T_1 \\ T_2 \\ T_3 \end{Bmatrix} = \begin{bmatrix} -0.756 & 1.422 & 0.0 \\ 1.422 & -1.512 & 1.422 \\ 0.0 & 1.422 & -0.756 \end{bmatrix} \begin{Bmatrix} 25.0 \\ 25.0 \\ 25.0 \end{Bmatrix} + \begin{Bmatrix} 3.0 \\ 6.0 \\ 3.0 \end{Bmatrix}.$$

$$\tag{6.46}$$

Simplification and application of boundary condition lead to

$$\begin{bmatrix} 3.448 & -0.938 \\ -0.938 & 1.724 \end{bmatrix} \begin{Bmatrix} T_2 \\ T_3 \end{Bmatrix} = \begin{Bmatrix} 133.4 \\ 19.6 \end{Bmatrix}. \tag{6.47}$$

Solution to the above system gives $T_3 = 38.05\,^\circ C$ and $T_2 = 49.04\,^\circ C$. Similarly at time $t = 40\,s$, we arrive at the following values:

$$\begin{bmatrix} 1.724 & -0.938 & 0.0 \\ -0.938 & 3.448 & -0.938 \\ 0.0 & -0.938 & 1.724 \end{bmatrix} \begin{Bmatrix} T_1 \\ T_2 \\ T_3 \end{Bmatrix} = \begin{bmatrix} -0.756 & 1.422 & 0.0 \\ 1.422 & -1.512 & 1.422 \\ 0.0 & 1.422 & -0.756 \end{bmatrix} \begin{Bmatrix} 100.0 \\ 49.04 \\ 38.05 \end{Bmatrix} + \begin{Bmatrix} 3.0 \\ 6.0 \\ 3.0 \end{Bmatrix}. \tag{6.48}$$

Solution of the above system results in $T_3 = 71.03\,^\circ C$ and $T_2 = 83.69\,^\circ C$. The above process of time stepping may be continued until desired time is reached.

In the above example, it has been demonstrated how the transient solution is calculated. In the following example, a similar case is considered using an explicit computer program (see Chapter 15).

Example 6.4.2   *A rod of 1 unit width and 20 units in length is initially assumed to be at $0\,^\circ C$. The left-hand side of the domain is subjected to a uniform heat flux of 1 and all other sides are assumed to be insulated as shown in Figure 6.5. Assume all other properties are equal to unity and compute the temperature distribution and compare with a known analytical solution.*
The analytical solution for this problem is given by Carslaw and Jaeger (1959) as

$$T(x,t) = 2(t/\pi)^{1/2} \left[\exp(-x^2/4t) - (1/2)x\sqrt{\frac{\pi}{t}}erfc\left(\frac{x}{2\sqrt{t}}\right) \right]. \tag{6.49}$$

Figure 6.6 shows the two different meshes used in the calculations. Figure 6.6(a) is coarse mesh with 122 nodes and 158 elements, and Figure 6.6(b) shows a mesh of 2349 nodes and 4276 elements. This is a one-dimensional problem, which is solved using a two-dimensional forward difference (explicit) computer program.
Figure 6.7 shows the temperature contours at a time of unity. As seen, the results generated from both meshes are very similar. The temperature variation along the length of the rod is shown in Figure 6.8. The results of both meshes indicate excellent agreement with the analytical solution.

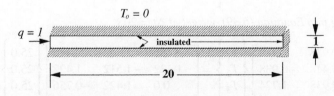

Figure 6.5   One-dimensional transient heat conduction analysis in a rod.

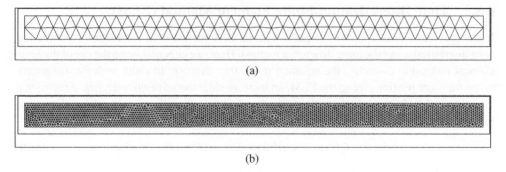

(a)

(b)

Figure 6.6   Linear triangular element meshes. (a) Coarse finite element mesh, 122 nodes and 158 elements. (b) Fine finite element mesh, 2349 nodes and 4276 elements.

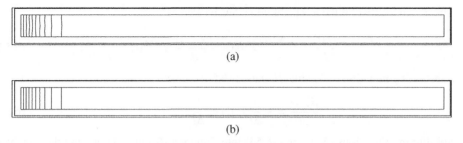

(a)

(b)

Figure 6.7   Temperature distribution at $t = 1$. (a) Temperature distribution on the coarse mesh, $T_{max} = 1.12$ at the right-hand face. (b) Temperature distribution on the fine mesh, $T_{max} = 1.128$ at the left hand face.

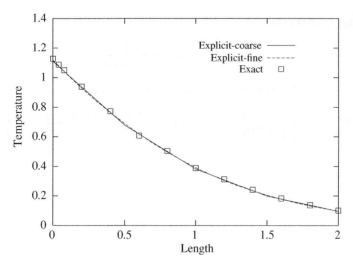

Figure 6.8   Temperature distribution along the length of the rod at $t = 1$.

6.4.2   Time Discretization-Finite Element Method (FEM)

In the previous subsection, the temporal term in the transient heat conduction equation has been discretized using the finite difference method. Here, we concentrate on the use of the finite element method to discretize the equation in the time domain. In order to derive the appropriate transient relations using the FEM, let us reconsider the semi-discrete, one-dimensional Equation (6.17). In this equation, the temperature is now discretized over a time element as (refer to Figure 6.9),

$$T_e(t) = N^n(t)\mathbf{T}^n(t) + N^{n+1}(t)\mathbf{T}^{n+1}(T). \tag{6.50}$$

where the linear shape functions $N_i(t)$ and $N_j(t)$ are given as

$$N^n(t) = 1 - \frac{t}{\Delta t}; \quad N^{n+1}(t) = \frac{t}{\Delta t}. \tag{6.51}$$

The time derivative of the temperature is thus written as

$$\frac{dT(t)}{dt} = \frac{dN^n(t)}{dt}\mathbf{T}^n(t) + \frac{dN^{n+1}(t)}{dt}\mathbf{T}^{n+1}(t). \tag{6.52}$$

Substituting Equation (6.51) into Equation (6.52) we get

$$\frac{dT(t)}{dt} = -\frac{1}{\Delta t}\mathbf{T}^n(t) + \frac{1}{\Delta t}\mathbf{T}^{n+1}(t). \tag{6.53}$$

Substituting Equations (6.50) and (6.53) into Equation (6.17) and applying the weighted residual principle (Galerkin method), we obtain for a time interval of Δt,

$$\int_{\Delta t} \left\{ \begin{array}{c} N^n(t) \\ N^{n+1}(t) \end{array} \right\} \left[[\mathbf{C}] \left(-\frac{\mathbf{T}^n(t)}{\Delta t} + \frac{\mathbf{T}^{n+1}(t)}{\Delta t} \right) + [\mathbf{K}] \left(N^n(t)\mathbf{T}^n(t) + N^{n+1}(t)\mathbf{T}^{n+1}(t) \right) - \{\mathbf{f}\} \right] dt = 0. \tag{6.54}$$

Employing (see Appendix C)

$$\int_{\Delta t} N^n(t)^a N^{n+1}(t)^b dt = \frac{a!b!}{(a+b+1)!}\Delta t, \tag{6.55}$$

we obtain the characteristic equation over the time interval Δt as

$$\frac{1}{\Delta t}[\mathbf{C}] \begin{bmatrix} -1 & 1 \\ -1 & 1 \end{bmatrix} \left\{ \begin{array}{c} \mathbf{T}^n(t) \\ \mathbf{T}^{n+1}(t) \end{array} \right\} + \frac{1}{3}[\mathbf{K}] \begin{bmatrix} 2 & 1 \\ 1 & 2 \end{bmatrix} \left\{ \begin{array}{c} \mathbf{T}^n(t) \\ \mathbf{T}^{n+1}(t) \end{array} \right\} = \left\{ \begin{array}{c} f^n \\ f^{n+1} \end{array} \right\} + 1. \tag{6.56}$$

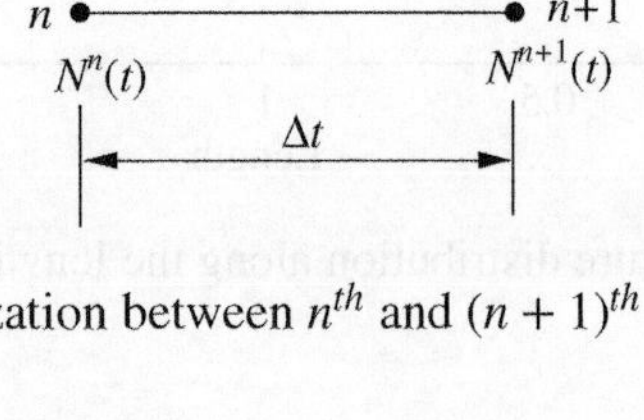

Figure 6.9   Time discretization between n^{th} and $(n+1)^{th}$ time levels (time element).

The above equation involves the temperature values at the nth and $(n + 1)^{th}$ level. A quadratic variation of temperature with respect to time may be derived in a similar fashion. It is also possible to take two time elements and assemble to obtain a different time discretization to that shown in Equation (6.56).

6.5   Stability

The stability of a numerical scheme may be obtained using a Fourier analysis (Hirsch 1988; Lewis *et al.* 2004). Here, we give a brief summary of the stability related issues of the time-stepping schemes discussed in this chapter.

Backward Euler: This is an implicit scheme with a backward difference approximation for the time term. This scheme is unconditionally stable and the accuracy of the scheme is governed by the size of the time step.

Forward Euler: This is an explicit scheme with a forward difference approximation to the time term. The scheme is conditionally stable and the stability limit for the time step is given as

$$\Delta t \leq \frac{l^2}{2\alpha},$$ (6.57)

where l is the element size and α is the thermal diffusivity.

Central difference: The central difference approximation of the time term, with an explicit treatment for the other terms, is unconditionally unstable and this scheme is not recommended.

Crank-Nicolson scheme (semi-implicit): Due to the oscillatory behavior of this semi-implicit scheme at larger time steps, it is often termed as a marginally stable scheme.

6.6   Multi-dimensional Transient Heat Conduction

A finite element solution for Multi-dimensional problems follows the same procedure as that for a one-dimensional case. However, the matrices $[\mathbf{C}], [\mathbf{K}]$ and $\{\mathbf{f}\}$ are different because of their multi-dimensional nature. For more details on the matrices, the reader should refer to Chapter 3. A numerical problem, using a two- and three-dimensional approximation, is solved in the following example.

Example 6.6.1   *A square plate and a cube are subjected to different thermal boundary conditions as shown in Figure 6.10. If the initial temperature of both the domains is 0°C, calculate the transient temperature distribution within these two geometries. Also, plot the temperature change with respect to time at a point (0.5,0.5) in the 2D geometry and at (0.5,0.5,0.5) in the three-dimensional geometry.*

The results from both the two- and three-dimensional geometries should be identical because of the insulated conditions on the two vertical sides of the cube.

Figure 6.11 shows the time evaluation of the temperature contours. The first two figures, that is, Figure 6.11(a) and (b), show a zero temperature value at the center of the plate. However, heat from the boundaries rapidly diffuses into the domain and the temperature

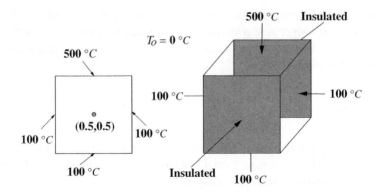

Figure 6.10   Square and cubical domains with thermal boundary conditions.

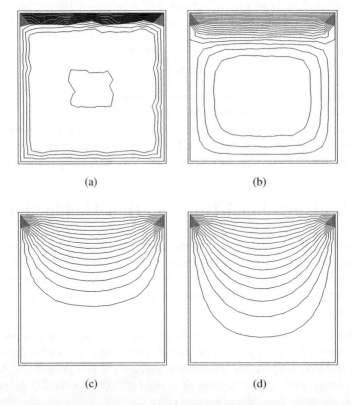

Figure 6.11   Transient temperature distribution in a 2D plane geometry. (a) Temperature distribution at $t = 0.001$ s, $T(0.5, 0.5) = 0.0\,°$C. (b) Temperature distribution at $t = 0.01$ s, $T(0.5, 0.5) = 0.0\,°$C. (c) Temperature distribution at $t = 0.1$ s, $T(0.5, 0.5) = 155.38\,°$C. (d) Temperature distribution at $t = 0.5$ s, $T(0.5, 0.5) = 200.40\,°$C.

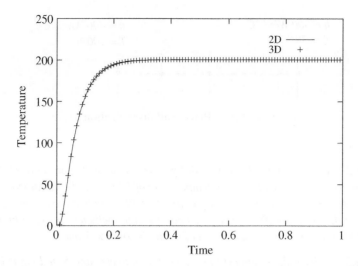

Figure 6.12   Temperature distribution at the center of square domain (cube in 3D) with respect to time.

reaches a steady value of 200.4 °C at the center by the time t = 0.5 s. In Figure 6.12 we show the temperature variation at the center point of both the two and three dimensional geometries with respect to time. It may be seen that both the results are identical. It should be noted that the temperature increases rapidly and reaches a value of 200.4 at about four seconds and thereafter it remains constant.

6.7   Summary

In this chapter, we have introduced the transient heat conduction problems and demonstrated solutions of such problems via many numerical examples. However, the problems discussed in this chapter are only the "tip of the iceberg." We recommend that the readers to formulate their own transient heat conduction problems and solve them using the transient computer programs available from the authors (see Chapter 15). For transient convection problems, readers should refer to Chapters 7.

6.8   Exercises

Exercise 6.8.1   *A large block of steel with a thermal conductivity of 40 W/m °C and a thermal diffusivity of 1.5×10^{-5} m^2/s is initially at a uniform temperature of 25 °C. The surface is exposed to (a) a heat flux of 3×10^5 W/m^2; (b) sudden rise in surface temperature of 200 °C Calculate the temperature at a depth of 1 cm after a time of 10 seconds for both cases. Verify the results with analytical results.*

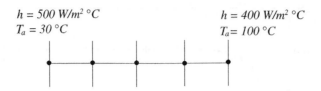

$h = 500 \, W/m^2 \, ^\circ C$
$T_a = 30 \, ^\circ C$

$h = 400 \, W/m^2 \, ^\circ C$
$T_a = 100 \, ^\circ C$

Figure 6.13   Plane wall discretization.

Exercise 6.8.2   *A fin of length 1 cm is initially at the ambient temperature of 30°C. If the base temperature is suddenly raised to a temperature of 150 °C and maintained at that value, determine the temperature distribution in the fin after 30 seconds if the thermal diffusivity of the material of the fin is $1 \times 10^{-5} \, m^2/s$. The heat transfer coefficient between fin surface and ambient is $100 \, W/m^2 \, ^\circ C$. The cross-section of the fin is $6 \, mm \times 5 \, mm$.*

Exercise 6.8.3   *A short aluminum cylinder 2.5 cm in diameter and 5 cm long is initially at a uniform temperature of 100°C. It is suddenly subjected to a convection environment at 50 °C and $h = 400 \, W/m^2 \, ^\circ C$. Calculate the temperature at a distance of 0.5 cm from one end of the cylinder 10 seconds after exposure to the environment.*

Exercise 6.8.4   *A plane wall of thickness 4 mm has internal heat generation of 25 MW/m³ with thermal properties of $k = 20 \, W/m \, ^\circ C$, $p = 8000 \, kg/m^3$ and specific heat $C_p = 500 \, J/kg \, ^\circ C$. It is initially at a uniform temperature of 50°C and is suddenly subjected to heat generation and convective boundary condition as shown in Figure 6.13. Calculate the temperature at a location of 2 mm after 10 seconds.*

Exercise 6.8.5   *A stainless steel plate size $2 \, cm \times 1 \, cm$ is surrounded by an insulating block as shown in Figure 6.14 and is initially at a uniform temperature of 40 °C with a convection environment at 40 °C. The top side of the plate is suddenly exposed to a radiant flux of $15 \, kW/m^2$. Calculate the temperature at the center of the top surface and bottom surface after 10 s. Take the properties of the stainless steel as $k = 18 \, W/mK$, $p = 8000 \, kg/m^3$, $C_p = 0.46 \, kJ/kg \, ^\circ C$, and $h = 30 \, W/m^2 K$.*

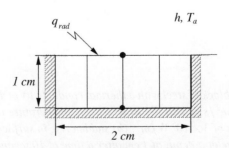

q_{rad}

h, T_a

1 cm

2 cm

Figure 6.14   Stainless steel plate.

References

Beck JV (1968) Surface heat flux determination using an integral method. *Nuclear Engineering Design*, **7**, 170–178.

Carslaw HS and Jaeger JC (1959) *Conduction of Heat in Solids*, 2nd Edition. Oxford University Press, Fairlawn, N.J.

Hirsch C (1988) *Numerical Computation of Internal and External Flows. Volume 1, Fundamentals of Numerical Discretization*. John Wiley & Sons, Inc., New York.

Lewis RW, Nithiarasu P and Seetharamu KN (2004) *Fundamentals of the Finite Element Method for Heat and Fluid Flow*. John Wiley & Sons, Inc., New York.

Ozisik MN (1968) *Boundary Value Problems of Heat Conduction*. International Text Book Company, Scranton, PA.

7

Laminar Convection Heat Transfer

7.1 Introduction

In the previous six chapters, the conduction mode of heat transfer has been discussed in detail. Occasionally, convective heat transfer boundary conditions were discussed in these chapters whenever appropriate. However, little information on fluid flow characteristics was given in any of the previous chapters. In the present chapter, the heat transfer mechanism due to a fluid motion is discussed in detail. This method of heat transfer, which is caused by fluid motion is referred to as "heat convection."

The study of fluid motion (fluid dynamics) is an important subject which has wide application in many engineering disciplines. Several industries use computer-based fluid dynamics analysis (Computational Fluid Dynamics or CFD) tools for both design and analysis. For instance aerospace applications, turbo-machines, weather forecasting, electronic cooling arrangements and flow in heat exchangers are merely a few examples. There has been a vast increase in the use of CFD tools in engineering industries in the last two decades due mainly to an ever-increasing computing power. In the 1980s a solution for a reasonably size three-dimensional fluid dynamics problem was rarely possible on a personal computer (PC). However, now it is very common for researchers to solve reasonably sized fluid dynamics problems in three dimensions using such computers.

There are several books written on the topic of computational fluid dynamics, which include texts explaining the basic solution scheme underlying a successful CFD software (Cheung, 2002; Donea and Huerta, 2003; Fletcher, 1988; Gresho and Sani 2000; Hirsch 1989; Lewis *et al.* 1996; Pironneau 1989; Zienkiewicz *et al.* 2005), or books on practical fluid dynamics calculations such as data structure and parallel computing (Lohner 2001). Several chapters could be written on the topic of CFD alone. However, our main interest is to give a practical

Fundamentals of the Finite Element Method for Heat and Mass Transfer, Second Edition.
P. Nithiarasu, R. W. Lewis, and K. N. Seetharamu.
© 2016 John Wiley & Sons, Ltd. Published 2016 by John Wiley & Sons, Ltd.

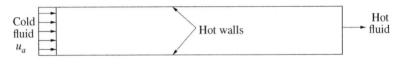

Figure 7.1   Flow and heat transport in a channel.

introduction to the role of fluid dynamics in heat transport. It is intended that this chapter will give a good starting point to pursue a further education and/or research in fluid dynamics assisted heat transport.

7.1.1   Types of Fluid motion assisted heat transport

The fluid motion assisted heat transfer (heat convection) may be classified into three different categories. In order to explain the different types, let us consider the fluid flow through a two-dimensional channel as shown in Figure 7.1. The inlet to the channel is at the left side and exit is at the right. Both the top and bottom walls of the channel are at higher temperatures than the invading fluid. The mechanism here is that the fluid, which is at a temperature lower than the wall temperature of the channel, comes into contact with the wall and removes heat by convection. Although this process is termed as being convective, there are aspects of the diffusion mode of heat transfer, which dominates very close to the hot walls.

It is obvious that flow with a higher incoming velocity will transport heat at a higher rate. The flow rate is often characterized by a quantity called the Reynolds number, which is defined as

$$Re = \frac{\rho_a u_a L}{\mu_a},\qquad(7.1)$$

where u_a is a reference velocity, for example average inlet velocity, L is a characteristic dimension, for example the width or height of the channel, ρ_a is a reference (inlet) density and μ_a is a reference (inlet) dynamic viscosity of the fluid. If the Reynolds number is small and below a certain critical value, the flow is laminar and if above this critical number then, the flow becomes turbulent. The critical Reynolds number for pipe and channel flows is approximately 2000, based on the diameter and height respectively.

In Figure 7.1, if the flow is forced into the channel by means of an external device, for example pump, then the convection process is referred to as "forced convection," and the Reynolds number is normally high (Jaluria 1986; Lewis *et al.* 1996, 1995b; Massarotti *et al.* 1998; Patnaik *et al.* 2001; Srinivas *et al.* 1994). In such situations, the fluid motion created by the density (or temperature) difference (buoyancy-driven motion) is negligibly small as compared to the forced motion of the fluid. However, at low and moderate Reynolds numbers, the motion created by the local density (or temperature) differences in the fluid is comparable to that of the forced flow. A situation where the forced and density difference driven motions are equally important is called "mixed convection" transport (Aung and Worku 1986a,b; Gowda *et al.* 1998). If the forced flow is suddenly stopped and the fluid is stagnant inside the channel, then the fluid motion will be entirely influenced by the local density (or temperature) differences until an equilibrium state is reached, that is, no local differences in density or temperature are present. Such a flow is often referred to as "natural, free or buoyancy driven convection" (de Vahl Davis 1983; Jaluria 1986; Jaluria and Torrance 1986; Nithiarasu *et al.* 1998; Zienkiewicz *et al.* 1996).

7.2 Navier-Stokes Equations

The mathematical model of any fundamental fluid dynamics problem is governed by the Navier-Stokes equations. These equations are important and represent the fluid as a continuum. The equations conserving mass, momentum and energy can be derived either following an integral or a differential approach. The integral form of the equations is derived using Reynolds Transport Theorem (RTT) and is discussed in many standard fluid mechanics texts (Shames 1982). The approach we follow in this book is the differential approach in which a differential control volume is considered in the fluid domain and the Taylor expansion is used to represent the variation of mass, momentum and energy.

7.2.1 Conservation of Mass or Continuity Equation

The conservation of mass equation ensures that the total mass is conserved, or in other words the total mass of a fluid system is completely accounted for. In order to derive a general conservation of the mass equation, consider the differential control volume as shown in Figure 7.2. The reader can assume the control volume to be infinitesimal for a typical flow problem, such as flow in a channel (Figure 7.1), flow over a flat plate or the temperature (or density) difference driven circulation of air inside a room as shown in Figure 7.3.

Let us assume that the mass flux entering the control volume (Figure 7.2) is ρu_1 in the x_1 direction and ρu_2 in the x_2 direction. It is also assumed that there is no reaction or mass production within the fluid domain. The Taylor series expansion may be used to express the mass flux exiting the control volume as (refer to Figure 7.2):

$$(\rho u_1)_{x_1+\Delta x_1} = (\rho u_1)_{x_1} + \frac{\Delta x_1}{1!}\frac{\partial(\rho u_1)}{\partial x_1} + \frac{{\Delta x_1}^2}{2!}\frac{\partial^2(\rho u_1)}{\partial x_1^2} + \dots \quad (7.2)$$

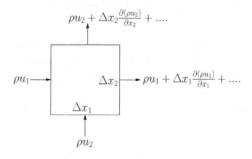

Figure 7.2   Infinitesimal control volume. Derivation of conservation of mass in a flow field.

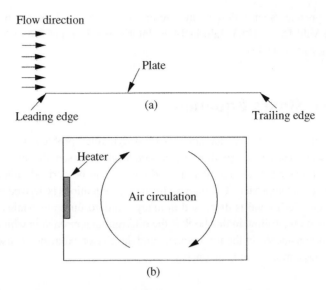

Figure 7.3   Forced flow over a flat plate and natural convection inside a room.

in the x_1 direction and

$$(\rho u_2)_{x_2+\Delta x_2} = (\rho u_2)_{x_2} + \frac{\Delta x_2}{1!}\frac{\partial(\rho u_2)}{\partial x_2} + \frac{\Delta x_2{}^2}{2!}\frac{\partial^2(\rho u_2)}{\partial x_2^2} + ...$$ (7.3)

in the x_2 direction. From an inspection of the control volume shown in Figure 7.2, we can write the difference between the total mass entering and exiting the control volume as

$$\Delta x_2\left[(\rho u_1)_{x_1} - (\rho u_1)_{x_1+\Delta x_1}\right] = -\Delta x_2\left[\frac{\Delta x_1}{1!}\frac{\partial(\rho u_1)}{\partial x_1} + \frac{\Delta x_1{}^2}{2!}\frac{\partial^2(\rho u_1)}{\partial x_1^2} + ...\right].$$ (7.4)

Similarly, in the x_2 direction

$$\Delta x_1\left[(\rho u_2)_{x_2} - (\rho u_2)_{x_2+\Delta x_2}\right] = -\Delta x_1\left[\frac{\Delta x_2}{1!}\frac{\partial(\rho u_2)}{\partial x_2} + \frac{\Delta x_2{}^2}{2!}\frac{\partial^2(\rho u_2)}{\partial x_2^2} + ...\right].$$ (7.5)

Note that the total mass is calculated as being the mass flux times the perpendicular area to the flow direction. For instance, the total mass entering the control volume in the x_1 direction is $\Delta x_2 \times 1 \times \rho u_1$. A unit thickness is assumed in the x_3 direction.

Adding the Equations (7.4) and (7.5) gives the total mass stored inside the control volume. Neglecting the second- and higher-order terms, the total mass stored inside the control volume is

$$-\Delta x_1\Delta x_2\left[\frac{\partial(\rho u_1)}{\partial x_1} + \frac{\partial(\rho u_2)}{\partial x_2}\right].$$ (7.6)

The above quantity, stored within the control volume, is equal to the rate of change of the total mass within the control volume, which is given as

$$\Delta x_1 \Delta x_2 \frac{\partial \rho}{\partial t}. \tag{7.7}$$

We can therefore write

$$\Delta x_1 \Delta x_2 \frac{\partial \rho}{\partial t} = -\Delta x_1 \Delta x_2 \left[\frac{\partial(\rho u_1)}{\partial x_1} + \frac{\partial(\rho u_2)}{\partial x_2} \right] \tag{7.8}$$

or

$$\frac{\partial \rho}{\partial t} + \frac{\partial(\rho u_1)}{\partial x_1} + \frac{\partial(\rho u_2)}{\partial x_2} = 0. \tag{7.9}$$

The above equation is known as the equation of conservation of mass, or the continuity equation for two-dimensional flows. In three dimensions, the continuity equation is

$$\frac{\partial \rho}{\partial t} + \frac{\partial(\rho u_1)}{\partial x_1} + \frac{\partial(\rho u_2)}{\partial x_2} + \frac{\partial(\rho u_3)}{\partial x_3} = 0. \tag{7.10}$$

If the density is assumed to be constant then the above equation is reduced to

$$\frac{\partial u_1}{\partial x_1} + \frac{\partial u_2}{\partial x_2} + \frac{\partial u_3}{\partial x_3} = 0. \tag{7.11}$$

Using vector notation, the above equation is written as (divergence free velocity field)

$$\nabla.\mathbf{u} = 0 \tag{7.12}$$

or, using an indicial notation

$$\frac{\partial u_i}{\partial x_i} = 0, \tag{7.13}$$

where $i = 1, 2$ for a two-dimensional case and $i = 1, 2, 3$ for three-dimensional flows.

7.2.2 Conservation of Momentum

The conservation of momentum equation can be derived in a similar fashion to the conservation of mass equation. Here, the momentum equations are derived based on the conservation of momentum principle, that is, the total force generated by the momentum transfer in each direction is balanced by the rate of change of momentum in each direction. The momentum equation has directional components and is therefore a vector equation. In order to derive the conservation of momentum equation let us consider the control volume shown in Figure 7.4.

The momentum entering the control volume in the x_1 direction is given as

$$(\rho u_1 \Delta x_2)u_1 = \left(\rho u_1^2\right)\Delta x_2. \tag{7.14}$$

Since the momentum equation is a vector equation, the momentum in the x_1 direction will also have a contribution in the x_2 direction. The momentum entering the bottom face in the x_1 direction is

$$(\rho u_2 \Delta x_1)u_1 = (\rho u_1 u_2)\Delta x_1. \tag{7.15}$$

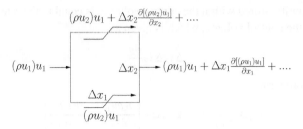

Figure 7.4   Infinitesimal control volume in a flow field. Derivation of conservation of momentum in x_1 direction. Rate of change of momentum.

A Taylor expansion is employed to work out the x_1 momentum leaving the control volume. In the x_1 direction we have

$$\rho u_1^2 \Delta x_2 + \Delta x_2 \frac{\partial \left(\rho u_1^2\right)}{\partial x_1} \Delta x_1. \tag{7.16}$$

Similarly, the x_1 momentum leaving the x_2 direction (top surface) is

$$\rho u_1 u_2 \Delta x_1 + \Delta x_1 \frac{\partial(\rho u_1 u_2)}{\partial x_2} \Delta x_2. \tag{7.17}$$

Note that the second- and higher-order terms in the Taylor expansion are neglected. The rate of change of momentum within the control volume due to the x_1 component is written as

$$\Delta x_1 \Delta x_2 \frac{\partial(\rho u_1)}{\partial t}. \tag{7.18}$$

The net momentum of the control volume is calculated as the "momentum exiting the control volume − momentum entering the control volume + rate of change of the momentum" which is

$$\Delta x_1 \Delta x_2 \left[\frac{\partial \left(\rho u_1^2\right)}{\partial x_1} + \frac{\partial(\rho u_1 u_2)}{\partial x_2} + \frac{\partial(\rho u_1)}{\partial t} \right]. \tag{7.19}$$

For equilibrium, the above net momentum should be balanced by the net force acting on the control volume. In order to derive the net force acting on the control volume, refer to Figure 7.5. From the figure, the total pressure force acting on the control volume in the x_1 direction is written as (positive in the positive x_1 direction and negative in the negative x_1 direction)

$$p \Delta x_2 - \left[p + \frac{\partial p}{\partial x_1} \Delta x_1 \right] \Delta x_2 = -\frac{\partial p}{\partial x_1} \Delta x_1 \Delta x_2. \tag{7.20}$$

Similarly, the total force due to the deviatoric stress (viscosity or friction) acting on the control volume in the x_1 direction is written as (see Figure 7.5):

$$\left[\tau_{11} + \frac{\partial \tau_{11}}{\partial x_1} \Delta x_1 \right] \Delta x_2 - \tau_{11} \Delta x_2 + \left[\tau_{12} + \frac{\partial \tau_{12}}{\partial x_2} \Delta x_2 \right] \Delta x_1 - \tau_{12} \Delta x_1. \tag{7.21}$$

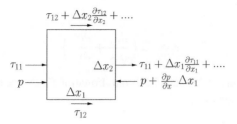

Figure 7.5   Infinitesimal control volume in a flow field. Derivation of conservation of momentum in x_1 direction. Viscous and pressure forces.

Simplifying, we obtain the net force due to the deviatoric stress as

$$\frac{\partial \tau_{11}}{\partial x_1} \Delta x_1 \Delta x_2 + \frac{\partial \tau_{12}}{\partial x_2} \Delta x_2 \Delta x_2. \tag{7.22}$$

The total force acting on the control volume in the x_1 direction is

$$\Delta x_1 \Delta x_2 \left[-\frac{\partial p}{\partial x_1} + \frac{\partial \tau_{11}}{\partial x_1} + \frac{\partial \tau_{12}}{\partial x_2} \right]. \tag{7.23}$$

As mentioned before, for equilibrium the net momentum in the x_1 direction should be equal to the total force acting on the control volume in the x_1 direction, that is,

$$\Delta x_1 \Delta x_2 \left[\frac{\partial(\rho u_1^2)}{\partial x_1} + \frac{\partial(\rho u_1 u_2)}{\partial x_2} + \frac{\partial(\rho u_1)}{\partial t} \right] = \Delta x_1 \Delta x_2 \left[-\frac{\partial p}{\partial x_1} + \frac{\partial \tau_{11}}{\partial x_1} + \frac{\partial \tau_{12}}{\partial x_2} \right]. \tag{7.24}$$

Simplifying, we obtain

$$\frac{\partial(\rho u_1)}{\partial t} + \frac{\partial(\rho u_1^2)}{\partial x_1} + \frac{\partial(\rho u_1 u_2)}{\partial x_2} = -\frac{\partial p}{\partial x_1} + \frac{\partial \tau_{11}}{\partial x_1} + \frac{\partial \tau_{12}}{\partial x_2}. \tag{7.25}$$

Note that the external and body forces (buoyancy) are not included in the above force balance. In the above equations, the deviatoric stresses τ_{ij} are expressed in terms of the velocity gradients and dynamic viscosity as

$$\tau_{ij} = \mu \left(\frac{\partial u_i}{\partial x_j} + \frac{\partial u_j}{\partial x_i} - \frac{2}{3} \frac{\partial u_k}{\partial x_k} \delta_{ij} \right), \tag{7.26}$$

where δ_{ij} is the Kroneker delta, which is equal to unity if $i = j$ and equal to zero if $i \neq j$. From the previous expression τ_{11} is expressed as

$$\tau_{11} = \mu \left(\frac{\partial u_1}{\partial x_1} + \frac{\partial u_1}{\partial x_1} - \frac{2}{3} \frac{\partial u_1}{\partial x_1} - \frac{2}{3} \frac{\partial u_2}{\partial x_2} \right). \tag{7.27}$$

Note that $i = j = 1$ in the above equation and $k = 1, 2$ for two-dimensional flow. The above equation may be simplified as follows:

$$\tau_{11} = \mu \left(\frac{4}{3} \frac{\partial u_1}{\partial x_1} - \frac{2}{3} \frac{\partial u_2}{\partial x_2} \right). \tag{7.28}$$

Similarly, τ_{12} is

$$\tau_{12} = \mu \left(\frac{\partial u_1}{\partial x_2} + \frac{\partial u_2}{\partial x_1} \right). \tag{7.29}$$

Substituting Equations (7.28) and (7.29) into Equation (7.25) we obtain the x_1 component of the momentum equation as

$$\frac{\partial(\rho u_1)}{\partial t} + \frac{\partial\left(\rho u_1^2\right)}{\partial x_1} + \frac{\partial(\rho u_1 u_2)}{\partial x_2} =$$
$$-\frac{\partial p}{\partial x_1} + \frac{\partial}{\partial x_1}\left[\mu\left(\frac{4}{3}\frac{\partial u_1}{\partial x_1} - \frac{2}{3}\frac{\partial u_2}{\partial x_2}\right)\right] + \frac{\partial}{\partial x_2}\left[\mu\left(\frac{\partial u_2}{\partial x_1} + \frac{\partial u_1}{\partial x_2}\right)\right]. \tag{7.30}$$

The momentum component in the x_2 direction can be derived by the following steps, which are similar to the derivation of the x_1 component of the momentum equation. The x_2 momentum equation is

$$\frac{\partial(\rho u_2)}{\partial t} + \frac{\partial(\rho u_1 u_2)}{\partial x_1} + \frac{\partial(\rho u_2^2)}{\partial x_2} =$$
$$-\frac{\partial p}{\partial x_2} + \frac{\partial}{\partial x_1}\left[\mu\left(\frac{\partial u_1}{\partial x_2} + \frac{\partial u_2}{\partial x_1}\right)\right] + \frac{\partial}{\partial x_2}\left[\mu\left(\frac{4}{3}\frac{\partial u_2}{\partial x_2} - \frac{2}{3}\frac{\partial u_1}{\partial x_1}\right)\right]. \tag{7.31}$$

For a constant density flow (incompressible flow), the momentum equations can be further reduced by taking the density term out of the differential signs. In addition, substitution of the conservation of mass equation (Equation (7.11)) into the momentum equation leads to a further simplification of the momentum equation. After simplification (see Appendix E for the detailed derivation), the momentum equations are

$$\rho\left(\frac{\partial u_1}{\partial t} + u_1\frac{\partial u_1}{\partial x_1} + u_2\frac{\partial u_1}{\partial x_2}\right) = -\frac{\partial p}{\partial x_1} + \mu\left[\frac{\partial^2 u_1}{\partial x_1^2} + \frac{\partial^2 u_1}{\partial x_2^2}\right] \tag{7.32}$$

in the x_1 direction and

$$\rho\left(\frac{\partial u_2}{\partial t} + u_1\frac{\partial u_2}{\partial x_1} + u_2\frac{\partial u_2}{\partial x_2}\right) = -\frac{\partial p}{\partial x_2} + \mu\left[\frac{\partial^2 u_2}{\partial x_1^2} + \frac{\partial^2 u_2}{\partial x_2^2}\right] \tag{7.33}$$

in the x_2 direction. In vector notation, the momentum equations can be written as

$$\rho\left[\frac{\partial \mathbf{u}}{\partial t} + \nabla.(\mathbf{u} \times \mathbf{u})\right] = \nabla.[-p\mathbf{I} + \tau] \tag{7.34}$$

or, in indicial form,

$$\rho\left(\frac{\partial u_i}{\partial t} + u_j\frac{\partial u_i}{\partial x_j}\right) = -\frac{\partial p}{\partial x_i} + \mu\left(\frac{\partial^2 u_i}{\partial x_i^2}\right). \tag{7.35}$$

Note that the above equation is applicable in any dimension.

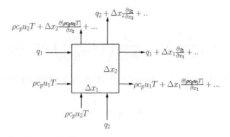

Figure 7.6   Infinitesimal control volume in a flow field. Derivation of conservation of energy.

7.2.3   Energy Equation

The energy equation can be derived by following a procedure, similar to the momentum equation derivation. However, the difference here is that the temperature, or energy equation, is a scalar equation. In order to derive this equation, let us consider the control volume as shown in Figure 7.6. The energy convected into the control volume in the x_1 direction is

$$\rho c_p(u_1 T)\Delta x_2. \tag{7.36}$$

Similarly, the energy convected into the control volume in the x_2 direction is

$$\rho c_p(u_2 T)\Delta x_1. \tag{7.37}$$

As before, a Taylor series expansion may be used to express the energy convected out of the control volume in both the x_1 and x_2 directions as

$$\rho c_p(u_1 T)\Delta x_2 + \rho c_p \frac{\partial(u_1 T)}{\partial x_1}\Delta x_1 \Delta x_2 \tag{7.38}$$

and

$$\rho c_p(u_2 T)\Delta x_1 + \rho c_p \frac{\partial(u_2 T)}{\partial x_2}\Delta x_2 \Delta x_1. \tag{7.39}$$

Note that the specific heat, c_p, and density, ρ, are assumed to be constants in deriving the above equation. The heat diffusion into and out of the control volume is also derived using the above approach. The heat diffusing into the domain in the x_1 direction (Fourier's law of heat conduction) is

$$q_1 \Delta x_2 = -k_{x_1} \frac{\partial T}{\partial x_1}\Delta x_2 \tag{7.40}$$

and the diffusion entering the control volume in the x_2 direction is

$$q_2 \Delta x_1 = -k_{x_2} \frac{\partial T}{\partial x_2}\Delta x_1. \tag{7.41}$$

Using a Taylor series expansion, the heat diffusing out of the control volume can be written as (see Figure 7.6):

$$-k_{x_1} \frac{\partial T}{\partial x_1}\Delta x_2 + \frac{\partial}{\partial x_1}\left(-k_{x_1} \frac{\partial T}{\partial x_1}\right)\Delta x_2 \Delta x_1 \tag{7.42}$$

in the x_1 direction and

$$-k_{x_2}\frac{\partial T}{\partial x_2}\Delta x_1 + \frac{\partial}{\partial x_2}\left(-k_{x_2}\frac{\partial T}{\partial x_2}\right)\Delta x_1 \Delta x_2 \tag{7.43}$$

in the x_2 direction. Finally, the rate of change of energy within the control volume is

$$\Delta x_1 \Delta x_2 \left(\rho c_p \frac{\partial T}{\partial t}\right). \tag{7.44}$$

Now, it is a simple matter of balancing the energy entering and exiting the control volume. The energy balance can be obtained as

"heat entering the control volume by convection + heat entering the control volume by diffusion = heat exiting the control volume by convection + heat exiting the control volume by diffusion + rate of change of energy within the control volume"

Following the above heat balance approach and rearranging, we get

$$\frac{\partial T}{\partial t} + \frac{\partial(u_1 T)}{\partial x_1} + \frac{\partial(u_2 T)}{\partial x_2} = \frac{1}{\rho c_p}\left[\frac{\partial}{\partial x_1}\left(k_{x_1}\frac{\partial T}{\partial x_1}\right) + \frac{\partial}{\partial x_2}\left(k_{x_2}\frac{\partial T}{\partial x_2}\right)\right]. \tag{7.45}$$

Differentiating the convection terms by parts and substituting Equation (7.11) (continuity) into Equation (7.45) we obtain the simplified energy equation in two dimensions as

$$\frac{\partial T}{\partial t} + u_1\frac{\partial T}{\partial x_1} + u_2\frac{\partial T}{\partial x_2} = \frac{1}{\rho c_p}\left[\frac{\partial}{\partial x_1}\left(k_{x_1}\frac{\partial T}{\partial x_1}\right) + \frac{\partial}{\partial x_2}\left(k_{x_2}\frac{\partial T}{\partial x_2}\right)\right]. \tag{7.46}$$

If the thermal conductivity is assumed to be constant and $k = k_{x_1} = k_{x_2}$, the energy equation is reduced to

$$\frac{\partial T}{\partial t} + u_1\frac{\partial T}{\partial x_1} + u_2\frac{\partial T}{\partial x_2} = \alpha\left(\frac{\partial^2 T}{\partial x_1^2} + \frac{\partial^2 T}{\partial x_2^2}\right). \tag{7.47}$$

where $\alpha = k/\rho c_p$ is called the thermal diffusivity. The energy equation in vector from is

$$\frac{\partial T}{\partial t} + \mathbf{u}.\nabla T = \alpha(\nabla^2 T) \tag{7.48}$$

and in indicial form

$$\frac{\partial T}{\partial t} + u_i\frac{\partial T}{\partial x_i} = \alpha\frac{\partial^2 T}{\partial x_i^2}. \tag{7.49}$$

The above equation is applicable in any space dimension.

7.3 Nondimensional Form of the Governing Equations

In the previous section, we discussed the derivation of the Navier-Stokes equations for an incompressible fluid. In many heat transfer applications, it is often easy to generate data by nondimensionalizing the equations using appropriate scales. To demonstrate the nondimensional form of the governing equations, let us consider the following two-dimensional incompressible flow equations in dimensional form

Continuity equation:

$$\frac{\partial u_1}{\partial x_1} + \frac{\partial u_2}{\partial x_2} = 0. \tag{7.50}$$

x_1 *momentum equation:*

$$\frac{\partial u_1}{\partial t} + u_1\frac{\partial u_1}{\partial x_1} + u_2\frac{\partial u_1}{\partial x_2} = -\frac{1}{\rho}\frac{\partial p}{\partial x_1} + v\left(\frac{\partial^2 u_1}{\partial x_1^2} + \frac{\partial^2 u_1}{\partial x_2^2}\right). \tag{7.51}$$

x_2 *momentum equation:*

$$\frac{\partial u_2}{\partial t} + u_1\frac{\partial u_2}{\partial x_1} + u_2\frac{\partial u_2}{\partial x_2} = -\frac{1}{\rho}\frac{\partial p}{\partial x_2} + v\left(\frac{\partial^2 u_2}{\partial x_1^2} + \frac{\partial^2 u_2}{\partial x_2^2}\right). \tag{7.52}$$

Energy equation:

$$\frac{\partial T}{\partial t} + u_1\frac{\partial T}{\partial x_1} + u_2\frac{\partial T}{\partial x_2} = \alpha\left(\frac{\partial^2 T}{\partial x_1^2} + \frac{\partial^2 T}{\partial x_2^2}\right), \tag{7.53}$$

where $v = \mu/\rho$ is the kinematic viscosity. To obtain a set of nondimensional equations, let us consider three different cases of convective heat transfer. We start with the forced convection problem followed by the "natural" and "mixed" convection problems. For each case, we discuss one set of nondimensional scales. There are several other ways of scaling the equations. Some of them are discussed in the latter part of the chapter and others can be found in various other publications listed at the end of this chapter.

Example 7.3.1   *Nondimensional form – forced convection*
   In forced convection problems the following nondimensional scales are normally employed.

$$x_1^* = \frac{x_1}{L}; \quad x_2^* = \frac{x_2}{L}; \quad t^* = \frac{tu_a}{L}; \quad u_1^* = \frac{u_1}{u_a};$$

$$u_2^* = \frac{u_2}{u_a}; \quad p^* = \frac{p}{\rho u_a^2}; \quad T^* = \frac{T - T_a}{T_w - T_a}, \tag{7.54}$$

*where * indicates a nondimensional quantity, L is a characteristic dimension (e.g., channel height), the subscript a indicates a constant reference value and T_w is a constant reference temperature, for example wall temperature. The density ρ and viscosity μ of the fluid are assumed to be constant everywhere.*
   Substitution of the above scales into the dimensional Equations (7.50) to (7.53) leads to the following nondimensional form of the equations.
   Continuity equation:

$$\frac{\partial u_1^*}{\partial x_1^*} + \frac{\partial u_2^*}{\partial x_2^*} = 0. \tag{7.55}$$

x_1 *momentum equation:*

$$\frac{\partial u_1^*}{\partial t^*} + u_1^* \frac{\partial u_1^*}{\partial x_1^*} + u_2^* \frac{\partial u_1^*}{\partial x_2^*} = -\frac{\partial p^*}{\partial x_1^*} + \frac{1}{Re} \left(\frac{\partial^2 u_1^*}{\partial x_1^{*2}} + \frac{\partial^2 u_1^*}{\partial x_2^{*2}} \right). \tag{7.56}$$

x_2 *momentum equation:*

$$\frac{\partial u_2^*}{\partial t^*} + u_1^* \frac{\partial u_2^*}{\partial x_1^*} + u_2^* \frac{\partial u_2^*}{\partial x_2^*} = -\frac{\partial p^*}{\partial x_2^*} + \frac{1}{Re} \left(\frac{\partial^2 u_2^*}{\partial x_1^{*2}} + \frac{\partial^2 u_2^*}{\partial x_2^{*2}} \right). \tag{7.57}$$

Energy equation:

$$\frac{\partial T^*}{\partial t^*} + u_1^* \frac{\partial T^*}{\partial x_1^*} + u_2^* \frac{\partial T^*}{\partial x_2^*} = \frac{1}{RePr} \left(\frac{\partial^2 T^*}{\partial x_1^{*2}} + \frac{\partial^2 T^*}{\partial x_2^{*2}} \right). \tag{7.58}$$

where Re is the Reynolds number defined as

$$Re = \frac{u_a L}{\nu} \tag{7.59}$$

and Pr is the Prandtl number given as

$$Pr = \frac{\nu}{\alpha}. \tag{7.60}$$

Once again, note that the density, kinematic viscosity and thermal conductivity are assumed to be constant in deriving the above nondimensional equations. Appropriate changes will be necessary if an appreciable variation in these quantities occurs in a flow field. Another nondimensional number, which is often employed in forced convection heat transfer calculations, is the Peclet number and is given as $Pe = RePr = u_a L/\alpha$. For buoyancy-driven natural convection problems, a different type of scale is necessary if there are no reference velocity values available. The following subsection gives the natural convection scales.

Example 7.3.2   *Nondimensional form – natural convection (buoyancy-driven convection)*
   Natural convection is generated by the density difference induced by the temperature differences within a fluid system. Because of the small density variations present in these type of flows, a general incompressible flow approximation is normally adopted. In most buoyancy-driven convection problems, flow is generated by either a temperature variation or a concentration variation in the fluid system, which leads to local density differences. Therefore, in such flows, a body force term needs to be added to the momentum equations to include the effect of local density differences. For temperature driven flows, the Boussenesq approximation is often employed, that is,

$$\frac{g(\rho - \rho_a)}{\rho_a} = g\beta(T - T_a), \tag{7.61}$$

where g is the acceleration due to gravity $(9.81 \ m/s^2)$ and β is the coefficient of thermal expansion. The above body force term is added to the momentum equations in the gravity direction. In a normal situation (refer to Figure 7.7), the body force is added to the x_2

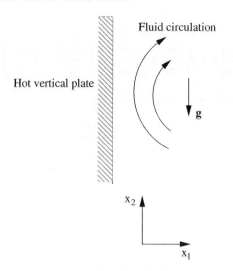

Figure 7.7   Natural convective flow near a hot vertical plate.

momentum (if the gravity direction is negative x_2), that is,

$$\frac{\partial u_2}{\partial t} + u_1 \frac{\partial u_2}{\partial x_1} + u_2 \frac{\partial u_2}{\partial x_2} = -\frac{1}{\rho}\frac{\partial p}{\partial x_2} + v\left(\frac{\partial^2 u_2}{\partial x_1^2} + \frac{\partial^2 u_2}{\partial x_2^2}\right) + g\beta(T - T_a). \qquad (7.62)$$

In practice, the following nondimensional scales are adopted for natural convection in the absence of a reference velocity value.

$$x_1^* = \frac{x_1}{L}; \quad x_2^* = \frac{x_2}{L}; \quad t^* = \frac{t\alpha}{L^2}; \quad u_1^* = \frac{u_1 L}{\alpha};$$

$$u_2^* = \frac{u_2 L}{\alpha}; \quad p^* = \frac{pL^2}{\rho\alpha^2}; \quad T^* = \frac{T - T_a}{T_w - T_a}. \qquad (7.63)$$

On introducing the above nondimensional scales into the governing equations we obtain the nondimensional form of the equations as follows:

Continuity equation:

$$\frac{\partial u_1^*}{\partial x_1^*} + \frac{\partial u_2^*}{\partial x_2^*} = 0. \qquad (7.64)$$

x_1 momentum equation:

$$\frac{\partial u_1^*}{\partial t^*} + u_1^* \frac{\partial u_1^*}{\partial x_1^*} + u_2^* \frac{\partial u_1^*}{\partial x_2^*} = -\frac{\partial p^*}{\partial x_1^*} + Pr\left(\frac{\partial^2 u_1^*}{\partial x_1^{*2}} + \frac{\partial^2 u_1^*}{\partial x_2^{*2}}\right). \qquad (7.65)$$

x_2 momentum equation:

$$\frac{\partial u_2^*}{\partial t^*} + u_1^* \frac{\partial u_2^*}{\partial x_1^*} + u_2^* \frac{\partial u_2^*}{\partial x_2^*} = -\frac{\partial p^*}{\partial x_2^*} + Pr\left(\frac{\partial^2 u_2^*}{\partial x_1^{*2}} + \frac{\partial^2 u_2^*}{\partial x_2^{*2}}\right) + GrPr^2 T^*. \qquad (7.66)$$

Energy equation

$$\frac{\partial T^*}{\partial t^*} + u_1^* \frac{\partial T^*}{\partial x_1^*} + u_2^* \frac{\partial T^*}{\partial x_2^*} = \left(\frac{\partial^2 T^*}{\partial x_1^{*2}} + \frac{\partial^2 T^*}{\partial x_2^{*2}} \right), \tag{7.67}$$

where Gr is the Grashof number given as

$$Gr = \frac{g\beta \Delta T L^3}{\nu^2}. \tag{7.68}$$

Often, another nondimensional number, called the Rayleigh number, is used in the calculations. This is given as

$$Ra = GrPr = \frac{g\beta \Delta T L^3}{\nu \alpha}. \tag{7.69}$$

On comparing the nondimensional equations of natural and forced convection, it is easy to identify the differences. If we substitute $1/Pr$ in place of the Reynolds number for the forced convection equations, we revert to a natural convection scaling. Obviously, the extra buoyancy term needs to be added to appropriate component(s) of the momentum equation for natural convection flows.

Example 7.3.3   *Nondimensional form – mixed convection*
   Mixed convection involves features from both forced and natural flow conditions. The buoyancy effects become comparable to the forced flow effects at small and moderate Reynolds numbers. Since the flow is partly forced, a reference velocity value is normally known (example: velocity at the inlet of a channel). Therefore, nondimensional scales of forced convection can be adapted here. However, in mixed convection problems, the buoyancy term needs to be added to the appropriate component of the momentum equation. If we replace $1/Pr$ with Re in the nondimensional natural convection equations of the previous subsection, we obtain the nondimensional equations for mixed convection flows. These equations are the same as for the forced convection flow problem, except for the body force term, which will be added to the momentum equation in the gravity direction. The body force term is

$$\frac{Gr}{Re^2} T^*. \tag{7.70}$$

Note that some times a nondimensional parameter referred to as the Richardson number (Gr/Re^2) is also used in the literature. Note also that the Peclit member reappears in the energy equation.

7.4   The Transient Convection-Diffusion Problem

An understanding of the fundamentals of the convection-diffusion equations is crucial in studying fluid dynamics assisted heat transfer. The equations governing the combined fluid flow and heat transfer mainly involve the convection and diffusion components. A typical scalar convection diffusion equation may be written as

$$\frac{\partial \phi}{\partial t} + u_i \frac{\partial \phi}{\partial x_i} + \phi \frac{\partial u_i}{\partial x_i} - \frac{\partial}{\partial x_i} \left(k \frac{\partial \phi}{\partial x_i} \right) + Q = 0, \tag{7.71}$$

where ϕ is a scalar variable, k is a diffusion coefficient (thermal conductivity if $\phi = T$), u_i are the convection velocity components and Q is a source term. In the above equation, the first term is a transient term, the second and third terms are convection terms and the fourth term is the diffusion term. For a one-dimensional problem, the above equation is reduced to

$$\frac{\partial \phi}{\partial t} + u_1 \frac{\partial \phi}{\partial x_1} + \phi \frac{\partial u_1}{\partial x_1} - \frac{\partial}{\partial x_1} \left(k \frac{\partial \phi}{\partial x_1} \right) + Q = 0. \tag{7.72}$$

If the convection velocity u_1 is assumed to be constant, we can rewrite Equation (7.72) as follows:

$$\frac{\partial \phi}{\partial t} + u_1 \frac{\partial \phi}{\partial x_1} - \frac{\partial}{\partial x_1} \left(k \frac{\partial \phi}{\partial x_1} \right) + Q = 0. \tag{7.73}$$

A one-dimensional convection equation without a source term is obtained by neglecting the diffusion and source terms as follows,

$$\frac{\partial \phi}{\partial t} + u_1 \frac{\partial \phi}{\partial x_1} = 0. \tag{7.74}$$

Note that an appropriate solution for the above equation is valid for any similar equations such as the energy equation.

7.4.1   Finite Element Solution to the Convection-Diffusion Equation

Unlike the conduction equation, a numerical solution for the convection equation has to deal with the convection part of the governing equation in addition to diffusion. For most of the conduction equations, the finite element solution is straightforward, as discussed in the previous chapters. However, if a similar Galerkin type approximation was used in the solution of convection equations, the results may be marked with spurious oscillations in space (see the example discussed later in this section) if certain parameters exceed a critical value (element Peclet number). This problem is not unique to finite elements as all other spatial discretization techniques have the same difficulties. In a finite difference formulation, the spatial oscillations are reduced, or suppressed, by a family of discretization methods called upwinding schemes (Fletcher 1988; Spalding 1972). In the finite element method, procedures such as Petrov-Galerkin (Zienkiewicz *et al.* 2005) and Streamline Upwind Galerkin (SUPG) (Brooks and Hughes 1982) are equivalent upwinding schemes with the specific purpose of eliminating spatial oscillations. In these methods, the basic shape function is modified to obtain the upwinding effect.

For time dependent equations, however, a different kind of approach is followed. The finite difference Lax-Wendroff (Hirsch 1989) scheme has an equivalent in the finite element method, which is referred to as the Taylor Galerkin (TG) scheme (Donea 1984). Another similar method, which is widely used, is known as the Characteristic Galerkin (CG) scheme

(Zienkiewicz *et al.* 2005). For scalar variables the CG and TG methods are identical (Lohner *et al.* 1984). In this book, we follow the Characteristic Galerkin (CG) approach to deal with spatial oscillations due to the discretization of the convection transport terms. In order to demonstrate the CG method, let us reconsider the simple convection diffusion equation in one dimension, viz.

$$\frac{\partial \phi}{\partial t} + u_1 \frac{\partial \phi}{\partial x_1} - \frac{\partial}{\partial x_1}\left(k\frac{\partial \phi}{\partial x_1}\right) = 0. \tag{7.75}$$

With negligible diffusion, the equation may be reduced to

$$\frac{\partial \phi}{\partial t} + u_1 \frac{\partial \phi}{\partial x_1} = 0. \tag{7.76}$$

Following example explains how the characteristic speed, dx_1/dt, for the above equation may be calculated.

Example 7.4.1   *Characteristic speed*
The total derivative of variable ϕ may be computed as

$$d\phi = \frac{\partial \phi}{\partial x_1}dx_1 + \frac{\partial \phi}{\partial t}dt. \tag{7.77}$$

Rearranging,

$$\frac{d\phi}{dt} = \frac{\partial \phi}{\partial x_1}\frac{dx_1}{dt} + \frac{\partial \phi}{\partial t}. \tag{7.78}$$

Comparison of Equations (7.78) and (7.76) gives

$$\frac{dx_1}{dt} = u_1 \tag{7.79}$$

is the characteristic speed. This speed can be represented in $x_1 - t$ plane as shown in Figure 7.8.

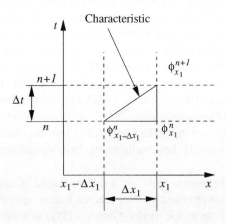

Figure 7.8   Characteristic in a space–time domain.

7.4.2 A Simple Characteristic Galerkin Method for Convection-Diffusion Equation

Let us consider a characteristic of the flow as shown in Figure 7.8 in the one-dimensional, time–space domain. The incremental time period covered by the flow is Δt from the nth time level to the $n + 1^{th}$ time level and the incremental distance covered during this time period is Δx_1, that is, from $(x_1 - \Delta x_1)$ to x_1. If a moving coordinate is assumed along the path of the characteristic wave with a speed of u_1, the convection term of Equation (7.75) disappears (as in a Lagrangian fluid dynamics approach). Although this approach eliminates the convection term responsible for spatial oscillation when discretized standard Galerkin method in space, the complication of a moving coordinate system x_1' is introduced, that is, Equation (7.75) becomes

$$\frac{\partial \phi}{\partial t}(x_1', t) - \frac{\partial}{\partial x_1'}\left(k\frac{\partial \phi}{\partial x_1'}\right) = 0. \tag{7.80}$$

The semi-discrete form of the above equation can be written as

$$\frac{\phi^{n+1}|_{x_1} - \phi^n|_{x_1-\Delta x_1}}{\Delta t} - \frac{\partial}{\partial x_1'}\left(k\frac{\partial \phi}{\partial x_1'}\right)^n\Big|_{x_1-\Delta x_1} = 0. \tag{7.81}$$

Note that the diffusion term is treated explicitly (a definition of explicit schemes has been given in Chapter 6 and later on in this chapter). It is possible to solve the above equation by adapting a moving coordinate strategy. However, a simple spatial Taylor series expansion in space avoids such a moving coordinate approach. With reference to Figure 7.8, we can write, using a Taylor series expansion, as follows:

$$\phi^n|_{x_1-\Delta x_1} = \phi^n|_{x_1} - \frac{\partial \phi}{\partial x_1}^n\frac{\Delta x_1}{1!} + \frac{\partial^2 \phi}{\partial x_1^2}\frac{\Delta x_1^2}{2!} - \ldots \tag{7.82}$$

Similarly, the diffusion term is expanded as

$$\frac{\partial}{\partial x_1'}\left(k\frac{\partial \phi}{\partial x_1'}\right)^n\Big|_{x_1-\Delta x_1} = \frac{\partial}{\partial x_1}\left(k\frac{\partial \phi}{\partial x_1}\right)^n\Big|_{x_1} - \frac{\partial}{\partial x_1}\left[\frac{\partial}{\partial x_1}\left(k\frac{\partial \phi}{\partial x_1}\right)^n\right]\Delta x. \tag{7.83}$$

On substituting Equations (7.82) and (7.83) into Equation (7.81) we obtain (third- and higher-order terms being neglected) the following expression:

$$\frac{\phi^{n+1} - \phi^n}{\Delta t} = -\frac{\Delta x}{\Delta t}\frac{\partial \phi}{\partial x_1}^n + \frac{\Delta x^2}{2\Delta t}\frac{\partial^2 \phi}{\partial x_1^2}^n + \frac{\partial}{\partial x_1}\left(k\frac{\partial \phi}{\partial x_1}\right)^n. \tag{7.84}$$

In this case all the terms are evaluated at the position x_1, and not at two positions as in Equation (7.81). If the flow velocity is u_1, we can write $\Delta x_1 = u_1\Delta t$. Substituting into Equation (7.84) we obtain the semi-discrete form as

$$\frac{\phi^{n+1} - \phi^n}{\Delta t} = -u_1\frac{\partial \phi}{\partial x_1}^n + u_1^2\frac{\Delta t}{2}\frac{\partial^2 \phi}{\partial x_1^2}^n + \frac{\partial}{\partial x_1}\left(k\frac{\partial \phi}{\partial x_1}\right)^n. \tag{7.85}$$

By carrying out a Taylor series expansion (see Figure 7.8), the convection term reappears in the equation along with an additional second-order term. This second-order term acts as a smoothing operator which reduces the oscillations arising from the Galerkin type spatial discretization of the convection terms. The equation is now ready for the Galerkin spatial approximation.

The following linear spatial approximation of the scalar variable ϕ in space is used to approximate Equation (7.85).

$$\tilde{\phi} = N_1\phi_1 + N_2\phi_2 + \ldots\ldots N_n\phi_n = \begin{bmatrix} N_1 & N_2 & \ldots & N_n \end{bmatrix} \begin{Bmatrix} \phi_1 \\ \phi_2 \\ \ldots \\ \ldots \\ \ldots \\ \phi_n \end{Bmatrix} = [\mathbf{N}]\{\boldsymbol{\phi}\}. \tag{7.86}$$

where $N_1, N_2 \ldots N_n$ are the shape functions, $\phi_1, \phi_2 \ldots \phi_n$ are the nodal values of ϕ, n is the total number of nodes and ˜ indicates an approximate value. On employing the Galerkin weighting to Equation (7.85), we obtain

$$\int_\Omega [\mathbf{N}]^T \left(\frac{\tilde{\phi}^{n+1} - \tilde{\phi}^n}{\Delta t} \right) d\Omega + \int_\Omega [\mathbf{N}]^T \left(u_1 \frac{\partial \tilde{\phi}}{\partial x_1} \right)^n d\Omega$$

$$- \frac{\Delta t}{2} \int_\Omega [\mathbf{N}]^T \left(u_1^2 \frac{\partial^2 \tilde{\phi}}{\partial x_1^2} \right)^n d\Omega - \int_\Omega [\mathbf{N}]^T \frac{\partial}{\partial x_1} \left(k \frac{\partial \tilde{\phi}}{\partial x_1} \right) d\Omega = 0, \tag{7.87}$$

where

$$[\mathbf{N}]^T = \begin{bmatrix} N_1 \\ \ldots \\ \ldots \\ \ldots \\ N_n \end{bmatrix}. \tag{7.88}$$

On substituting Equation (7.86) into (7.87), we get

$$\int_\Omega [\mathbf{N}]^T [\mathbf{N}] \frac{\{\boldsymbol{\phi}^{n+1} - \boldsymbol{\phi}^n\}}{\Delta t} d\Omega = -u_1 \int_\Omega [\mathbf{N}]^T \frac{\partial}{\partial x_1} ([\mathbf{N}]\{\boldsymbol{\phi}\})^n d\Omega$$

$$+ \frac{\Delta t}{2} u_1^2 \int_\Omega [\mathbf{N}]^T \frac{\partial^2}{\partial x_1^2} ([\mathbf{N}]\{\boldsymbol{\phi}\})^n d\Omega + \int_\Omega [\mathbf{N}]^T \frac{\partial^2}{\partial x_1^2} ([\mathbf{N}]\{\boldsymbol{\phi}\})^n d\Omega, \tag{7.89}$$

where the definition of $[\mathbf{N}]$ and $\{\boldsymbol{\phi}\}$ are given in Equation (7.86). Before utilizing the linear integration formulae, we apply Green's lemma to some of the integrals in the above equation. Green's lemma is given as follows:

$$\int_\Omega \alpha \frac{\partial \beta}{\partial x_1} d\Omega = - \int_\Omega \frac{\partial \alpha}{\partial x_1} \beta d\Omega + \int_\Gamma \alpha \beta n_1 d\Gamma, \tag{7.90}$$

where n_1 is the direction cosine of the outward normal $\boldsymbol{n}$, Ω is the domain and Γ is the domain boundary. The second-order derivatives can also be similarly expressed (see Appendix B).

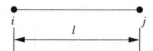

Figure 7.9   One-dimensional linear element.

Applying Green's lemma to the second-order terms of Equation (7.89) we obtain

$$\int_\Omega [\mathbf{N}]^T [\mathbf{N}] \frac{\{\boldsymbol{\phi}^{n+1}\} - \{\boldsymbol{\phi}^n\}}{\Delta t} d\Omega = -u_1 \int_\Omega [\mathbf{N}]^T \frac{\partial}{\partial x_1}([\mathbf{N}]\{\boldsymbol{\phi}\})^n d\Omega$$

$$- \frac{\Delta t}{2} u_1^2 \int_\Omega \frac{\partial [\mathbf{N}]^T}{\partial x_1} \frac{\partial [\mathbf{N}]}{\partial x_1} \{\boldsymbol{\phi}\} d\Omega + \frac{\Delta t}{2} u_1^2 \int_\Gamma [\mathbf{N}]^T \frac{\partial [\mathbf{N}]}{\partial x_1} \{\boldsymbol{\phi}\} n_1 d\Gamma$$

$$- \int_\Omega \frac{\partial [\mathbf{N}]^T}{\partial x_1} k \frac{\partial [\mathbf{N}]}{\partial x_1} \{\boldsymbol{\phi}\} d\Omega + \int_\Gamma [\mathbf{N}]^T k \frac{\partial [\mathbf{N}]}{\partial x_1} \{\boldsymbol{\phi}\} n_1 d\Gamma. \tag{7.91}$$

The first-order convection term can either be directly integrated or integrated via Green's lemma. In this section, the convection term is integrated directly without applying Green's lemma. However, integration of the first derivatives, by parts is useful for the solution of Navier Stokes equations, as demonstrated in the Section 7.6. The integration of the above equation can now be carried out locally element by element by introducing element matrices. The element matrices and vectors for a linear element shown in Figure 7.9 may be written as

$$[\mathbf{N}]_e = \begin{bmatrix} N_i & N_j \end{bmatrix} \quad \text{and} \quad \{\boldsymbol{\phi}\}_e = \begin{Bmatrix} \phi_i \\ \phi_j \end{Bmatrix}. \tag{7.92}$$

For one-dimensional linear elements of length l we may use the integration formula

$$\int_{\Omega_e} N_i^a N_j^b d\Omega = \frac{a! b! l}{(a + b + 1)!} \tag{7.93}$$

to derive the element matrices for all the terms in Equation (7.91). Note that Ω_e here indicates an element subdomain. The term on the left-hand side for a single element (see Figure 7.9) is

$$\int_{\Omega_e} [\mathbf{N}]_e^T [\mathbf{N}]_e \frac{\{\boldsymbol{\phi}^{n+1}\}_e - \{\boldsymbol{\phi}^n\}_e}{\Delta t} d\Omega = \int_{\Omega_e} \begin{bmatrix} N_i \\ N_j \end{bmatrix} \begin{bmatrix} N_i & N_j \end{bmatrix} \begin{Bmatrix} \frac{\phi_i^{n+1} - \phi_i^n}{\Delta t} \\ \frac{\phi_j^{n+1} - \phi_j^n}{\Delta t} \end{Bmatrix} d\Omega$$

$$= \int_{\Omega_e} \begin{bmatrix} N_i^2 & N_i N_j \\ N_j N_i & N_j^2 \end{bmatrix} \begin{Bmatrix} \frac{\phi_i^{n+1} - \phi_i^n}{\Delta t} \\ \frac{\phi_j^{n+1} - \phi_j^n}{\Delta t} \end{Bmatrix} d\Omega$$

$$= \frac{l}{6} \begin{bmatrix} 2 & 1 \\ 1 & 2 \end{bmatrix} \begin{Bmatrix} \frac{\phi_i^{n+1} - \phi_i^n}{\Delta t} \\ \frac{\phi_j^{n+1} - \phi_j^n}{\Delta t} \end{Bmatrix}$$

$$= [M]_e \frac{\Delta \{\boldsymbol{\phi}\}_e}{\Delta t}, \tag{7.94}$$

where $[\mathbf{M}]_e$ is the mass matrix. For a single element, the mass matrix is given as

$$[\mathbf{M}]_e = \frac{l}{6}\begin{bmatrix} 2 & 1 \\ 1 & 2 \end{bmatrix}. \tag{7.95}$$

The above mass matrix for a single element will have to be utilized in an assembly procedure for a fluid domain containing many elements. In Equation (7.94)

$$\Delta\{\boldsymbol{\phi}\}_e = \left\{ \begin{array}{c} \dfrac{\phi_i^{n+1}-\phi_i^n}{\Delta t} \\[2mm] \dfrac{\phi_j^{n+1}-\phi_j^n}{\Delta t} \end{array} \right\}. \tag{7.96}$$

In a similar fashion all other terms can be integrated, for example, the convection term is given by

$$
u_1 \int_{\Omega_e} [\mathbf{N}]_e^T \frac{\partial[\mathbf{N}]_e}{\partial x_1}\{\boldsymbol{\phi}\}_e^n d\Omega = u_1 \int_{\Omega_e} \begin{bmatrix} N_i \\ N_j \end{bmatrix}\begin{bmatrix} \dfrac{\partial N_i}{\partial x_1} & \dfrac{\partial N_j}{\partial x_1} \end{bmatrix}\left\{ \begin{array}{c} \phi_i \\ \phi_j \end{array} \right\}^n d\Omega
$$

$$
= u_1 \int \begin{bmatrix} \dfrac{l}{2}\dfrac{\partial N_i}{\partial x_1} & \dfrac{l}{2}\dfrac{\partial N_j}{\partial x_1} \\[2mm] \dfrac{l}{2}\dfrac{\partial N_i}{\partial x_1} & \dfrac{l}{2}\dfrac{\partial N_j}{\partial x_1} \end{bmatrix}\left\{ \begin{array}{c} \phi_i \\ \phi_j \end{array} \right\}^n
$$

$$
= \frac{u_1}{2}\begin{bmatrix} -1 & 1 \\ -1 & 1 \end{bmatrix}\left\{ \begin{array}{c} \phi_i \\ \phi_j \end{array} \right\}^n
$$

$$
= [\mathbf{C}]_e\{\boldsymbol{\phi}\}_e^n, \tag{7.97}
$$

where $[\mathbf{C}]_e$ is the elemental convection matrix, that is,

$$[\mathbf{C}]_e = \frac{u_1}{2}\begin{bmatrix} -1 & 1 \\ -1 & 1 \end{bmatrix}. \tag{7.98}$$

Note that the values of derivatives of the shape functions

$$\frac{dN_i}{dx_1} = -\frac{1}{l}; \quad \text{and} \quad \frac{dN_j}{dx} = \frac{1}{l} \tag{7.99}$$

are substituted to derive $[\mathbf{C}]_e$ matrix. The diffusion term within an element subdomain Ω_e is integrated as

$$
\int_{\Omega_e} \frac{\partial[\mathbf{N}]_e^T}{\partial x_1} k \frac{\partial[\mathbf{N}]_e}{\partial x_1}\{\boldsymbol{\phi}\}_e^n d\Omega = \int_{\Omega_e} \begin{bmatrix} \dfrac{\partial N_i}{\partial x_1} \\[2mm] \dfrac{\partial N_j}{\partial x_1} \end{bmatrix} k \begin{bmatrix} \dfrac{\partial N_i}{\partial x_1} & \dfrac{\partial N_j}{\partial x_1} \end{bmatrix}\left\{ \begin{array}{c} \phi_i \\ \phi_j \end{array} \right\}^n d\Omega
$$

$$
= \int_{\Omega_e} k \begin{bmatrix} \dfrac{\partial N_i}{\partial x_1}\dfrac{\partial N_i}{\partial x_1} & \dfrac{\partial N_i}{\partial x_1}\dfrac{\partial N_j}{\partial x_1} \\[2mm] \dfrac{\partial N_j}{\partial x_1}\dfrac{\partial N_i}{\partial x_1} & \dfrac{\partial N_j}{\partial x_1}\dfrac{\partial N_j}{\partial x_1} \end{bmatrix}\left\{ \begin{array}{c} \phi_i \\ \phi_j \end{array} \right\}^n d\Omega
$$

$$
= \frac{k}{l}\begin{bmatrix} 1 & -1 \\ -1 & 1 \end{bmatrix}\left\{ \begin{array}{c} \phi_i \\ \phi_j \end{array} \right\}^n
$$

$$
= [\mathbf{K}]_e\{\boldsymbol{\phi}\}_e^n, \tag{7.100}
$$

where $[\mathbf{K}]_e$ is the elemental diffusion matrix and given as

$$[\mathbf{K}]_e = \frac{k}{l}\begin{bmatrix} 1 & -1 \\ -1 & 1 \end{bmatrix}. \qquad (7.101)$$

The characteristic Galerkin term within the element subdomain Ω_e is integrated as

$$u_1^2 \frac{\Delta t}{2} \int_{\Omega_e} \frac{\partial [\mathbf{N}]_e^T}{\partial x_1} \frac{\partial [\mathbf{N}]_e}{\partial x_1} \{\boldsymbol{\phi}\}^n d\Omega = u_1^2 \frac{\Delta t}{2} \int_{\Omega_e} \begin{bmatrix} \frac{\partial N_i}{\partial x_1} \\ \frac{\partial N_j}{\partial x_1} \end{bmatrix} \begin{bmatrix} \frac{\partial N_i}{\partial x_1} & \frac{\partial N_j}{\partial x_1} \end{bmatrix} \left\{ \begin{matrix} \phi_i \\ \phi_j \end{matrix} \right\}^n d\Omega$$

$$= u_1^2 \frac{\Delta t}{2} \int_{\Omega_e} \begin{bmatrix} \frac{\partial N_i}{\partial x_1} \frac{\partial N_i}{\partial x_1} & \frac{\partial N_i}{\partial x_1} \frac{\partial N_j}{\partial x_1} \\ \frac{\partial N_j}{\partial x_1} \frac{\partial N_i}{\partial x_1} & \frac{\partial N_j}{\partial x_1} \frac{\partial N_j}{\partial x_1} \end{bmatrix} \left\{ \begin{matrix} \phi_i \\ \phi_j \end{matrix} \right\} d\Omega$$

$$= u_1^2 \frac{\Delta t}{2} \frac{1}{l} \begin{bmatrix} 1 & -1 \\ -1 & 1 \end{bmatrix} \left\{ \begin{matrix} \phi_i \\ \phi_j \end{matrix} \right\}^n$$

$$= [\mathbf{K}_s]_e \{\boldsymbol{\phi}_e\}^n, \qquad (7.102)$$

where $[\mathbf{K}_s]_e$ is a stabilization matrix,

$$[\mathbf{K}_s]_e = u_1^2 \frac{\Delta t}{2} \frac{1}{l} \begin{bmatrix} 1 & -1 \\ -1 & 1 \end{bmatrix}. \qquad (7.103)$$

The boundary term from the diffusion operator is integrated by assuming that i is a boundary node, as follows:

$$\int_{\Gamma_e} [\mathbf{N}]_e^T k \frac{\partial [\mathbf{N}]_e}{\partial x_1} \{\boldsymbol{\phi}\}_e^n d\Gamma = \int_{\Gamma_e} \begin{bmatrix} N_i \\ 0 \end{bmatrix} k \begin{bmatrix} \frac{\partial N_i}{\partial x_1} & \frac{\partial N_j}{\partial x_1} \end{bmatrix} \left\{ \begin{matrix} \phi_i \\ \phi_j \end{matrix} \right\}^n d\Gamma$$

$$= \int_{\Gamma_e} k \begin{bmatrix} N_i \frac{\partial N_i}{\partial x_1} & N_i \frac{\partial N_j}{\partial x_1} \\ 0 & 0 \end{bmatrix} \left\{ \begin{matrix} \phi_i \\ \phi_j \end{matrix} \right\}^n d\Gamma$$

$$= k \begin{bmatrix} -\frac{1}{l} & \frac{1}{l} \\ 0 & 0 \end{bmatrix} \left\{ \begin{matrix} \phi_i \\ \phi_j \end{matrix} \right\}^n$$

$$= \{\mathbf{f}\}_e, \qquad (7.104)$$

where $\{\mathbf{f}\}_e$ is the forcing vector due to the diffusion term, that is,

$$\{\mathbf{f}\}_e = k \left\{ \begin{matrix} \frac{\phi_j - \phi_i}{l} \\ 0 \end{matrix} \right\}^n. \qquad (7.105)$$

The boundary integral from the characteristic Galerkin term is integrated, again by assuming that i is a boundary node, as

$$\int_{\Gamma_e} u_1^2 \frac{\Delta t}{2} [\mathbf{N}]_e^T \frac{\partial [\mathbf{N}]_e}{\partial x_1} \{\boldsymbol{\phi}\}_e^n d\Gamma = u_1^2 \frac{\Delta t}{2} \int_{\Gamma_e} \begin{bmatrix} N_i \\ 0 \end{bmatrix} \begin{bmatrix} \frac{\partial N_i}{\partial x_1} & \frac{\partial N_j}{\partial x_1} \end{bmatrix} \left\{ \begin{matrix} \phi_i \\ \phi_j \end{matrix} \right\}^n d\Gamma$$

$$= u_1^2 \frac{\Delta t}{2} \int_{\Gamma_e} \begin{bmatrix} N_i \frac{\partial N_i}{\partial x_1} & N_i \frac{\partial N_j}{\partial x_1} \\ 0 & 0 \end{bmatrix} \left\{ \begin{matrix} \phi_i \\ \phi_j \end{matrix} \right\}^n d\Gamma$$

$$= u_1^2 \frac{\Delta t}{2} \begin{bmatrix} -\frac{1}{l} & \frac{1}{l} \\ 0 & 0 \end{bmatrix} \left\{ \begin{matrix} \phi_i \\ \phi_j \end{matrix} \right\}^n$$

$$= \{\mathbf{f}_s\}_e, \tag{7.106}$$

where $\{\mathbf{f}_s\}_e$ is the forcing vector due to the stabilization term,

$$\{\mathbf{f}_s\}_e = u_1^2 \frac{\Delta t}{2} \left\{ \begin{matrix} \frac{\phi_j - \phi_i}{l} \\ 0 \end{matrix} \right\}^n. \tag{7.107}$$

The forcing vectors are formulated by assuming that the node i is a boundary node. Because of the opposite signs of the outward normals at the interface between any two elements within the domain, these forcing vector terms vanish for all nodes other than the boundary nodes. The remaining terms will have a value only at the domain boundaries. Also, the boundary terms due to the CG stabilizing operator (Equation (7.107)) can be neglected during the calculations without any loss in accuracy due to the fact that the residual vanishes on the boundaries.

For a one-dimensional domain with more than one element, all the matrices and vectors need to be assembled in order to obtain the global matrices. Once assembled, the discretized one-dimensional equation becomes

$$[\mathbf{M}] \frac{\Delta\{\boldsymbol{\phi}\}}{\Delta t} + [\mathbf{C}]\{\boldsymbol{\phi}\}^n + [\mathbf{K}]\{\boldsymbol{\phi}\}^n + [\mathbf{K}_s]\{\boldsymbol{\phi}\}^n = \{\mathbf{f}\}_{n_1}^n + \{\mathbf{f}_s\}_{n_1}^n. \tag{7.108}$$

Note that all the matrices and vectors in the above equation are assembled.

Example 7.4.2   *Convection-diffusion problem*

Let us now consider a simple one-dimensional convection problem, as given in Figure 7.10 to demonstrate the effect of a discretization with and without CG scheme.

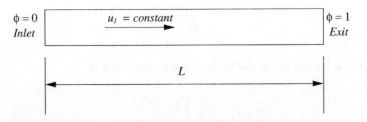

Figure 7.10   Convection-diffusion of a scalar variable. Problem setup.

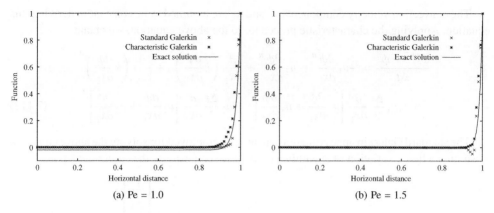

Figure 7.11   Steady-state spatial variation of a function, ϕ, in one-dimensional space for different element Peclet numbers.

The scalar variable value at the inlet is $\phi = 0$ and at the exit its value is 1.0. This scalar variable is transported in the direction of the velocity as shown in Figure 7.10. Note that the convection velocity u_1 is constant. The element Peclet number for this problem is defined as

$$Pe = \frac{u_1 l}{2k}, \tag{7.109}$$

where l is the element size in the flow direction, which in one dimension is the local element length. Figure 7.11 shows the comparison between a solution with the CG discretization scheme and one without it. Only two Peclet numbers are shown in these diagrams to demonstrate the spatial oscillations without the CG discretization. As seen, both discretizations give no spatial oscillations at a Pe of unity. However, at a Pe value of 1.5, the CG discretization is accurate and stable while the discretization without CG term becomes oscillatory. The exact solution to this problem is given as follows (Brooks and Hughes 1982):

$$\phi = \frac{1 - e^{\frac{u_1 x_1}{k}}}{1 - e^{\frac{u_1 L}{k}}}. \tag{7.110}$$

In this equation, L is the total length of the domain and x_1 is the local length of the domain.

7.4.3   Extension to Multi-dimensions

An approximate extension of the characteristic Galerkin scheme to a multi-dimensional scalar convection-diffusion equation is straightforward and follows the previous procedure as discussed for a one-dimensional case. The two-dimensional convection-diffusion equation without the source term is

$$\frac{\partial \phi}{\partial t} + u_1 \frac{\partial \phi}{\partial x_1} + u_2 \frac{\partial \phi}{\partial x_2} = \frac{\partial}{\partial x_1}\left(k\frac{\partial \phi}{\partial x_1}\right) + \frac{\partial}{\partial x_1}\left(k\frac{\partial \phi}{\partial x_2}\right). \tag{7.111}$$

The convection velocity components u_1 and u_2 are assumed to be constant in deriving this equation. Applying the characteristic procedure to the above equation, we obtain

$$\frac{\phi^{n+1} - \phi^n}{\Delta t} = -u_1 \frac{\partial \phi}{\partial x_1}^n - u_2 \frac{\partial \phi}{\partial x_2}^n + \frac{\partial}{\partial x_1} \left(k \frac{\partial \phi}{\partial x_1} \right)^n + \frac{\partial}{\partial x_2} \left(k \frac{\partial \phi}{\partial x_2} \right)^n$$

$$+ u_1 \frac{\Delta t}{2} \frac{\partial}{\partial x_1} \left[u_1 \frac{\partial \phi}{\partial x_1} + u_2 \frac{\partial \phi}{\partial x_2} \right]^n + u_2 \frac{\Delta t}{2} \frac{\partial}{\partial x_2} \left[u_1 \frac{\partial \phi}{\partial x_1} + u_2 \frac{\partial \phi}{\partial x_2} \right]^n. \qquad (7.112)$$

The standard Galerkin approximation can now be employed for solving the above equation. Assuming a linear variation of ϕ within a domain we can express the variation of ϕ as

$$\tilde{\phi} = N_1 \phi_1 + N_2 \phi_2 + \dots N_n \phi_n = \begin{bmatrix} N_1 & N_2 & \dots\dots & N_n \end{bmatrix} \begin{Bmatrix} \phi_1 \\ \phi_2 \\ \dots \\ \dots \\ \phi_n \end{Bmatrix} = [\mathbf{N}]\{\boldsymbol{\phi}\}. \quad (7.113)$$

Employing the Galerkin weighting, we obtain the global system as

$$\int_\Omega [\mathbf{N}]^T \frac{\tilde{\phi}^{n+1} - \tilde{\phi}^n}{\Delta t} d\Omega = -\int_\Omega [\mathbf{N}]^T u_1 \frac{\partial \tilde{\phi}}{\partial x_1}^n d\Omega - \int_\Omega [\mathbf{N}]^T u_2 \frac{\partial \tilde{\phi}}{\partial x_2}^n d\Omega + \int_\Omega [\mathbf{N}]^T \frac{\partial}{\partial x_1} \left(k \frac{\partial \tilde{\phi}}{\partial x_1} \right)^n d\Omega$$

$$+ \int_\Omega [\mathbf{N}]^T \frac{\partial}{\partial x_2} \left(\frac{\partial \tilde{\phi}}{\partial x_2} \right)^n d\Omega + \frac{\Delta t}{2} u_1 \int_\Omega [\mathbf{N}]^T \frac{\partial}{\partial x_1} \left[u_1 \frac{\partial \tilde{\phi}}{\partial x_1} + u_2 \frac{\partial \tilde{\phi}}{\partial x_2} \right]^n d\Omega$$

$$+ \frac{\Delta t}{2} u_2 \int_\Omega [\mathbf{N}]^T \frac{\partial}{\partial x_2} \left[u_1 \frac{\partial \tilde{\phi}}{\partial x_1} + u_2 \frac{\partial \tilde{\phi}}{\partial x_2} \right]^n d\Omega. \qquad (7.114)$$

The above equation is valid globally. On substituting the global spatial approximation for the scalar variable ϕ into the above equation we obtain

$$\int_\Omega [\mathbf{N}]^T [\mathbf{N}] \frac{\{\boldsymbol{\phi}\}^{n+1} - \{\boldsymbol{\phi}\}^n}{\Delta t} d\Omega = -u_1 \int_\Omega [\mathbf{N}]^T \frac{\partial [\mathbf{N}]}{\partial x_1} \{\boldsymbol{\phi}\}^n d\Omega - u_2 \int_\Omega [\mathbf{N}]^T \frac{\partial [\mathbf{N}]}{\partial x_2} \{\boldsymbol{\phi}\}^n d\Omega$$

$$+ \int_\Omega [\mathbf{N}]^T \frac{\partial}{\partial x_1} \left(k \frac{\partial [\mathbf{N}]}{\partial x_1} \right) \{\boldsymbol{\phi}\}^n d\Omega + \int_\Omega [\mathbf{N}]^T \frac{\partial}{\partial x_2} \left(k \frac{\partial [\mathbf{N}]}{\partial x_2} \right) \{\boldsymbol{\phi}\}^n d\Omega$$

$$+ \frac{\Delta t}{2} u_1 \int_\Omega \left[\frac{\partial}{\partial x_1} \left(u_1 \frac{\partial [\mathbf{N}]}{\partial x_1} \{\boldsymbol{\phi}\}^n + u_2 \frac{\partial [\mathbf{N}]}{\partial x_2} \{\boldsymbol{\phi}\}^n \right) \right] d\Omega$$

$$+ \frac{\Delta t}{2} u_2 \int_\Omega \left[\frac{\partial}{\partial x_2} \left(u_1 \frac{\partial [\mathbf{N}]}{\partial x_1} \{\boldsymbol{\phi}\}^n + u_2 \frac{\partial [\mathbf{N}]}{\partial x_2} \{\boldsymbol{\phi}\}^n \right) \right] d\Omega. \qquad (7.115)$$

The above global system can now be reduced to element systems by introducing the approximation over an element as shown in Figure 7.12 as

$$\phi_e = N_i \phi_i + N_j \phi_j + N_k \phi_k = \begin{bmatrix} N_i & N_j & N_k \end{bmatrix} \begin{Bmatrix} \phi_i \\ \phi_j \\ \phi_k \end{Bmatrix} = [\mathbf{N}]_e \{\boldsymbol{\phi}\}_e. \qquad (7.116)$$

Figure 7.12   Two-dimensional linear triangular element.

The elemental matrices should be assembled before the system of simultaneous equations can be solved. The elemental matrices are derived by applying the following formula for integration over linear triangular elements

$$\int_{\Omega} N_i^a N_j^b N_k^c d\Omega = \frac{a!b!c!2A}{(a+b+c+2)!} \tag{7.117}$$

and for the line integral

$$\int_{\Gamma} N_i^a N_i^b N_k^c d\Gamma = \frac{a!b!c!l}{(a+b+c+1)!}, \tag{7.118}$$

where A is the area of a triangular element and l is the length of a boundary edge. Applying the above formulae along with the elemental approximations to Equation (7.115), we obtain the element characteristic equations as follows.

The mass matrix is

$$[\mathbf{M}]_e = \int_{\Omega_e} [\mathbf{N}]_e^T [\mathbf{N}]_e d\Omega = \frac{A}{12} \begin{bmatrix} 2 & 1 & 1 \\ 1 & 2 & 1 \\ 1 & 1 & 2 \end{bmatrix}, \tag{7.119}$$

The convection matrix is

$$\begin{aligned}
[\mathbf{C}]_e &= \int_{\Omega_e} [\mathbf{N}]_e^T \left(u_1 \frac{\partial [\mathbf{N}]_e}{\partial x_1} + u_2 \frac{\partial [\mathbf{N}]_e}{\partial x_2} \right) d\Omega \\
&= \frac{u_1}{6} \begin{bmatrix} b_i & b_j & b_k \\ b_i & b_j & b_k \\ b_i & b_j & b_k \end{bmatrix} + \frac{u_2}{6} \begin{bmatrix} c_i & c_j & c_k \\ c_i & c_j & c_k \\ c_i & c_j & c_k \end{bmatrix},
\end{aligned} \tag{7.120}$$

where

$$\begin{aligned}
b_i &= y_j - y_k; & c_i &= x_k - x_j \\
b_j &= y_k - y_i; & c_j &= x_i - x_k \\
b_k &= y_i - y_j; & c_k &= x_j - x_i.
\end{aligned} \tag{7.121}$$

As before, the diffusion term can be integrated after applying Green's lemma. The diffusion matrix for the elements inside the domain is

$$[\mathbf{K}]_e = \int_{\Omega_e} \left(\frac{\partial[\mathbf{N}]_e^T}{\partial x_1} k \frac{\partial[\mathbf{N}]_e}{\partial x_1} + \frac{\partial[\mathbf{N}]_e^T}{\partial x_2} k \frac{\partial[\mathbf{N}]_e}{\partial x_2} \right) d\Omega$$

$$= \frac{k}{4A} \begin{bmatrix} b_i^2 & b_i b_j & b_i b_k \\ b_j b_i & b_j^2 & b_j b_k \\ b_k b_i & b_k b_j & b_k^2 \end{bmatrix} + \frac{k}{4A} \begin{bmatrix} c_i^2 & c_i c_j & c_i c_k \\ c_j c_i & c_j^2 & c_j c_k \\ c_k c_i & c_k c_j & c_k^2 \end{bmatrix}. \qquad (7.122)$$

The stabilization matrix is

$$[\mathbf{K}_s]_e = u_1 \frac{\Delta t}{2} \left[u_1 \int_\Omega \frac{\partial[\mathbf{N}]_e^T}{\partial x_1} \frac{\partial[\mathbf{N}]_e}{\partial x_1} d\Omega + u_2 \int_\Omega \frac{\partial[\mathbf{N}]_e^T}{\partial x_1} \frac{\partial[\mathbf{N}]_e}{\partial x_2} d\Omega \right]$$

$$+ u_2 \frac{\Delta t}{2} \left[u_1 \int_\Omega \frac{\partial[\mathbf{N}]_e^T}{\partial x_2} \frac{\partial[\mathbf{N}]_e}{\partial x_1} d\Omega + u_2 \int_\Omega \frac{\partial[\mathbf{N}]_e^T}{\partial x_2} \frac{\partial[\mathbf{N}]_e}{\partial x_2} d\Omega \right]$$

$$= \frac{u_1}{4A} \frac{\Delta t}{2} \begin{bmatrix} u_1 b_i^2 + u_2 b_i c_i & u_1 b_i b_j + u_2 b_i c_j & u_1 b_i b_k + u_2 b_i c_k \\ u_1 b_j b_i + u_2 b_j c_i & u_1 b_j^2 + u_2 b_j c_j & u_1 b_j b_k + u_2 b_j c_k \\ u_1 b_k b_i + u_2 b_k c_i & u_1 b_k b_j + u_2 b_k c_j & u_1 b_k^2 + u_2 b_k c_k \end{bmatrix}$$

$$+ \frac{u_2}{4A} \frac{\Delta t}{2} \begin{bmatrix} u_1 c_i b_i + u_2 c_i^2 & u_1 c_i b_j + u_2 c_i c_j & u_1 c_i b_k + u_2 c_i c_k \\ u_1 c_j b_i + u_2 c_j c_i & u_1 c_j b_j + u_2 c_j^2 & u_1 c_j b_k + u_2 c_j c_k \\ u_1 c_k b_i + u_2 c_k c_i & u_1 c_k b_j + u_2 c_k c_j & u_1 c_k b_k + u_2 c_k^3. \end{bmatrix}. \qquad (7.123)$$

The forcing vectors along the boundary edges are (assuming ij as the boundary edge)

$$[\mathbf{f}]_e = k \int_{\Gamma_e} \begin{bmatrix} N_i \\ N_j \\ 0 \end{bmatrix} \begin{bmatrix} \frac{\partial N_i}{\partial x_1} & \frac{\partial N_j}{\partial x_1} & \frac{\partial N_k}{\partial x_1} \end{bmatrix} \{\phi\}_e^n d\Gamma n_1$$

$$+ k \int_{\Gamma_e} \begin{bmatrix} N_i \\ N_j \\ 0 \end{bmatrix} \begin{bmatrix} \frac{\partial N_i}{\partial x_2} & \frac{\partial N_j}{\partial x_2} & \frac{\partial N_k}{\partial x_2} \end{bmatrix} \{\phi\}_e d\Gamma n_2$$

$$= \frac{l_{ij}}{4A} k \begin{bmatrix} b_i \phi_i + b_j \phi_j + b_k \phi_k \\ b_i \phi_i + b_j \phi_j + b_k \phi_k \\ 0 \end{bmatrix} n_1$$

$$+ \frac{l_{ij}}{4A} k \begin{bmatrix} c_i \phi_i + c_j \phi_j + c_k \phi_k \\ c_i \phi_i + c_j \phi_j + c_k \phi_k \\ 0 \end{bmatrix} n_2. \qquad (7.124)$$

$$[\mathbf{f}_s]_e = u_1 \frac{\Delta t}{2} \int_{\Gamma_e} u_1 \begin{bmatrix} N_i \\ N_j \\ 0 \end{bmatrix} \begin{bmatrix} \frac{\partial N_i}{\partial x_1} & \frac{\partial N_j}{\partial x_1} & \frac{\partial N_k}{\partial x_1} \end{bmatrix} \{\phi\}_e^n$$

$$+ u_1 \frac{\Delta t}{2} \int_{\Gamma_e} u_2 \begin{bmatrix} N_i \\ N_j \\ 0 \end{bmatrix} \begin{bmatrix} \frac{\partial N_i}{\partial x_2} & \frac{\partial N_j}{\partial x_2} & \frac{\partial N_k}{\partial x_2} \end{bmatrix} \{\phi\}_e^n d\Gamma n_1$$

$$+ u_2 \frac{\Delta t}{2} \int_{\Gamma_e} u_1 \begin{bmatrix} N_i \\ N_j \\ 0 \end{bmatrix} \begin{bmatrix} \frac{\partial N_i}{\partial x_1} & \frac{\partial N_j}{\partial x_1} & \frac{\partial N_k}{\partial x_1} \end{bmatrix} \{\phi\}_e^n d\Gamma n_2$$

$$+ u_2 \frac{\Delta t}{2} \int_{\Gamma_e} u_2 \begin{bmatrix} N_i \\ N_j \\ 0 \end{bmatrix} \begin{bmatrix} \frac{\partial N_i}{\partial x_2} & \frac{\partial N_j}{\partial x_2} & \frac{\partial N_k}{\partial x_2} \end{bmatrix} \{\phi\}_e^n d\Gamma n_2$$

$$= \frac{u_1}{2A} \frac{\Delta t}{2} \frac{l_{ij}}{2} \begin{bmatrix} u_1(b_i\phi_i + b_j\phi_j + b_k\phi_k) + u_2(c_i\phi_i + c_j\phi_j + c_k\phi_k) \\ u_1(b_i\phi_i + b_j\phi_j + b_k\phi_k) + u_2(c_i\phi_i + c_j\phi_j + c_k\phi_k) \\ 0 \end{bmatrix}^n n_1$$

$$+ \frac{u_2}{2A} \frac{\Delta t}{2} \frac{l_{ij}}{2} \begin{bmatrix} u_1(b_i\phi_i + b_j\phi_j + b_k\phi_k) + u_2(c_i\phi_i + c_j\phi_j + c_k\phi_k) \\ u_1(b_i\phi_i + b_j\phi_j + b_k\phi_k) + u_2(c_i\phi_i + c_j\phi_j + c_k\phi_k) \\ 0 \end{bmatrix}^n n_2. \quad (7.125)$$

where l_{ij} is the length of edge ij. The assembled equation for a two-dimensional analysis takes an identical form to the one-dimensional Equation (7.108). Once again, the boundary terms from Equation (7.125) may be neglected in the calculations.

Example 7.4.3 *Steady-state 2D convection-diffusion problem*
    The example considered here involves the convection of a discontinuous inlet data at an angle θ^o to the horizontal axis, over a square domain as shown in Figure 7.13(a). The convective velocity is unidirectional and constant, with a magnitude of unity in the direction of the flow. The left and bottom sides make up the inlet boundary and are defined as shown in Figure 7.13(a). On the outlet boundary (top and right sides), natural boundary conditions are considered (Donea and Huerta 2003). For this example a skew angle $\theta = 45^o$ is chosen and a 39×39 uniform-structured mesh is used. The results for this case are given in Figure 7.13(b). The stabilized procedure used here is the characteristic Galerkin method.

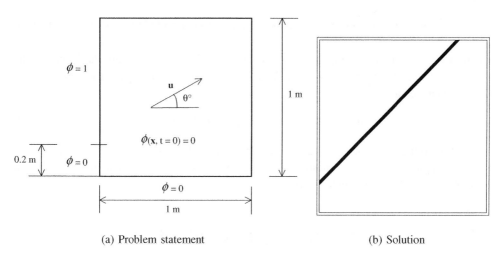

(a) Problem statement   (b) Solution

Figure 7.13   Convection of discontinuous inlet data at at angle θ^o skew to the horizontal axis. Problem statement and contours of the scalar variable distribution.

7.5 Stability Conditions

The stability conditions for a given time discretization may be derived using a Von Neumann or Fourier analysis for either the convection or the convection-diffusion equations. However, for more complicated equations such as the Navier-Stokes equations, the derivation of the stability limit is not straightforward. A detailed discussion on stability criteria is not within the scope of this book and readers are asked to refer to the relevant text books and papers for details (Hirsch 1989; Zienkiewicz and Codina 1995). A stability analysis will give some idea about the time-step restrictions of any numerical scheme.

In general, for fluid dynamics problems the time-step magnitude is controlled by two wave speeds. The first one due to the convection velocity and the second to the real diffusion introduced by the equations. In the case of a convection diffusion equation, the convection velocity is $\sqrt{u_i u_i}$ which is $\sqrt{u_1^2 + u_2^2} = |\mathbf{u}|$. The diffusion velocity is $2\,k/h$ where h is the local element size. The time-step restrictions are calculated as the ratio of the local element size and the local wave speed. It is therefore correct to write that the time step is calculated as

$$\Delta t = min(\Delta t_c, \Delta t_d), \tag{7.126}$$

where Δt_c and Δt_d are the convection and diffusion time-step limits respectively, which are

$$\Delta t_c = \frac{h}{|u|}$$

$$\Delta t_d = \frac{h^2}{2k}. \tag{7.127}$$

Often it may be necessary to multiply the time step Δt by a safety factor due to different methods of element size calculations. A simple procedure to calculate the element size in two dimensions is

$$h = min(\frac{2Area_i}{l_i}), i = 1, \text{number of elements connected to the node}, \tag{7.128}$$

where $Area_i$ are the area of the elements connected to the node and l_i are the length of the opposite sides as shown in Figure 7.14. For the node shown in this figure, the local element area is calculated as

$$h = min(A_1/l_1, A_2/l_2, A_3/l_3, A_4/l_4, A_5/l_5). \tag{7.129}$$

In three dimensions, the term $2Area_i$ is replaced by $3Volume_i$ and l_i is replaced by the face area opposite the node in question. Although more expensive, element size calculated in streamline direction gives a better accuracy to convection time step.

7.6 Characteristic Based Split (CBS) Scheme

It is essential to understand the approximate characteristic Galerkin (CG) procedure, discussed in the previous section for the convection-diffusion equation, in order to apply the concept

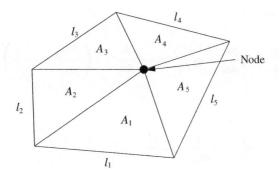

Figure 7.14   Two-dimensional linear triangular element.

to solve the real heat convection equations. Unlike the convection diffusion equation, the momentum equation, which is part of a set of heat convection equations, is a vector equation. A direct extension of the CG scheme to solve the momentum equation is difficult. In order to apply the characteristic Galerkin approach to the momentum equations, we have to introduce two steps. In the first step, the pressure term from the momentum equation will be dropped and an intermediate velocity field will be calculated. In the second step, the intermediate velocities will be corrected. This two-step procedure for the treatment of the momentum equations has two advantages. The first advantage is that, without the pressure terms, each component of the momentum equation is similar to that of a convection diffusion equation and the CG procedure can be readily applied. The second advantage is that removing the pressure term from the momentum equations enhances the pressure stability and allows the use of arbitrary interpolation functions for both velocity and pressure. In other words, the well-known Babuska-Brezzi condition is satisfied. Due to the split introduced in the equations, the method is referred to as the Characteristic Based Split (CBS) scheme.

The CG procedure may be applied to the individual momentum components without removing the pressure term provided the pressure term is treated as a source term. However, such a procedure will lose the advantages mentioned in the previous paragraph.

For more mathematical details, readers are directed to earlier publications on the method (Zienkiewicz and Codina 1995; Zienkiewicz *et al.* 2005) and for recent developments, references (Nithiarasu 2003; Nithiarasu *et al.* 2006; Zienkiewicz *et al.* 2013) are recommended. In order to apply the CG procedure, the governing equations in two dimensions (note that body forces are not included for simplicity), may be written as follows:

Continuity equation:

$$\frac{\partial u_1}{\partial x_1} + \frac{\partial u_2}{\partial x_2} = 0. \tag{7.130}$$

x_1 momentum equation:

$$\frac{\partial u_1}{\partial t} + u_1 \frac{\partial u_1}{\partial x_1} + u_2 \frac{\partial u_1}{\partial x_2} = -\frac{1}{\rho} \frac{\partial p}{\partial x_1} + \nu \left(\frac{\partial^2 u_1}{\partial x_1^2} + \frac{\partial^2 u_1}{\partial x_2^2} \right). \tag{7.131}$$

x_2 momentum equation:

$$\frac{\partial u_2}{\partial t} + u_1 \frac{\partial u_2}{\partial x_1} + u_2 \frac{\partial u_2}{\partial x_2} = -\frac{1}{\rho} \frac{\partial p}{\partial x_2} + v \left(\frac{\partial^2 u_2}{\partial x_1^2} + \frac{\partial^2 u_2}{\partial x_2^2} \right). \tag{7.132}$$

Energy equation:

$$\frac{\partial T}{\partial t} + u_1 \frac{\partial T}{\partial x_1} + u_2 \frac{\partial T}{\partial x_2} = \alpha \left(\frac{\partial^2 T}{\partial x_1^2} + \frac{\partial^2 T}{\partial x_2^2} \right). \tag{7.133}$$

From the governing equations, it is obvious that the application of the CG scheme is not straightforward. However, by implementing the following steps, it is possible to obtain a solution to the convection heat transfer equation.

Step 1 Intermediate velocity or momentum field: This step is carried out by removing the pressure terms from Equations (7.131) and (7.132). The intermediate velocity component equations, in their semi-discrete form, are:

Intermediate x_1 momentum equation:

$$\frac{u_1^* - u_1^n}{\Delta t} + u_1 \frac{\partial u_1}{\partial x_1}^n + u_2 \frac{\partial u_1}{\partial x_2}^n = v \left(\frac{\partial^2 u_1}{\partial x_1^2} + \frac{\partial^2 u_1}{\partial x_2^2} \right)^n. \tag{7.134}$$

Intermediate x_2 momentum equation:

$$\frac{u_2^* - u_2^n}{\Delta t} + u_1 \frac{\partial u_2}{\partial x_1}^n + u_2 \frac{\partial u_2}{\partial x_2}^n = v \left(\frac{\partial^2 u_2}{\partial x_1^2} + \frac{\partial^2 u_2}{\partial x_2^2} \right)^n, \tag{7.135}$$

where u_1^* and u_2^* are the intermediate momentum variables. It is obvious that the CG scheme can now be applied as the above equations are very similar to the convection-diffusion equations of the previous section. If the characteristic procedure is applied to the above equations, a semi-discrete from of the equations is obtained, viz.

Intermediate x_1 momentum equation:

$$\frac{u_1^* - u_1^n}{\Delta t} = -u_1 \frac{\partial u_1}{\partial x_1}^n - u_2 \frac{\partial u_1}{\partial x_2}^n + v \left(\frac{\partial^2 u_1}{\partial x_1^2} + \frac{\partial^2 u_1}{\partial x_2^2} \right)^n$$

$$+ u_1 \frac{\Delta t}{2} \frac{\partial}{\partial x_1} \left[u_1 \frac{\partial u_1}{\partial x_1}^n + u_2 \frac{\partial u_1}{\partial x_2}^n \right]$$

$$+ u_2 \frac{\Delta t}{2} \frac{\partial}{\partial x_2} \left[u_1 \frac{\partial u_1}{\partial x_1}^n + u_2 \frac{\partial u_1}{\partial x_2}^n \right]. \tag{7.136}$$

Intermediate x_2 momentum equation:

$$\frac{u_2^* - u_2^n}{\Delta t} = -u_1 \frac{\partial u_2}{\partial x_1}^n - u_2 \frac{\partial u_2}{\partial x_2}^n + v \left(\frac{\partial^2 u_2}{\partial x_1^2} + \frac{\partial^2 u_2}{\partial x_2^2} \right)^n$$

$$+ u_1 \frac{\Delta t}{2} \frac{\partial}{\partial x_1} \left[u_1 \frac{\partial u_2}{\partial x_1}^n + u_2 \frac{\partial u_2}{\partial x_2}^n \right]$$

$$+ u_2 \frac{\Delta t}{2} \frac{\partial}{\partial x_2} \left[u_1 \frac{\partial u_2}{\partial x_1}^n + u_2 \frac{\partial u_2}{\partial x_2}^n \right]. \tag{7.137}$$

Step 2 Pressure calculation: The pressure field is calculated from a pressure equation of the Poisson type. The pressure equation is derived from the fact that the intermediate velocities at the first step need to be corrected. If the pressure terms are not removed from the momentum equations, then the correct velocities are obtained but with the loss of some advantages. If the semi-discrete form of the momentum equations are written, without removing the pressure terms, then,

Semi-discrete x_1 momentum equation:

$$\frac{u_1^{n+1} - u_1^n}{\Delta t} = -u_1 \frac{\partial u_1}{\partial x_1}^n - u_2 \frac{\partial u_1}{\partial x_2}^n + v \left(\frac{\partial^2 u_1}{\partial x_1^2} + \frac{\partial^2 u_1}{\partial x_2^2} \right)^n - \frac{1}{\rho} \frac{\partial p}{\partial x_1}^n$$

$$+ u_1 \frac{\Delta t}{2} \frac{\partial}{\partial x_1} \left[u_1 \frac{\partial u_1}{\partial x_1}^n + u_2 \frac{\partial u_1}{\partial x_2}^n + \frac{1}{\rho} \frac{\partial p}{\partial x_1}^n \right]$$

$$+ u_2 \frac{\Delta t}{2} \frac{\partial}{\partial x_2} \left[u_1 \frac{\partial u_1}{\partial x_1}^n + u_2 \frac{\partial u_1}{\partial x_2}^n + \frac{1}{\rho} \frac{\partial p}{\partial x_1}^n \right]. \tag{7.138}$$

Semi-discrete x_2 momentum equation:

$$\frac{u_2^{n+1} - u_2^n}{\Delta t} = -u_1 \frac{\partial u_2}{\partial x_1}^n - u_2 \frac{\partial u_2}{\partial x_2}^n + v \left(\frac{\partial^2 u_2}{\partial x_1^2} + \frac{\partial^2 u_2}{\partial x_2^2} \right)^n - \frac{1}{\rho} \frac{\partial p}{\partial x_2}^n$$

$$+ u_1 \frac{\Delta t}{2} \frac{\partial}{\partial x_1} \left[u_1 \frac{\partial u_2}{\partial x_1}^n + u_2 \frac{\partial u_2}{\partial x_2}^n + \frac{1}{\rho} \frac{\partial p}{\partial x_2}^n \right]$$

$$+ u_2 \frac{\Delta t}{2} \frac{\partial}{\partial x_2} \left[u_1 \frac{\partial u_2}{\partial x_1}^n + u_2 \frac{\partial u_2}{\partial x_2}^n + \frac{1}{\rho} \frac{\partial p}{\partial x_2}^n \right]. \tag{7.139}$$

The real velocity field may be directly obtained if the above equations are utilized. Subtracting Equation (7.136) from (7.138) and (7.137) from (7.139) results in the following two equations:

$$\frac{u_1^{n+1} - u_1^*}{\Delta t} = -\frac{1}{\rho} \frac{\partial p}{\partial x_1}^n + u_1 \frac{\Delta t}{2} \frac{\partial}{\partial x_1} \left(\frac{1}{\rho} \frac{\partial p}{\partial x_1} \right)^n + u_2 \frac{\Delta t}{2} \frac{\partial}{\partial x_2} \left(\frac{1}{\rho} \frac{\partial p}{\partial x_1} \right)^n$$

$$\frac{u_2^{n+1} - u_2^*}{\Delta t} = -\frac{1}{\rho} \frac{\partial p}{\partial x_2}^n + u_1 \frac{\Delta t}{2} \frac{\partial}{\partial x_1} \left(\frac{1}{\rho} \frac{\partial p}{\partial x_1} \right)^n + u_2 \frac{\Delta t}{2} \frac{\partial}{\partial x_2} \left(\frac{1}{\rho} \frac{\partial p}{\partial x_2} \right)^n. \tag{7.140}$$

It is obvious that if the pressure terms can be calculated from another source, the intermediate velocities of Step 1 can be corrected using Equation (7.140). However, an independent pressure equation is required in order to substitute the pressure values into the above equation. In order to do this, let us introduce a relaxation parameter θ into the continuity equation, that is,

$$\frac{\partial u_1}{\partial x_1}^{n+\theta} + \frac{\partial u_2}{\partial x_2}^{n+\theta} = \theta \left(\frac{\partial u_1}{\partial x_1} + \frac{\partial u_2}{\partial x_2} \right)^{n+1} + (1 - \theta) \left(\frac{\partial u_1}{\partial x_1} + \frac{\partial u_2}{\partial x_2} \right)^{n} = 0. \quad (7.141)$$

The relaxation parameter θ must be nonzero (less than unity) to ensure pressure stability (Zienkiewicz *et al.* 2013). The numerical value option chosen is 0.5. Since $n + 1$ values are not available at step 2, we require to eliminate these values in Equation (7.141) with step 3, Equation (7.140), that is, Equation (7.141) becomes

$$\theta \left(\frac{\partial u_1^*}{\partial x_1} + \frac{\partial u_2^*}{\partial x_2} \right) - \theta \Delta t \frac{1}{\rho} \left(\frac{\partial^2 p}{\partial x_1^2} + \frac{\partial^2 p}{\partial x_2^2} \right)^{n} + (1 - \theta) \left(\frac{\partial u_1}{\partial x_1} + \frac{\partial u_2}{\partial x_2} \right)^{n} = 0. \quad (7.142)$$

Note that third- and higher-order terms are neglected. Substituting conservation of mass (Equation 7.130) and simplifying,

$$\frac{1}{\rho} \left(\frac{\partial^2 p}{\partial x_1^2} + \frac{\partial^2 p}{\partial x_2^2} \right)^{n} = \frac{1}{\Delta t} \left(\frac{\partial u_1^*}{\partial x_1} + \frac{\partial u_2^*}{\partial x_2} \right). \quad (7.143)$$

It should be noted that there are no transient or convection terms present in the above equation. Although this equation does not require any special treatment in order to stabilize the oscillations, the absence of a transient term makes it compulsory to build a matrix and simultaneous solution. However, it is possible to introduce an artificial compressibility formulation to avoid a full matrix. This is discussed in a later section.

Step 3 Velocity or momentum correction: The velocity correction has already been derived in the previous step (Equation (7.140)). This involves the pressure and intermediate velocity field and is written as

$$\frac{u_1^{n+1} - u_1^*}{\Delta t} = -\frac{1}{\rho} \frac{\partial p}{\partial x_1}^{n} + u_1 \frac{\partial}{\partial x_1} \left(\frac{1}{\rho} \frac{\partial p}{\partial x_1} \right)^{n} + u_2 \frac{\partial}{\partial x_2} \left(\frac{1}{\rho} \frac{\partial p}{\partial x_1} \right)^{n}$$

$$\frac{u_2^{n+1} - u_2^*}{\Delta t} = -\frac{1}{\rho} \frac{\partial p}{\partial x_2}^{n} + u_1 \frac{\partial}{\partial x_1} \left(\frac{1}{\rho} \frac{\partial p}{\partial x_1} \right)^{n} + u_2 \frac{\partial}{\partial x_2} \left(\frac{1}{\rho} \frac{\partial p}{\partial x_2} \right)^{n}. \quad (7.144)$$

Step 4 Temperature calculation: Applying the characteristic procedure to the temperature equation, we get

$$\frac{T^{n+1} - T^n}{\Delta t} = -u_1 \frac{\partial T}{\partial x_1}^{n} - u_2 \frac{\partial T}{\partial x_2}^{n} + \alpha \left(\frac{\partial^2 T}{\partial x_1^2} + \frac{\partial^2 T}{\partial x_2^2} \right)^{n}$$

$$+ u_1 \frac{\Delta t}{2} \frac{\partial}{\partial x_1} \left[u_1 \frac{\partial T}{\partial x_1}^{n} + u_2 \frac{\partial T}{\partial x_2}^{n} \right]$$

$$+ u_2 \frac{\Delta t}{2} \frac{\partial}{\partial x_2} \left[u_1 \frac{\partial T}{\partial x_1}^{n} + u_2 \frac{\partial T}{\partial x_2}^{n} \right]. \quad (7.145)$$

All four semi-discrete steps of the CBS scheme for convection heat transfer may now be summarized

Step 1: Intermediate velocity

Intermediate x_1 momentum equation:

$$\frac{u_1^* - u_1^{\,n}}{\Delta t} = -u_1 \frac{\partial u_1}{\partial x_1}^n - u_2 \frac{\partial u_1}{\partial x_2}^n + v \left(\frac{\partial^2 u_1}{\partial x_1^2} + \frac{\partial^2 u_1}{\partial x_2^2} \right)^n$$

$$+ u_1 \frac{\Delta t}{2} \frac{\partial}{\partial x_1} \left[u_1 \frac{\partial u_1}{\partial x_1} + u_2 \frac{\partial u_1}{\partial x_2} \right]^n$$

$$+ u_2 \frac{\Delta t}{2} \frac{\partial}{\partial x_2} \left[u_1 \frac{\partial u_1}{\partial x_1} + u_2 \frac{\partial u_1}{\partial x_2} \right]^n . \tag{7.146}$$

Intermediate x_2 momentum equation:

$$\frac{u_2^* - u_2^{\,n}}{\Delta t} = -u_1 \frac{\partial u_2}{\partial x_1}^n - u_2 \frac{\partial u_2}{\partial x_2}^n + v \left(\frac{\partial^2 u_2}{\partial x_1^2} + \frac{\partial^2 u_2}{\partial x_2^2} \right)^n$$

$$+ u_1 \frac{\Delta t}{2} \frac{\partial}{\partial x_1} \left[u_1 \frac{\partial u_2}{\partial x_1} + u_2 \frac{\partial u_2}{\partial x_2} \right]^n$$

$$+ u_2 \frac{\Delta t}{2} \frac{\partial}{\partial x_2} \left[u_1 \frac{\partial u_2}{\partial x_1} + u_2 \frac{\partial u_2}{\partial x_2} \right]^n . \tag{7.147}$$

Step 2: Pressure calculation

$$\frac{1}{\rho} \left(\frac{\partial^2 p}{\partial x_1^2} + \frac{\partial^2 p}{\partial x_2^2} \right)^n = \frac{1}{\Delta t} \left(\frac{\partial u_1^*}{\partial x_1} + \frac{\partial u_2^*}{\partial x_2} \right) . \tag{7.148}$$

Step 3: Velocity correction

$$\frac{u_1^{n+1} - u_1^*}{\Delta t} = -\frac{1}{\rho} \frac{\partial p}{\partial x_1}^n + u_1 \frac{\Delta t}{2} \frac{\partial}{\partial x_1} \left(\frac{1}{\rho} \frac{\partial p}{\partial x_1} \right)^n + u_2 \frac{\Delta t}{2} \frac{\partial}{\partial x_2} \left(\frac{1}{\rho} \frac{\partial p}{\partial x_1} \right)^n$$

$$\frac{u_2^{n+1} - u_2^*}{\Delta t} = -\frac{1}{\rho} \frac{\partial p}{\partial x_2}^n + u_1 \frac{\Delta t}{2} \frac{\partial}{\partial x_1} \left(\frac{1}{\rho} \frac{\partial p}{\partial x_1} \right)^n + u_2 \frac{\Delta t}{2} \frac{\partial}{\partial x_2} \left(\frac{1}{\rho} \frac{\partial p}{\partial x_2} \right)^n . \tag{7.149}$$

Step 4: Temperature calculation

$$\frac{T^{n+1} - T^n}{\Delta t} = -u_1 \frac{\partial T}{\partial x_1}^n - u_2 \frac{\partial T}{\partial x_2}^n + \alpha \left(\frac{\partial^2 T}{\partial x_1^2} + \frac{\partial^2 T}{\partial x_2^2} \right)^n$$

$$+ u_1 \frac{\Delta t}{2} \frac{\partial}{\partial x_1} \left[u_1 \frac{\partial T}{\partial x_1} + u_2 \frac{\partial T}{\partial x_2} \right]^n$$

$$+ u_2 \frac{\Delta t}{2} \frac{\partial}{\partial x_2} \left[u_1 \frac{\partial T}{\partial x_1} + u_2 \frac{\partial T}{\partial x_2} \right]^n . \tag{7.150}$$

The temporal discretization of the CBS scheme has now been completed and the following subsection gives the spatial discretization procedure.

7.6.1   Spatial Discretization

The Galerkin approximation and spatial discretization of the four steps discussed previously follow the same procedure as given for the convection-diffusion equation in Section 7.4.3. On assuming linear interpolation functions for all the variables, then the spatial variation for a linear triangular element may be written as (refer to Figure 7.12):

$$
\begin{aligned}
u_{1_e} &= N_i u_{1i} + N_j u_{1j} + N_k u_{1k} = [\mathbf{N}]_e \{\mathbf{u_1}\}_e \\
u_{2_e} &= N_i u_{2i} + N_j u_{2j} + N_k u_{2k} = [\mathbf{N}]_e \{\mathbf{u_2}\}_e \\
p_e &= N_i p_i + N_j p_j + N_k p_k = [\mathbf{N}]_e \{\mathbf{p}\}_e \\
T_e &= N_i T_i + N_j T_j + N_k T_k = [\mathbf{N}]_e \{\mathbf{T_e}\}.
\end{aligned}
\tag{7.151}
$$

The elemental convection, diffusion and other matrices are very similar to the one discussed for the convection-diffusion equation. However, the difference here is that the convection velocities are not constant. Also, a nonlinearity is introduced in the convection terms of the momentum equation. The following element matrices arose from the CBS scheme after spatial discretization:

Elemental mass matrix

$$
[\mathbf{M}]_e = \frac{A}{12}
\begin{bmatrix}
2 & 1 & 1 \\
1 & 2 & 1 \\
1 & 1 & 2
\end{bmatrix}.
\tag{7.152}
$$

Elemental convection matrix

$$
\begin{aligned}
[\mathbf{C}]_e = \frac{1}{24} &
\begin{bmatrix}
(usu + u_{1i})b_i & (usu + u_{1i})b_j & (usu + u_{1i})b_k \\
(usu + u_{1j})b_i & (usu + u_{1j})b_j & (usu + u_{1j})b_k \\
(usu + u_{1k})b_i & (usu + u_{1k})b_j & (usu + u_{1k})b_k
\end{bmatrix} \\
+ \frac{1}{24} &
\begin{bmatrix}
(vsu + u_{2i})c_i & (vsu + u_{2i})c_j & (vsu + u_{2i})c_k \\
(vsu + u_{2j})c_i & (vsu + u_{2j})c_j & (vsu + u_{2j})c_k \\
(vsu + u_{2k})c_i & (vsu + u_{2k})c_j & (vsu + u_{2k})c_k
\end{bmatrix},
\end{aligned}
\tag{7.153}
$$

where

$$
\begin{aligned}
usu &= u_{1i} + u_{1j} + u_{1k} \\
vsu &= u_{2i} + u_{2j} + u_{2k}
\end{aligned}
\tag{7.154}
$$

and definition of b_i, b_j, b_k, c_i, c_j and c_k are given in Equation (7.121) and in Chapter 3.

The differences in the above convection matrix from that of the convection matrix discussed in the Section 7.4.3 are due to the variable and nonlinear velocity field. The diffusion matrix is the same as the convection diffusion equation but k is replaced with the kinematic viscosity v for the momentum equation. Two diffusion matrices are required for convection heat transfer problems, one for the momentum equation and another for the temperature equation. These are

$$
[\mathbf{K_m}]_e = \frac{v}{4A}
\begin{bmatrix}
b_i^2 & b_i b_j & b_i b_k \\
b_j b_i & b_j^2 & b_j b_k \\
b_k b_i & b_k b_j & b_k^2
\end{bmatrix}
+ \frac{v}{4A}
\begin{bmatrix}
c_i^2 & c_i c_j & c_i c_k \\
c_j c_i & c_j^2 & c_j c_k \\
c_k c_i & c_k c_j & c_k^2
\end{bmatrix}
\tag{7.155}
$$

for the momentum diffusion and

$$[\mathbf{K_t}]_e = \frac{k}{4A}\begin{bmatrix} b_i^2 & b_ib_j & b_ib_k \\ b_jb_i & b_j^2 & b_jb_k \\ b_kb_i & b_kb_j & b_k^2 \end{bmatrix} + \frac{k}{4A}\begin{bmatrix} c_i^2 & c_ic_j & c_ic_k \\ c_jc_i & c_j^2 & c_jc_k \\ c_kc_i & c_kc_j & c_k^2 \end{bmatrix} \tag{7.156}$$

for the heat diffusion. The elemental stabilization matrix is

$$[\mathbf{K_s}]_e = \left(\frac{u_{1av}}{48A}\right)\begin{bmatrix} b_i^2 & b_ib_j & b_ib_k \\ b_jb_i & b_j^2 & b_jb_k \\ b_kb_i & b_kb_j & b_k^2 \end{bmatrix} + \left(\frac{u_{12av}}{48A}\right)\begin{bmatrix} b_ic_i & b_ic_j & b_ic_k \\ b_jc_i & b_jc_j & b_jc_k \\ b_kc_i & b_kc_j & b_kc_k \end{bmatrix}$$

$$+ \left(\frac{u_{12av}}{48A}\right)\begin{bmatrix} c_ib_i & c_ib_j & c_ib_k \\ c_jb_i & c_jb_j & c_jb_k \\ c_kb_i & c_kb_j & c_kb_k \end{bmatrix} + \left(\frac{u_{2av}}{48A}\right)\begin{bmatrix} c_i^2 & c_ic_j & c_ic_k \\ c_jc_i & c_j^2 & c_jc_k \\ c_kc_i & c_kc_j & c_k^2 \end{bmatrix}, \tag{7.157}$$

where u_{1av} and u_{2av} are given as

$$u_{1av} = u_{1i}(u_{1i} + usu) + u_{1j}(u_{1j} + usu) + u_{1k}(u_{1k} + usu)$$

$$u_{12av} = u_{1i}(u_{2i} + vsu) + u_{1j}(u_{2j} + vsu) + u_{1k}(u_{2k} + vsu)$$

$$u_{2av} = u_{2i}(u_{2i} + vsu) + u_{2j}(u_{2j} + vsu) + u_{2k}(u_{2k} + vsu)$$

The discretization of the CBS steps requires three more matrices and four forcing vectors to complete the process. The matrix from the discretized second-order terms for Step 2 is

$$[\mathbf{K}]_e = \frac{1}{4A\rho}\begin{bmatrix} b_i^2 & b_ib_j & b_ib_k \\ b_jb_i & b_j^2 & b_jb_k \\ b_kb_i & b_kb_j & b_k^2 \end{bmatrix} + \frac{1}{4A\rho}\begin{bmatrix} c_i^2 & c_ic_j & c_ic_k \\ c_jc_i & c_j^2 & c_jc_k \\ c_kc_i & c_kc_j & c_k^2 \end{bmatrix}. \tag{7.158}$$

The first gradient matrix in the x_1 direction is

$$[\mathbf{G_1}]_e = \frac{1}{6}\begin{bmatrix} b_i & b_j & b_k \\ b_i & b_j & b_k \\ b_i & b_j & b_k \end{bmatrix} \tag{7.159}$$

and the second gradient matrix in the x_2 direction is

$$[\mathbf{G_2}]_e = \frac{1}{6}\begin{bmatrix} c_i & c_j & c_k \\ c_i & c_j & c_k \\ c_i & c_j & c_k \end{bmatrix}. \tag{7.160}$$

The forcing terms are the result of the application of Green's lemma to the second-order derivatives of the differential equations. This issue has been previously discussed in the context of the discretization of the convection diffusion equations. However, one important change is that it will be assumed that the boundary integral values of the stabilization terms are equal to zero on the boundaries and will be ignored. This is an appropriate assumption as these terms will be equal to zero because the residual of the discrete equations are zero on the boundaries (Zienkiewicz *et al.* 2005). However, the forcing terms resulting from the discretization of the other second-order terms (and first-order terms, if integrated by parts) are important and need to be taken into account. The forcing vector of the x_1 component of momentum equation is

$$\{\mathbf{f}_1\}_e = \frac{l_{ij}}{4A}v\begin{bmatrix}b_iu_{1i}+b_ju_{1j}+b_ku_{1k}\\b_iu_{1i}+b_ju_{1j}+b_ku_{1k}\\0\end{bmatrix}^n n_1 + \frac{l_{ij}}{4A}v\begin{bmatrix}c_iu_{1i}+c_ju_{1j}+c_ku_{1k}\\c_iu_{1i}+c_ju_{1j}+c_ku_{1k}\\0\end{bmatrix}^n n_2. \quad (7.161)$$

Note that ij is assumed as being the boundary edge of an element. The forcing vector of the x_2 component of the momentum equation is

$$\{\mathbf{f}_2\}_e = \frac{l_{ij}}{4A}v\begin{bmatrix}b_iu_{2i}+b_ju_{2j}+b_ku_{2k}\\b_iu_{2i}+b_ju_{2j}+b_ku_{2k}\\0\end{bmatrix}^n n_1 + \frac{l_{ij}}{4A}v\begin{bmatrix}c_iu_{2i}+c_ju_{2j}+c_ku_{2k}\\c_iu_{2i}+c_ju_{2j}+c_ku_{2k}\\0\end{bmatrix}^n n_2. \quad (7.162)$$

The forcing vector from the discretization of the second-order pressure terms in Step 2 is

$$\{\mathbf{f}_3\}_e = \frac{l_{ij}}{4A\rho}\begin{bmatrix}b_ip_i+b_jp_j+b_kp_k\\b_ip_i+b_jp_j+b_kp_k\\0\end{bmatrix}^n n_1 + \frac{l_{ij}}{4A\rho}\begin{bmatrix}c_ip_i+c_jp_j+c_kp_k\\c_ip_i+c_jp_j+c_kp_k\\0\end{bmatrix}^n n_2. \quad (7.163)$$

The above forcing vector has often been ignored in the past, which is not an unreasonable assumption. Finally, the forcing term due to the discretization of the second-order terms in the energy equation is

$$\{\mathbf{f}_4\}_e = \frac{l_{ij}}{4A}k\begin{bmatrix}b_iT_i+b_jT_j+b_kT_k\\b_iT_i+b_jT_j+b_kT_k\\0\end{bmatrix}^n n_1 + \frac{l_{ij}}{4A}k\begin{bmatrix}c_iT_i+c_jT_j+c_kT_k\\c_iT_i+c_jT_j+c_kT_k\\0\end{bmatrix}^n n_2. \quad (7.164)$$

The four steps of the CBS scheme may now be written in matrix form. The above elemental equations need to be assembled before they can be used in the steps. It will be assumed that the matrices without the subscript e are already assembled and therefore the steps in terms of the assembly (discrete form) can now be written as

Step 1: Intermediate velocity calculation x_1 component

$$[\mathbf{M}]\frac{\Delta\{\mathbf{u}_1{}^*\}}{\Delta t} = -[\mathbf{C}]\{\mathbf{u}_1\}^n - [\mathbf{K}_m]\{\mathbf{u}_1\}^n - \frac{\Delta t}{2}[\mathbf{K}_s]\{\mathbf{u}_1\}^n + \{\mathbf{f}_1\} \quad (7.165)$$

and for the x_2 component

$$[\mathbf{M}]\frac{\Delta\{\mathbf{u}_2{}^*\}}{\Delta t} = -[\mathbf{C}]\{\mathbf{u}_2\}^n - [\mathbf{K}_m]\{\mathbf{u}_2\}^n - \frac{\Delta t}{2}[\mathbf{K}_s]\{\mathbf{u}_2\}^n + \{\mathbf{f}_2\}. \quad (7.166)$$

Step 2: Pressure calculation

$$[\mathbf{K}]\{\mathbf{p}\}^n = -\frac{1}{\Delta t}\left[[\mathbf{G}_1]\{\mathbf{u}_1{}^*\} + [\mathbf{G}_2]\{\mathbf{u}_2{}^*\}\right] + \{\mathbf{f}_3\}. \quad (7.167)$$

Step 3: Velocity correction

$$[\mathbf{M}]\{\mathbf{u}_1\}^{n+1} = [\mathbf{M}]\{\mathbf{u}_1{}^*\} - \Delta t[\mathbf{G}_1]\{\mathbf{p}\}^n. \quad (7.168)$$

$$[\mathbf{M}]\{\mathbf{u}_2\}^{n+1} = [\mathbf{M}]\{\mathbf{u}_2{}^*\} - \Delta t[\mathbf{G}_2]\{\mathbf{p}\}^n. \quad (7.169)$$

Step 4: Temperature calculation

$$[\mathbf{M}]\frac{\Delta\{\mathbf{T}\}}{\Delta t} = -[\mathbf{C}]\{\mathbf{T}\}^n - [\mathbf{K}_t]\{\mathbf{T}\}^n - \frac{\Delta t}{2}[\mathbf{K}_s]\{\mathbf{T}\}^n + \{\mathbf{f}_4\}. \tag{7.170}$$

The above four steps are the cornerstone of the CBS scheme for the solution of the heat convection equations. An extension of the above steps for solving the conservation and three-dimensional forms is straightforward. Interested readers should consult some of the appropriate publications (Nithiarasu 2003; Zienkiewicz *et al.* 1999).

The mass matrix [**M**] used in the above steps may be "lumped" to simplify the solution procedure. This is an approximation but a worth while and time-saving approximation. Mass lumping will eliminate the need for the matrix solution procedure necessary for consistent mass matrices. The lumped mass matrix for a linear triangular element is constructed by summing the rows and placing on the diagonals. The elemental lumped mass matrix of a linear triangular element is

$$[\mathbf{M_L}]_e = \frac{A}{12}\begin{bmatrix} 4 & 0 & 0 \\ 0 & 4 & 0 \\ 0 & 0 & 4 \end{bmatrix} = \frac{A}{3}\begin{bmatrix} 1 & 0 & 0 \\ 0 & 1 & 0 \\ 0 & 0 & 1 \end{bmatrix}. \tag{7.171}$$

If the above mass lumping procedure is introduced into the CBS steps, some small errors may occur in the transient solution. For steady-state solutions, however, no errors are introduced. However, for transient problems an accurate solution can still be obtained by appropriate mesh refinement.

7.6.2 Time-step Calculation

The time-step restrictions are very similar to the convection-diffusion equation (Equation (7.126)). The local time step at each and every node can be computed as follows:

$$\Delta t = min(\Delta t_c, \Delta t_d). \tag{7.172}$$

The convection time step Δt_c is identical to that of Equation (7.127). The diffusion time steps contain two parts. One due to the kinematic viscosity and another to the thermal diffusivity of the fluid. The diffusion time step may be expressed as

$$\Delta t_d = min(\frac{h^2}{2v}, \frac{h^2}{2\alpha}), \tag{7.173}$$

where v is the kinematic viscosity and α is the thermal diffusivity. The local element size, may be calculated using the same procedure as that discussed in Section 7.5. However, a more advanced method of the calculation of element size, for example, an element size in the streamline direction, is possible and readers are referred to the appropriate publication (Tezduyar *et al.* 2000).

7.6.3 Boundary and Initial Conditions

The two main boundary conditions prevalent in heat convection problems are the prescribed temperature, pressure and velocity (Dirichlet conditions) and flux boundary conditions (Neumann conditions). Other possibilities may be derived from these conditions.

Prescribed values If a value of the velocity components, temperature or pressure is given at a boundary node, the value will be "forced" at these nodes. To implement this the corresponding discrete nodal equations are removed and any associated load from the prescribed value is moved to the load vector.

Flux conditions In a heat transfer calculation, it is possible to have prescribed heat flux conditions, which are normally given as

$$-k\frac{\partial T}{\partial x_1}n_1 - k\frac{\partial T}{\partial x_2}n_2 = -k\frac{\partial T}{\partial n} = \bar{q},$$
(7.174)

where n is the normal direction to the surface on which the prescribed flux boundary is imposed and n_1 and n_2 are the components of outward pointing normals on a surface. The heat flux condition is imposed by rearranging $\{\mathbf{f_4}\}$ (Equation (7.164)) as follows:

$$\{\mathbf{f_4}\}_e = \frac{l_{ij}}{2}\bar{q}\begin{bmatrix} 1 \\ 1 \\ 0 \end{bmatrix},$$
(7.175)

assuming that the flux is applied along the edge ij of an element e. In the above equation $\bar{q}$ is assumed to be entering the domain. If the flux is leaving a domain then a negative $\bar{q}$ should be used. Note that often, symmetry (or zero flux) boundary conditions are employed in convection heat transfer calculations. In such cases, the forcing vector terms disappear.

In many industrial heat transfer applications, convection heat transfer boundary conditions are common. If a boundary, as shown in Figure 7.15, is convecting to the atmosphere, then the

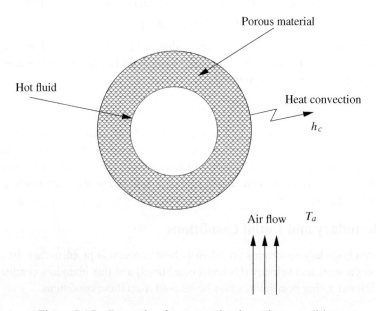

Figure 7.15   Example of a convection boundary condition.

boundary condition on this wall can be expressed as

$$-k\frac{\partial T}{\partial x_1}n_1 - k\frac{\partial T}{\partial x_2}n_2 = -k\frac{\partial T}{\partial n} = h_c(T - T_a), \tag{7.176}$$

where the wall temperature T is unknown. The implementation along a side ij of an element may be carried out for the CBS scheme as

$$\{\mathbf{f_4}\}_e = \int_{\Gamma_e} \begin{bmatrix} N_i \\ N_j \\ 0 \end{bmatrix} h_c(T_e^n - T_a)d\Gamma, \tag{7.177}$$

with $T_e = N_iT_i + N_jT_j + N_kT_k$, integration of the above equation over edge ij becomes

$$\{\mathbf{f_4}\}_e = \frac{h_c l_{ij}}{6} \begin{bmatrix} 2 & 1 & 0 \\ 1 & 2 & 0 \\ 0 & 0 & 0 \end{bmatrix} \begin{Bmatrix} T_i \\ T_j \\ 0 \end{Bmatrix}^n - \frac{h_c l_{ij}}{2} \begin{bmatrix} 1 \\ 1 \\ 0 \end{bmatrix} T_a. \tag{7.178}$$

The initial conditions, which describe the initial state of the fluid (temperature, pressure, velocity and properties), are employed at the onset of heat convection calculations. These conditions are problem dependent and are discussed for various applications in the latter sections of this chapter.

7.6.4 Steady and Transient Solution

A steady-state solution for a problem can be obtained, using the CBS scheme, by time stepping to achieve a steady-state. This may be done by fixing a tolerance criteria as follows

$$\sum_{i=1}^{no.nodes} \frac{\phi_i^{n+1} - \phi_i^n}{\Delta t} \leq \epsilon. \tag{7.179}$$

where ϕ_i is any heat convection variable at a node, *nnodes* is the total number of nodes and ϵ is a prescribed tolerance, which will tend toward zero as the solution approaches steady-state.

A transient solution can be of two types. The first type is the "real" time variation of the solution for problems where a steady-state solution exists. The second category is one which has no real steady-state, for example vortex shedding behind a cylinder or Rayleigh-Bénard convection. In the first type, the calculations commence with prescribed initial conditions and progress with a suitable time-stepping algorithm until a steady-state is reached. The time history of the variables need to be stored and monitored as the transient solution progresses in order to study the behavior of the solution. In the second type of problems, that is, Rayleigh-Bénard convection and vortex shedding, the steady-state tolerance of Equation (7.179) is not applicable and steady-state is never reached. The time history of these type of problems needs to be followed as long as the user is interested in the solution.

7.7 Artificial Compressibility Scheme

As mentioned before, convection heat transfer calculations can be carried out using a fully explicit Artificial Compressibility (AC) scheme. In AC schemes an artificial compressibility is introduced at Step 2 of the CBS scheme, that is,

$$\frac{1}{\beta^2 \rho} \frac{\partial p}{\partial t} - \Delta t \left(\frac{\partial^2 p}{\partial x_1^2} + \frac{\partial^2 p}{\partial x_2^2} \right) = -\frac{\partial u_1^*}{\partial x_1} - \frac{\partial u_2^*}{\partial x_2}, \tag{7.180}$$

where β is an artificial compressibility parameter. The above equation can be derived by assuming a density variation in the continuity equation by substituting

$$\frac{\partial \rho}{\partial t} \approx \frac{1}{c^2} \frac{\partial p}{\partial t}, \tag{7.181}$$

where c the speed of sound, which for incompressible flows approaches infinity. However, c can be replaced by an artificial compressibility parameter β, as given in Equation (7.180), for the purpose of introducing an explicit scheme. In the artificial compressibility based CBS scheme, Step 2 will be replaced with

$$\frac{1}{\beta^2 \rho} [\mathbf{M}] \frac{\{\mathbf{\Delta p}\}}{\mathbf{\Delta t}} + [\mathbf{K}]\{\mathbf{p}\}^n = -\frac{1}{\Delta t} \left[[\mathbf{G_1}]\{\mathbf{u_1}^*\} + [\mathbf{G_2}]\{\mathbf{u_2}^*\} \right] + \{\mathbf{f_3}\}, \tag{7.182}$$

where $\{\mathbf{\Delta p}\} = \{\mathbf{p^{n+1}} - \mathbf{p^n}\}$. The artificial compressibility parameter can be chosen as

$$\beta = max(c_o, u_{conv}, u_{diff}, u_{therm}), \tag{7.183}$$

where c_o is a small constant (between 0.1 to 0.5) and u_{conv}, u_{diff} and u_{therm} are respectively the convection, diffusion and thermal velocities, which may be defined as

$$u_{conv} = \sqrt{u_1^2 + u_2^2}$$
$$u_{diff} = \frac{2v}{h}$$
$$u_{therm} = \frac{2\alpha}{h}. \tag{7.184}$$

All other steps of the CBS scheme remain the same. However, for the solution of transient problems, a dual time-stepping procedure has to be introduced. In this dual time-stepping procedure, a transient problem is split into several instantaneous steady-states and integrated via a real global time step (Malan *et al.* 2002a,b; Nithiarasu 2003; Zienkiewicz *et al.* 2013). This can be achieved by adding a source term to step 3 of the CBS scheme, that is,

$$[\mathbf{M}]\{\mathbf{u_1}\}^{n+1} = [\mathbf{M}]\{\mathbf{u_1}^*\} - \Delta t[\mathbf{G_1}]\{\mathbf{p}\}^n - \Delta t[\mathbf{M}]\frac{\Delta \mathbf{u_1}}{2\Delta \tau}$$

and

$$[\mathbf{M}]\{\mathbf{u_2}\}^{n+1} = [\mathbf{M}]\{\mathbf{u_2}^*\} - \Delta t[\mathbf{G_2}]\{\mathbf{p}\}^n - \Delta t[\mathbf{M}]\frac{\Delta \mathbf{u_2}}{2\Delta \tau}. \tag{7.185}$$

where τ is the real time and here t becomes an iterative pseudo time. For a second-order real-time accuracy

$$\Delta \mathbf{u}_1 = 3\mathbf{u}_1{}^{m+1} - 4\mathbf{u}_1{}^m + \mathbf{u}_1{}^{m-1}$$
$$\Delta \mathbf{u}_2 = 3\mathbf{u}_2{}^{m+1} - 4\mathbf{u}_2{}^m + \mathbf{u}_2{}^{m-1}, \tag{7.186}$$

with m being the real-time tag.

7.8 Nusselt Number, Drag and Stream Function

The two important quantities of interest in many heat transfer applications are the rate of heat transfer (Nusselt number) and the flow resistance offered by a surface (drag). The stream function is often used to draw streamlines in order to understand the flow pattern of a problem. In this section a brief summary of how to calculate these quantities is given.

7.8.1 Nusselt Number

The Nusslet number is derived as follows. Let us assume that a hot surface is cooled by a cold fluid stream. The heat from the hot surface, which is maintained at a constant temperature, is diffused through a boundary layer and convected away by the cold stream. This phenomenon is normally defined by Newton's law of cooling per unit surface area as

$$h_c(T_w - T_f) = -k\frac{\partial T}{\partial n}, \tag{7.187}$$

where h_c is the heat transfer coefficient, k is an average thermal conductivity of the fluid, T_f is the free stream temperature of the fluid and n is the normal direction to the heat transfer surface. The above equation can be rewritten as

$$\frac{h_c L}{k} = -\left(\frac{1}{T_w - T_f}\right)\frac{\partial T}{\partial n}L, \tag{7.188}$$

where L is any characteristic dimension. The quantity on the left-hand side of the above equation is the Nusselt number. If we apply nondimensional scales ($T^* = (T - T_f)/(T_w - T_f)$ and $n^* = n/L$), as discussed in Section 7.3, we can rewrite the above equation as

$$Nu = \frac{h_c L}{k} = -\frac{\partial T^*}{\partial n^*}, \tag{7.189}$$

where Nu is the local Nusselt number. It should be observed that the local Nusselt number is equal to the local, nondimensional, normal temperature gradient. The above definition of the Nusselt number is valid for any heat transfer problem as long as the surface temperature is constant, or the reference wall temperature is known. However, for prescribed heat flux conditions, taking a different approach is required to derive the Nuselt number. Let us assume a surface subjected to an uniform heat flux $\bar{q}$. We can write locally

$$\bar{q} = -k\frac{\partial T}{\partial n} = h_c(T_w - T_f), \tag{7.190}$$

where T_w is not a constant. The Nusselt number relation can be obtained by multiplying the RHS of the previous equations by L/k, that is,

$$\frac{h_c L}{k}(T_w - T_f) = \frac{\bar{q}L}{k}.$$ (7.191)

Rearranging, we obtain

$$Nu = \frac{\frac{\bar{q}L}{k}}{(T_w - T_f)}.$$ (7.192)

When a wall is subjected to heat flux boundary conditions, the temperature scale is qL/k which nondimensionlises the temperature. Therefore, the above equation can be rewritten in non-dimensional form as

$$Nu = \frac{1}{T_w^* - T_f^*}.$$ (7.193)

The equation is simpler than that derived for a constant wall temperature and is limited to the calculation of local nondimensional wall temperatures (assuming T_f is constant). Therefore, the calculation of the Nusselt number on a wall subjected to a constant heat flux is straightforward in any numerical method. However, in the Nusselt number calculation for a surface subjected to a constant temperature, it is necessary to calculate the normal temperature gradient. This calculation is simple using a finite element discretization, where the normal gradient can be directly calculated from the boundary terms arising due to the discretization of the second-order temperature terms, that is,

$$\frac{\partial T}{\partial n} = \frac{\partial T}{\partial x_1}n_1 + \frac{\partial T}{\partial x_2}n_2 + \frac{\partial T}{\partial x_3}n_3,$$ (7.194)

where n_1, n_2 and n_3 are the direction cosines of the surface normal. All the above discussed quantities are local (on the surface nodes or elements). However, it is often necessary to have an average Nusselt number for a heat transfer problem. The average Nusselt number can be easily calculated by integrating the local Nusselt number over a length (in two dimensions) or over a surface (in three dimensions). For example, in two dimensions,

$$Nu_{av} = \frac{1}{L}\int_L Nu\, dl = \frac{1}{L}\sum_{i=1}^{no.\ wall\ elements} Nu_i\ dl_i,$$ (7.195)

where L is the total length of the wall, i indicates a single incremental length of a one-dimensional element on the wall on which the Nusselt number is calculated and *no.wallelements* indicates the total number of one-dimensional elements on the wall. In order to use the above formula, the local Nusselt number over an incremental length (dl_i) is assumed to be constant.

7.8.2 Drag Calculation

The drag force is the resistance offered by a body which is equal to the force exerted by the flow on the body at equilibrium conditions. The drag force arises from two different sources.

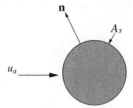

Figure 7.16   Normal gradient of velocity close to the wall.

One is from the pressure p acting in the flow direction on the surface of the body (form drag) and the second is due to the force caused by effects of friction in the flow direction. In general the drag force is characterized by a drag coefficient defined as

$$C_d = \frac{D}{A_f \frac{1}{2} \rho_a u_a^2},$$
(7.196)

where D is the drag force, A_f is the frontal area in the flow direction and the subscript a indicates the free stream value. The drag force D contains the contributions from both the influence of pressure and friction, that is,

$$D = D_p + D_f,$$
(7.197)

where D_p is the pressure drag force and D_f is the friction drag force in the flow direction. The pressure drag, or form drag, is calculated from the nodal pressure values. For a two-dimensional problem, the solid wall may be a curve or a line and the boundary elements on the solid wall are one-dimensional with two nodes if linear elements are used. The pressure may be averaged over each one-dimensional element to calculate the average pressure over the boundary element. If this average pressure is multiplied by the length of the element, then the normal pressure acting on the boundary element is obtained. If the pressure force is multiplied by the direction cosine in the flow direction, we obtain the local pressure drag force in the flow direction. Integration of these forces over the solid boundary gives the drag force due to pressure D_p.

The viscous drag force D_f is calculated by integrating the viscous traction in the flow direction, over the surface area. The relation for the total drag force in x_1 direction may be written for a two-dimensional case as

$$D_{x_1} = \int_{A_s} [(-p + \tau_{11})n_1 + \tau_{12}n_2] dA_s,$$
(7.198)

where n_1 and n_2 are components of the surface normal $\mathbf{n}$ as shown in Figure 7.16.

7.8.3   Stream Function

In most fluid dynamics and convection heat transfer problems, it is often easier to understand the flow results if the streamlines are plotted. In order to plot these streamlines, or flow pattern, it is first necessary to calculate the stream function values at the nodes. A line with a constant stream function value is referred to as a streamline. The stream function is defined by the following relationships:

$$u_1 = \frac{\partial \psi}{\partial x_2}$$

$$u_2 = -\frac{\partial \psi}{\partial x_1},$$ (7.199)

where ψ is the stream function. If we differentiate the first relation with respect to x_2 and the second with respect to x_1 and then sum, we get the differential equation for the stream function as

$$\frac{\partial^2 \psi}{\partial x_2^2} + \frac{\partial^2 \psi}{\partial x_1^2} = \frac{\partial u_2}{\partial x_1} - \frac{\partial u_1}{\partial x_2}.$$ (7.200)

The solution to the above equation is straightforward for any numerical procedure. This equation is similar to Step 2 of the CBS scheme and an implicit procedure immediately gives the solution.

7.9   Mesh Convergence

All numerical schemes are, by their nature, approximate and the CBS scheme is no exception. However, if a scheme is said to be convergent, the approximate solution should approach the exact answer as the mesh is refined. A converged solution is one which is nearly independent of meshing errors. A very coarse mesh would give a very approximate solution, which is far from reality. As the mesh is refined by reducing the size of the elements, the solution slowly approaches the exact solution. It should be noted that, in theory, the solution will not be exact until the mesh size is zero, which is obviously impossible. However, it is possible to fix a tolerance to the solution error and this can be achieved by solving the problem on several meshes.

In order to insure that the solution obtained is as close as possible to reality, solutions should be obtained from several meshes starting with a very coarse mesh and finishing with a very fine mesh. Once these solutions are available, many key quantities can be compared and plotted against mesh densities (or number of points) as shown in Figure 7.17. If the difference between two consecutive meshes (or number of nodes) is less than a fixed tolerance (arrow showing "converged" in Figure 7.17), the coarser mesh among the two is normally accepted as a suitable mesh for the analysis.

For two-dimensional problems, it is not difficult to carry out a detailed mesh convergence study for different parameters or cases. However, in large three-dimensional problems, it is often difficult to carry out a complete mesh convergence study. In such situations, it is customary to compare the results with analytical, or experimental, data if available. The past experience of the user also helps in obtaining an accurate solution for complicated problems. An adaptive refinement strategy is another way of obtaining better accuracy (see Chapter 14).

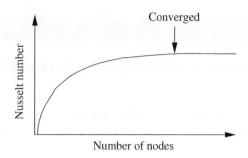

Figure 7.17 Typical convergence study.

7.10 Laminar Isothermal Flow

In this section examples of a steady and unsteady-state isothermal flow problems are discussed. The isothermal solution procedure is obtained by neglecting the temperature, or energy, equation from the governing set of equations. In other words, Step 4 of the CBS scheme is neglected thereby assuming isothermal flow.

Example 7.10.1 *Steady flow through a rectangular channel*

The problem considered here is a simple two-dimensional developing flow in a rectangular channel as shown in Figure 7.18. The "CBSflow" code is used to solve this problem. The steps employed are as discussed in Section 7.6. However, the "CBSflow" code is written using a nondimensional form of the governing equations. Therefore, the steps of the scheme have to undergo appropriate changes. The nondimensional scaling, discussed in Section 7.3, should be reflected in the geometry. The nondimensional geometry used is shown in Figure 7.18. The defined inlet Reynolds number is based on the inlet height and is therefore equal to unity in the nondimensional form. The length of the channel was assumed to be 15 times that of the height.

Based on the characteristic analysis discussed in many books (e.g., Hirsch 1989), a subsonic, incompressible two-dimensional isothermal flow problem requires two boundary conditions at the inlet and one boundary condition at the exit. It is normal practice to impose the velocity components at the inlet and pressure at the exit. In order that pressure may be

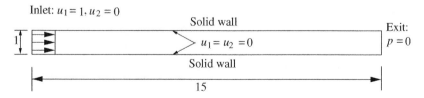

Figure 7.18 Flow through a two-dimensional rectangular channel. Geometry and boundary conditions.

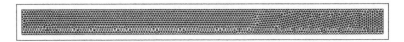

Figure 7.19   Flow through a two-dimensional rectangular channel. Finite element mesh.

imposed at the exit, it is necessary that the flow does not undergo any appreciable variation close to the exit. In other words, the channel length should be much greater than the height.

The boundary conditions may be summarized as follows:

- *Inlet: uniform velocity component u_1 of a nondimensional value of unity and the velocity component u_2 equal to zero.*

- *Exit: Nondimensional pressure value equal to constant. Here the value is prescribed as being zero.*

- *Walls: Both velocity components are forced to zero (no-slip condition)*

- *Initial conditions: Zero velocities and pressure at all points within the domain.*

Figure 7.19 shows the unstructured mesh used for the calculations. It is a uniform mesh with 3242 linear triangular elements and 1782 nodes.

The inlet Reynolds number of the flow is assumed to be 100, which is well within the laminar range. Figure 7.20 shows the velocity profiles along the length of the channel. This solution is a steady-state solution generated by an artificial compressibility form of the CBS scheme. The momentum boundary layer develops as the flow travels downstream. Figure 7.21 shows a comparison of the velocity profiles for nondimensional distances between 0 and 6. It may be seen that the parabolic profile is developed close to a distance of 4.0. The analytical solution obtained from boundary layer theory (Schlichting 1968) gives an approximate relation for the nondimensional developing length as

$$l_e = 0.04Re, \tag{7.201}$$

which gives a $l_e = 4.0$ for a Reynolds number of 100. It should be noted that the velocity profile is continuously changing in the downstream direction. A completely unchanged u_1 velocity profile can only be obtained by extending the length of the channel further (Schlichting 1968). Also, more accurate velocity profiles can be obtained by either employing a structured mesh

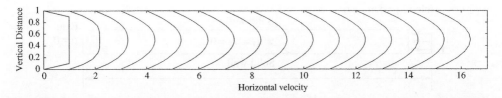

Figure 7.20   Flow through a two-dimensional rectangular channel. Velocity profiles at different sections.

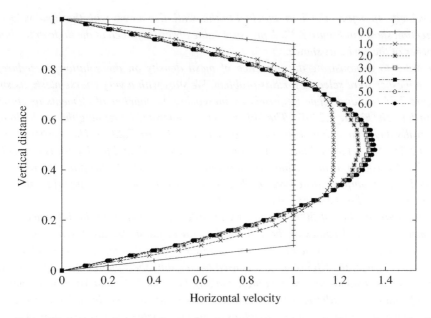

Figure 7.21 Flow through a two-dimensional rectangular channel. Comparison of velocity profiles at various distances.

or using a finer unstructured mesh. The interested reader is advised to carry out a mesh convergence study on this type of a problem.

Example 7.10.2 *Steady-flow inside a lid-driven cavity*

Flow in a lid-driven cavity is one of the most widely used benchmark problems to test steady-state incompressible fluid dynamics codes. Our interest will be to present this problem as a benchmark problem for the steady-state solution. The definition of the problem is given in Figure 7.22. The geometry is a simple square enclosure with solid walls on all four sides. All the walls, except for the top one, are fixed. The top wall is assumed to be moving with a

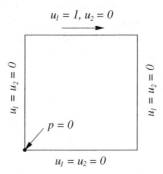

Figure 7.22 Incompressible isothermal flow in a lid-driven cavity. Geometry and boundary conditions.

given velocity, therefore, the fluid attached to this wall also moves with the same velocity in the direction shown in Figure 7.22. A pressure value of zero is forced at the node at the bottom left corner of the cavity as shown.

In order to demonstrate the influence of mesh density on the solution procedure, six different meshes were selected for this problem. We start with a very coarse mesh, as shown in Figure 7.23 (a), and refine uniformly by increasing the number of elements as shown in the fourth mesh (Figure 7.23(d)). The fifth mesh was generated by refining the mesh along the cavity walls and coarsening at the center as shown in Figure 7.23(e). The meshes shown in Figures 7.23(a) to (e) are all unstructured in nature. The sixth and final mesh is a structured mesh of 100×100 uniform divisions as shown in Figure 7.23(f). At this point the readers are reminded that an uniform structured mesh gives better accuracy as compared to an uniform unstructured mesh for the same number of nodes.

A Reynolds number of 5000 is selected in order to demonstrate the influence of mesh refinement. The initial values of the velocities at all inside nodes are taken as $u_1 = 1$ and $u_2 = 0$. The pressure is assumed to be equal to zero at the beginning of the computation. The semi-implicit form of the CBS scheme was used to calculate the solution in time for all the six meshes. Nondimensional time-step values, ranging between 10^{-3} and 10^{-2}, were employed in the calculations. In order to achieve a steady-state solution, the calculation was continued until the maximum difference of the variables u_1, u_2 and p between two consecutive time steps became less than 10^{-6}.

In Figure 7.24, the pressure contours generated from all meshes are shown. As seen, the pressure contours are distinguished by large oscillations when the mesh was relatively coarse (Figures 7.24(a) and (b)). These oscillations disappear from most of the domain as the mesh was progressively refined. The last two meshes (Figure 7.24(e) and (f)) result in much smoother contours than for the other meshes. However, even the fine meshes give oscillatory solution close to the singular point at the top left corner of the cavity.

The stream traces of meshes five and six are shown in Figure 7.25. At a Reynolds number of 5000, a secondary vortex appeared close to the bottom right-hand corner. In general it is difficult to predict this vortex and very fine meshes are necessary if this is to be achieved. Due to the small size of the secondary vortex, the first four meshes failed to produce its occurrence. However, the last two meshes (Figures 7.24(e) and (f)) were capable of predicting the secondary vortex as shown in Figure 7.25. In addition to this small secondary vortex, the figure also shows the recirculating vortices at both bottom corners and close to the top-left corner.

The quantitative result selected for this study was the horizontal velocity component distribution at the mid-vertical plane of the cavity. The horizontal velocity components of all the meshes have been calculated and plotted as shown in Figure 7.26. It is obvious that the first and second meshes result in inaccurate solutions due to insufficient mesh resolution. However, from the third mesh onwards sensible solutions were obtained. The comparison of the computed solution with the available benchmark data shows that the results obtained by the sixth mesh agreed excellently with the fine mesh solution of Ghia et al. (1982). The third, fourth and fifth meshes also give solutions which were close to that of Ghia et al. but were not identical.

The stream traces and pressure contours for Reynolds numbers of 400 and 1000 are shown in Figure 7.27. These results were generated using the sixth mesh. A comparison of the velocity profiles for the steady-state solution is shown in Figure 7.28. As seen, the comparison between

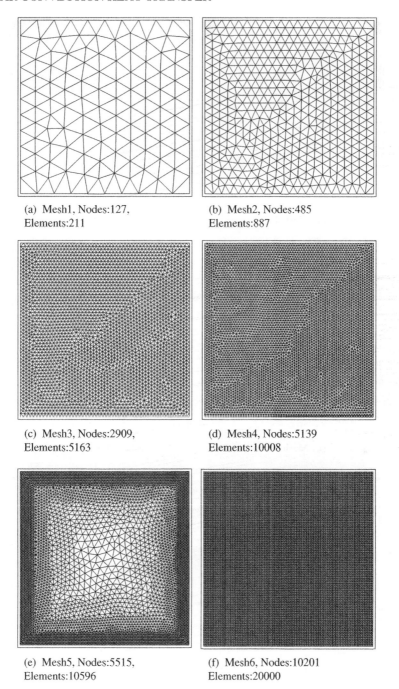

(a) Mesh1, Nodes:127,
Elements:211

(b) Mesh2, Nodes:485
Elements:887

(c) Mesh3, Nodes:2909,
Elements:5163

(d) Mesh4, Nodes:5139
Elements:10008

(e) Mesh5, Nodes:5515,
Elements:10596

(f) Mesh6, Nodes:10201
Elements:20000

Figure 7.23   Linear triangular element meshes, (a-e) unstructured meshes, (f) 100x100 structured mesh.

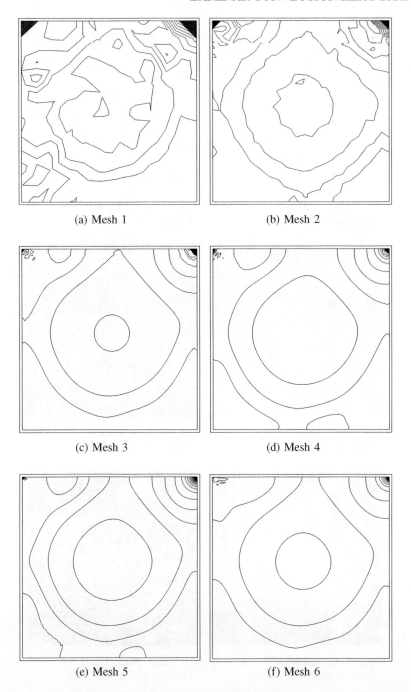

(a) Mesh 1   (b) Mesh 2

(c) Mesh 3   (d) Mesh 4

(e) Mesh 5   (f) Mesh 6

Figure 7.24   Isothermal flow in a lid-driven cavity. Pressure contours at Re = 5000.

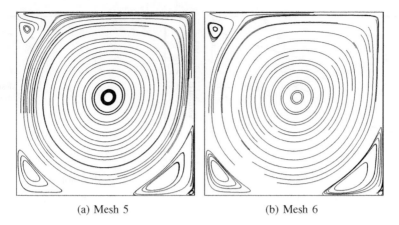

(a) Mesh 5   (b) Mesh 6

Figure 7.25   Isothermal flow in a lid-driven cavity. Stream traces at Re = 5000.

the present solution and the benchmark solution of Ghia et al. (1982) indicates excellent agreement. Further details may be obtained from Lewis et al. (1995); Malan et al. (2002a,b); Nithiarasu (2003) and the readers are encouraged to compute results for other Reynolds numbers.

Example 7.10.3   *Flow past a backward-facing step*

Another typical benchmark example is the flow past a backward facing step which is widely employed by researchers in validating flow solvers. In addition to the available numerical solutions, experimental data is also available for flow past a backward facing step.

The problem definition is shown in Figure 7.29. The inlet is situated at a distance of 4L upstream of the step, where L is the height of the step as shown in Figure 7.29. The inlet section is twice as high as the step. The total length of the channel is taken to be equal to 40 times the height of the step. Apart from the inlet and exit, all the other boundaries are assumed to be solid walls, where no slip boundary conditions are assumed to prevail. At the inlet to the channel, a nearly parabolic velocity profile of u_1 was assumed. The reason why a perfect parabolic velocity profile was not taken is that the experimental data was not available on a perfectly parabolic velocity profile. In order to compare the numerical results with the available experimental data, we imposed the experimental inlet velocity profile from Denham and Patrik (1974), which was not perfectly parabolic. The u_2 velocity at the inlet was assumed to be equal to zero at all times. The exit of the problem was situated at a distance of 36 times the step height in order to make sure that the disturbance created by the recirculation in the vicinity of the step was stabilized by the time flow reached the exit. At the exit, the pressure was prescribed as being equal to zero.

The Reynolds number, based on the average inlet velocity and step height, was taken to be equal to 229 in order to compare the velocity profiles with the available experimental velocity profile. The flow was assumed to be laminar and the computation was started with an initial value of u_1 equal to unity and u_2 equal to zero. In addition to the velocity values, an initial pressure value of zero was assumed on all nodal points.

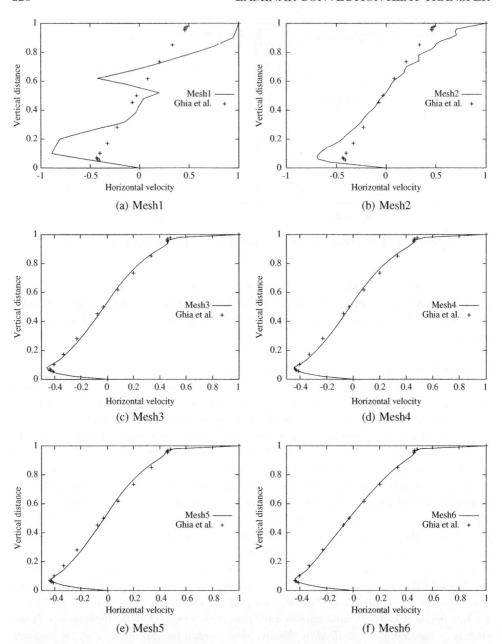

Figure 7.26   Incompressible isothermal flow in a lid-driven cavity. u_1 velocity profile along the mid-vertical line. Comparison with the benchmark steady-state results of Ghia *et al.*, (1982). *Source*: Data from Ghia *et al.* 1982.

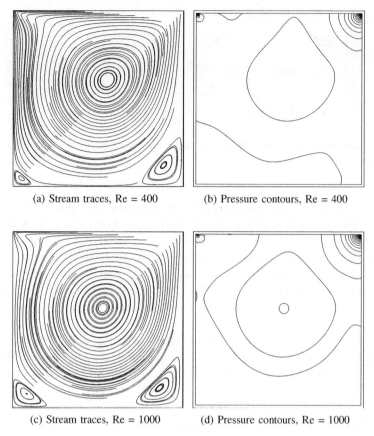

(a) Stream traces, Re = 400   (b) Pressure contours, Re = 400

(c) Stream traces, Re = 1000   (d) Pressure contours, Re = 1000

Figure 7.27   Isothermal flow in a lid-driven cavity. Stream traces and pressure contours for different Reynolds numbers.

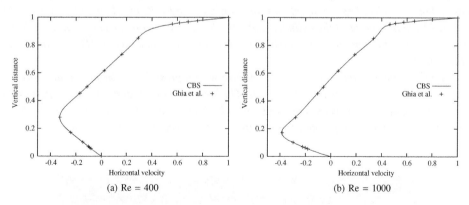

(a) Re = 400   (b) Re = 1000

Figure 7.28   Isothermal flow in a lid-driven cavity. Comparison of mid-vertical plane u_1 velocity profiles for different Reynolds numbers with Ghia *et al.*, (1982). *Source*: Data from Ghia *et al.* 1982.

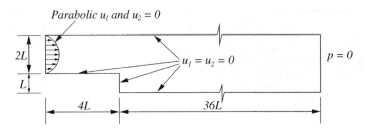

Figure 7.29   Incompressible isothermal flow past a backward facing step. Problem definition and boundary conditions.

The unstructured mesh employed in the calculation is shown in Figure 7.30. In Figure 7.31 the results are shown, which are produced by the CBS scheme in its fully explicit form (artificial compressibility form). Here, the use of local time-stepping techniques accelerated the solution towards the steady-state as compared to a fixed global time step (Malan et al. 2002a,b; Nithiarasu 2003). The u_1 velocity contours and pressure contours generated are given in Figures 7.31(a) and (b). In Figure 7.31 (c) the velocity profiles generated at different sections of the geometry, are compared with the experimental data of Denham et al. (Denham and Patrik 1974).

The u_1 velocity contours (Figures 7.31(a) is marked with the recirculation pattern downstream of the step. This was the expected pattern in a problem of this nature. The pressure contours are marked with minor oscillations, which was due to relatively coarse unstructured mesh used.

Example 7.10.4   *Transient flow past a circular cylinder*

In this section, a widely used transient benchmark problem of periodic vortex shedding behind a circular cylinder is briefly considered. The problem definition is simple and is shown in Figure 7.32. A circular cylinder of diameter D is placed in a fluid stream with a uniform approaching velocity. The computational domain inlet and exit are placed at lengths of 4D upstream from the center of the cylinder and 12D downstream from the center of the cylinder respectively. The top and bottom boundaries are situated at a distance of 4D from the center of the cylinder.

The inlet velocity was assumed to be uniform with a prescribed nonzero value of u_1 and a zero value for the u_2 velocity components. On both the bottom and top sides, the normal velocity component, u_2, was assumed to be equal to zero and u_1 was not prescribed. On the cylinder surface, the no-slip condition of zero velocity components was applied. At the exit,

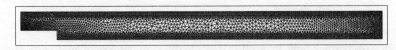

Figure 7.30   Incompressible isothermal flow past a backward facing step. Finite element mesh, Nodes:4656, Elements:8662.

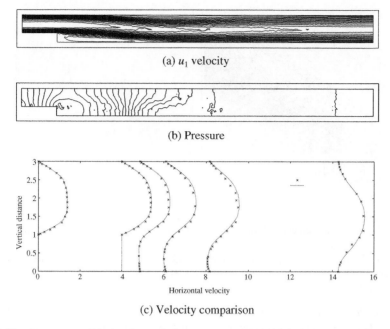

(a) u_1 velocity

(b) Pressure

(c) Velocity comparison

Figure 7.31   Incompressible isothermal flow past a backward facing step: (a) velocity contours; (b) pressure contours; (c) comparison of velocity profiles with experimental data, Re =
229.

*the pressure value was assumed to be constant. In this study, a zero value for pressure was
assumed at the exit. The inlet Reynolds number was defined based on the free stream inlet
velocity and the diameter D of the cylinder.*

   *A three-dimensional mesh was used in the vortex shedding calculations. For three-
dimensional flow calculations, two additional boundary conditions are necessary on the two*

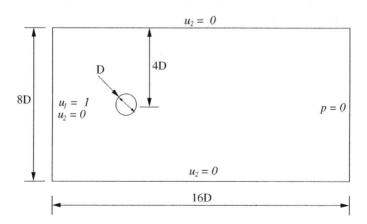

Figure 7.32   Isothermal flow past a circular cylinder. Geometry and boundary conditions.

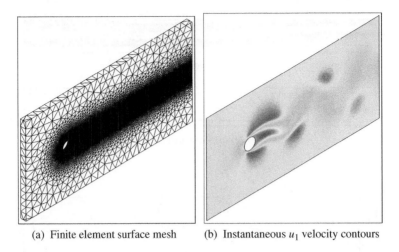

(a) Finite element surface mesh   (b) Instantaneous u_1 velocity contours

Figure 7.33   Isothermal flow past a circular cylinder. Three dimensional finite element mesh and an instantaneous u_1 velocity contour, Re = 100.

additional surfaces at the front and back (see Figure 7.33). The two additional surfaces were assumed to have no flow in the direction normal to the surfaces. Since the two-dimensional problem was solved in three dimensions, by introducing a third dimension, the width of the domain in the third dimension is arbitrary. The smaller the size of the domain in the third dimension, then the smaller will be the number of elements in the mesh. For the three-dimensional computations carried out here the length in the third dimension was assumed to be equal to 0.5 D.

The three-dimensional surface mesh is shown in Figure 7.33(a). The volume mesh used within the domain was generated using linear tetrahedral elements. A total number of approximately 600 000 elements were used in the calculations. As may be observed, the mesh is very fine behind the cylinder, along the expected von Karman vortex street. This is essential in order to accurately predict the flow. A mesh convergence study in three dimensions is time-consuming and difficult, and it is advisable to analyze many meshes in order to prove the convergence of the results. Alternatively, if the problem has existing benchmark results then a comparison with these will give confidence in the results generated. Here we chose the alternative approach and compare our results with the existing data.

The calculation was carried out using the artificial compressibility form of the CBS scheme (Nithiarasu 2003). The initial values of u_1 and u_2 were assumed to be equal to unity and zero respectively. Note that these values are nondimensional. All the velocity values are nondimensionalised using the reference inlet velocity value (see Section 7.3 for details). Similarly, the distances are scaled with respect to the diameter of the cylinder. These scalings result in a nondimensional inlet velocity value of unity and a cylinder diameter of unity in the nondimensional space. The initial values of pressure were assumed to be zero everywhere in the domain.

As mentioned previously, the solution to this problem is known to be periodic with respect to time. Once the solution reaches a steady periodic state, the periodic vortex shedding continues indefinitely. This process consists of vortex formation behind the cylinder and shedding.

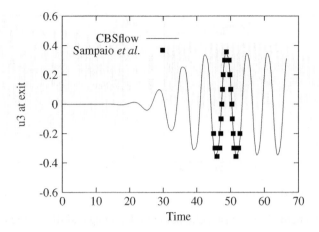

Figure 7.34 Isothermal flow past a circular cylinder. Comparison of u_1 velocity variation at an exit point, Re = 100.

In Figure 7.33(b), we only show a "snapshot" of the u_1 velocity distribution at a certain nondimensional time. Several such "snapshots" can be plotted but for the sake of brevity only one sample solution is given. We, however, provide the distribution of u_1 with respect to time at an exit point of the domain in Figure 7.34. The exit point is selected at the domain horizontal center line on the exit plane. As anticipated, the velocity at the selected exit point undergoes a steady periodic change with respect to time after establishing a steady periodic pattern. The initial period of the solution process (up to a nondimensional time of about 20) is marked with no sign of any periodic behavior of the velocity at the exit. The periodic behavior starts between nondimensional times of 20 and 30 and establishes a steady periodic pattern between the nondimensional time of 40 and 50. The peak values remain the same after establishing a steady pattern. The initial flow pattern depends heavily on the initial values of the variables, time steps and mesh used. However, once a steady periodic pattern is established the results should agree with other solutions as shown in Figure 7.34. The solution used in the comparison was generated from an adaptive analysis in two dimensions by de Sampaio et al. (de Sampaio et al. 1993). We also show the drag and lift co-efficient distribution with time in Figure 7.35. As seen, both conservation and nonconservation form of equations produce periodic patterns, though the results differ slightly between the two formulations (Nithiarasu and Zienkiewicz 2006).

7.11   Laminar Nonisothermal Flow

In this section, some examples of nonisothermal flow problems are discussed. In the previous section, the temperature effects are ignored, but they are included in this section in order to study some heat convection problems. The categories of forced convection, buoyancy-driven convection and mixed convection, are discussed in the following subsections.

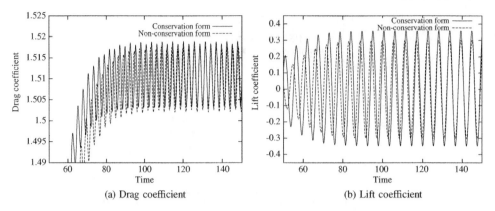

(a) Drag coefficient   (b) Lift coefficient

Figure 7.35   Isothermal flow past a circular cylinder. Drag and lift coefficients, Re = 100.

7.11.1   Forced Convection Heat Transfer

Forced convection heat transfer is induced by forcing a liquid, or gas, over a hot body or surface. Three forced convection problems will be studied in this section. The first problem is the extension of flow through a two-dimensional channel as discussed in the previous section, second is the forced convection over a sphere and the third is the problem of a backward facing step.

Example 7.11.1   *Forced convection in a rectangular channel*

The difference between the problem studied here and the one in the previous section is that the top and bottom walls are at a higher temperature than that of the air flowing into the channel. The nondimensional temperature scale employed is

$$T^* = \frac{T - T_a}{T_w - T_a}. \tag{7.202}$$

Since the CBSflow code is based on nondimensional governing equations, the above nondimensional scaling needs to be employed. This scale will give a temperature value of unity on the walls (T = T_w) and zero at the inlet (T = T_a). Dirichlet boundary conditions for temperature are not necessary at the exit as the no-flux conditions are assume across exit. For a steady-state solution, all four steps of the CBS scheme can either be solved simultaneously, or, firstly a steady-flow solution is obtained, then using these results a temperature distribution can be established independently. The Reynolds number is again assumed to be equal to 100, and the velocity distribution is the same as shown in Figure 7.20. The temperature profile distribution is as shown in Figure 7.36. As may be seen, a parabolic temperature profile is achieved at around the same distance from the entrance as that for the parabolic velocity profile. It should also be noted that as the length of the channel increases, the average temperature of the fluid also increases and approaches that of the wall temperature.

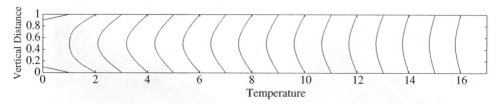

Figure 7.36   Forced convection flow through a two-dimensional rectangular channel. Temperature profiles at various distances.

Example 7.11.2   *Forced convection from a hot sphere*

The second problem considered is a three-dimensional flow over a hot sphere. The heat transfer aspects of the hot sphere are studied as it is exposed to a cold air stream. The problem definition is different from that of the channel flow, which is an internal flow but in the case of flow past a sphere, the flow is external to the sphere as shown in Figure 7.37.

In the problem discussed here, an outer boundary is fixed in such a way that the inlet is at a distance of five diameters from the center of the sphere and the exit is at 20 diameters from downstream of the center of the sphere (Nithiarasu et.al. 2004). The side boundaries are also at a distance of 5 diameters away from the center of the sphere. It is possible to imagine the sphere being placed inside a three-dimensional channel, which is 25 diameters in length and has 10 diameter long sides. However, the difference from the previous channel problem is that there is no solid outer wall in this case.

The boundary conditions are simple as in the previous problem. The inlet has a nondimensional velocity of unity and a nondimensional temperature of zero. The surface of the sphere is subjected to a no-slip velocity boundary condition and a nondimensional temperature of unity. All the side walls are subjected to a zero heat flux and a zero normal velocity value. At the exit the insulated conditions are assumed.

It is obvious that a three-dimensional mesh is required and for the problem under consideration linear tetrahedral elements were used. Three dimensional meshes were generated using an efficient mesh generator as reported by Morgan et al. 1999. The total number of elements used in the computation was approximately a million. The sphere and a cross-sectional side view along the axis are shown in Figure 7.38.

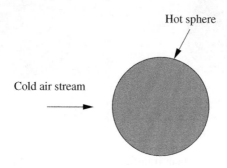

Figure 7.37   Forced convection flow past a sphere.

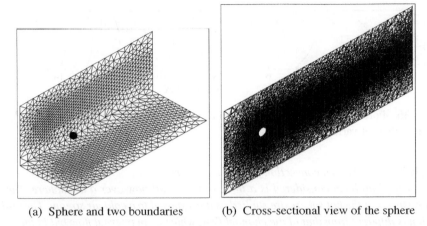

(a) Sphere and two boundaries   (b) Cross-sectional view of the sphere

Figure 7.38   Forced convection heat transfer from a sphere. Three dimensional mesh.

The temperature contours in the vicinity of the sphere are shown in Figure 7.39 for inlet Reynolds numbers of 100 and 200. As mentioned previously, the temperature on the surface of the sphere is unity. This diagram shows a cut view along the axis in the direction of the flow. Therefore, the temperature values close to the surface of the sphere are near to unity, which reduce in value away from the sphere and finally reach zero value (except in the wake), in the free air stream. In the downstream direction, however, the temperatures are greater than that of the free stream temperature all the way to the exit (see Figure 7.40). This indicates that the cold air stream removes heat from the sphere, which is then transported to the exit.

The values of drag coefficient and average Nusselt numbers are given in Tables 7.1 and 7.2 respectively. In Table 7.1, the quantity inside the brackets is the pressure drag coefficient.

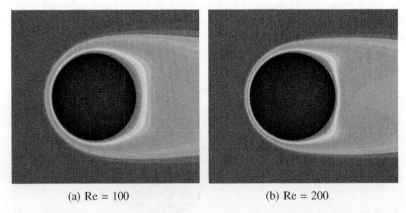

(a) Re = 100   (b) Re = 200

Figure 7.39   Forced convection heat transfer from a sphere. Temperature distribution in the vicinity of the sphere.

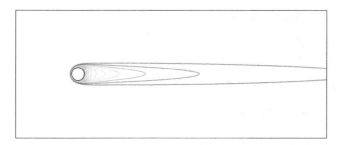

Figure 7.40   Forced convection flow past a sphere. Temperature contours, Re = 100.

Example 7.11.3   *Forced convection downstream of a backward facing step*
   The problem definition is similar to the isothermal flow past a backward facing step, as discussed in the previous section. The difference being that additional boundary conditions are prescribed for the temperature field. The boundary conditions discussed by (Kondoh et al. 1993) will be adopted. The solid bottom wall was assumed to be at a higher temperature than the fluid (results presented here are for air with Pr = 0.71) entering the channel. All other solid walls were assumed to be insulated. Zero flux conditions were prescribed at the exit. All other boundary conditions for the velocity and pressure are the same as the ones discussed for the isothermal problem in the previous section and are repeated in Figure 7.41.
   Three different meshes have been employed to make sure that the solutions presented were accurate. The first mesh used was the mesh shown in Figure 7.30. The second and third meshes are finer than the first mesh and are shown in Figure 7.42.

Table 7.1   Comparison of coefficient of drag with existing literature

Author references	100	200
Clift *et al.* 1978	1.087	-
Lee 2000	1.096 (0.512)	-
Gulcat and Aslan 1997	1.07	0.78
Rimon and Cheng 1969	1.014	0.727
Le Clair *et al.* 1970	1.096 (0.590)	0.772 (0.372)
Magnaudet *et al.* 1995	1.092 (0.584)	0.765 (0.368)
CBS	1.105 (0.564)	0.7708 (0.347)

Table 7.2   Comparison of average Nusselt number.

Re	(Yuge 1960)	(Whitaker 1983)	(Feng *et al.* 2000)	CBS
50	5.4860	5.1764	5.4194	5.2176
100	6.9300	6.6151	6.9848	6.6589
200	8.9721	8.7219	9.1901	8.7599

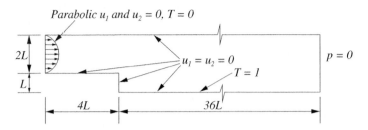

Figure 7.41   Forced convection heat transfer downstream of a backward facing step. Geometry and boundary conditions.

All three meshes were employed to study the heat transfer at a Reynolds number of 500. The local Nusselt number distribution on the hot wall downstream of the step is shown in Figure 7.43. As seen, the Nusselt number difference between all three meshes was very small. Therefore, the second mesh was used in all the calculations in order to save computational time, as the difference between the local Nusselt number distribution of the finest mesh (third mesh) and the second was very small. The small oscillations in the local Nusselt number distribution, especially on the first mesh, was generated by the coarseness of the unstructured mesh.

Figure 7.44 shows the temperature contours for all the different Reynolds numbers considered. Previous studies indicate that the maximum heat transfer occurred close to the reattachment point. The incompressible flow is attached to the wall from the inlet until it reaches the step. The flow is detached from the bottom wall and recirculation develops downstream of the step as shown previously for the nonisothermal case. The flow reattaches itself to the bottom wall after the recirculation in the downstream portion of the step. The location at which the reattachment takes place varies with the Reynolds number. The higher the Reynolds number, then the farther will be the reattachment point from the step. The reattachment distances from the step are given in Figure 7.44. These values are in close agreement with reported results (Kondoh et al. 1993).

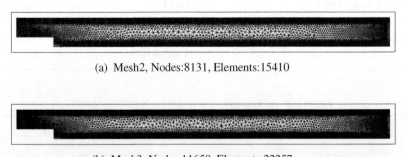

(a) Mesh2, Nodes:8131, Elements:15410

(b) Mesh3, Nodes:11659, Elements:22257

Figure 7.42   Forced convection heat transfer downstream of a backward facing step. Unstructured meshes.

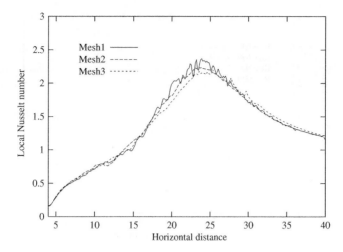

Figure 7.43   Forced convection heat transfer downstream of a backward facing step. Local Nusselt number distribution on the hot wall for a Reynolds number of 500 on different meshes.

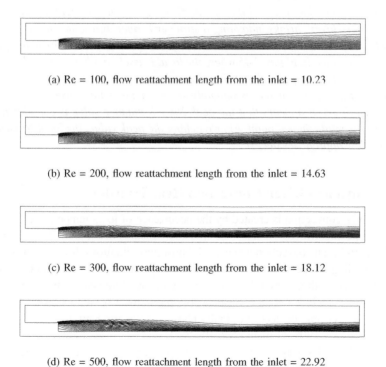

(a) Re = 100, flow reattachment length from the inlet = 10.23

(b) Re = 200, flow reattachment length from the inlet = 14.63

(c) Re = 300, flow reattachment length from the inlet = 18.12

(d) Re = 500, flow reattachment length from the inlet = 22.92

Figure 7.44   Forced convection heat transfer downstream of a backward facing step. Temperature contours at different Reynolds numbers.

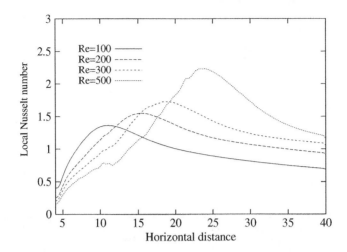

Figure 7.45   Forced convection heat transfer downstream of a backward facing step. Local Nusselt number distribution on the hot wall for different Reynolds numbers.

The thermal action predominantly takes place downstream of the step in the bottom portion of the channel. It may be observed that as the flow approaches the reattachment point, the thermal boundary layer shrinks indicating a stronger temperature gradient in the vicinity of the reattachment point and thus a higher heat transfer rate taking place close to this point. This is clearly demonstrated in Figure 7.45 where the local Nusselt number is plotted along the hot wall downstream of the step. The local Nusselt number starts with an almost zero value at the corner close to the step and increases smoothly to a maximum value close to the reattachment point and then drops. It appears that the peak Nusselt number value is calculated close to, but just after the reattachment point. After reaching the peak value, the local Nusselt number drops as the flow approaches the exit.

7.11.2   Buoyancy-driven Convection Heat Transfer

Buoyancy-driven convection is created by the occurrence of local temperature differences in a fluid. This type of convection can also be created by local concentration differences within a fluid. Buoyancy-driven convection is present in most flow situations; however, its significance can vary according to the situation. For instance in a situation where a hot surface and a cold fluid interact, without any other external force, then buoyancy-driven convection will develop. Examples include radiators inside a cold room, most solar appliances, some cooling applications of electronic devices and finally phase change applications (Lewis *et al.* 1995a; Ravindran and Lewis 1998; Usmani *et al.* 1992a,b).

The principles of buoyancy-driven convection are simple. A local temperature difference creates a local density difference within the fluid resulting in fluid motion because of the local density variation. Although the principles are simple, the development of an accurate numerical solution for such buoyancy-driven flows is far from simple. This is mainly due to the very slow flow rates involved, which are often marked with turbulence, which again complicates the numerical prediction.

Example 7.11.4 *Buoyancy-driven flow in a square enclosure*

In order to demonstrate buoyancy-driven convection, we shall consider the standard benchmark problem of natural convection within a two-dimensional square enclosure, as shown in Figure 7.46. The geometry is a two-dimensional square with a nondimensional size of unity. The walls are solid and subjected to no slip velocity boundary conditions (zero velocity components). One of the vertical walls is subjected to a higher temperature ($T = 1$) than the other vertical wall ($T = 0$). Both the top and bottom walls are assumed to be insulated (zero heat flux). The steady-state solution to this problem is sought herein.

In order to obtain a steady-state solution, the CBSflow is used in its semi-implicit form with zero initial velocity and temperature values and a small constant value for of pressure (0.1). A simple pressure boundary condition is essential in order to solve the pressure equations implicitly. One of the corner points has a fixed pressure value of zero all the time. The parameter varied in this problem is the Rayleigh number. The mesh employed in the calculations is a structured mesh and is shown in Figure 7.47. Unstructured meshes are equally valid but require a greater number of elements in order to obtain the same accuracy as structured meshes. The mesh shown in Figure 7.47 contains 5000 elements and 2601 nodes.

Figures 7.48 shows the temperature contours and streamlines for different Rayleigh numbers. The flow raises along side of the hot left-side wall, taking the heat with it and losing it along side of the right-side wall. As the Rayleigh number increases, the flow becomes stronger and is marked with a thinner flow regime and thermal boundary layers close to the vertical walls.

Table 7.3 reports various quantities, which has been calculated for the natural convection in a square cavity (Massarotti et al. 1998). In Table 7.3, ψ is the stream function, Nu_{av} is the average Nusselt number and u_{2max} is the maximum vertical velocity component. These values compare very well with the benchmark data available in the literature.

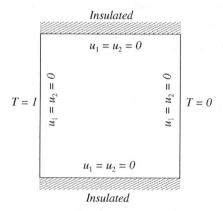

Figure 7.46   Buoyancy-driven flow in a square enclosure. Geometry and boundary conditions.

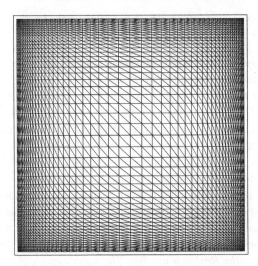

Figure 7.47   Buoyancy-driven flow in a square enclosure. Finite element mesh. Nodes: 2601, Elements: 5000.

7.11.3   Mixed Convection Heat Transfer

A mixed convection heat transfer mode has features of both forced convection and natural convection. The mixed convection solution to a heat transfer problem is necessary if the Reynolds number is small and the importance of the buoyancy contribution is significant. The equations solved are those of forced convection with an addition of a source term (Equation (7.70)) in the gravitational direction. If the direction of gravity is not aligned with either of the coordinate directions (x_1 and x_2), then appropriate components of the source term need to be added to the momentum equations. The effect of mixed convection can be measured by calculating the source term of Equation (7.70). If this term is close to zero then the buoyancy effects can be ignored and a forced convection solution is sufficient. However, if the value of the source term is far from being zero (either in the negative or positive sense), then a mixed convection solution is essential.

Example 7.11.5   *Mixed convection in a channel*

Here, we consider a simple mixed convection problem in a rectangular vertical channel as shown in Figure 7.49. In order to compare the results with the analytical solution for fully developed flow in a channel as given by Aung and Worku (1986a), then the nondimensional scales require changing. The scales used by Aung and Worku are

$$x_2^* = \frac{x_2}{ReL}; \quad u_1^* = \frac{u_1 L}{\nu}. \tag{7.203}$$

All other scales are the same as for the forced convection scale discussed in Section 7.3. The above scales lead to some changes in the nondimensional form of the mixed convection equation. The source term GrT^*/Re^2, in the mixed convection equation, will be GrT/Re and the Reynolds number at all other locations will disappear. The great advantage of applying

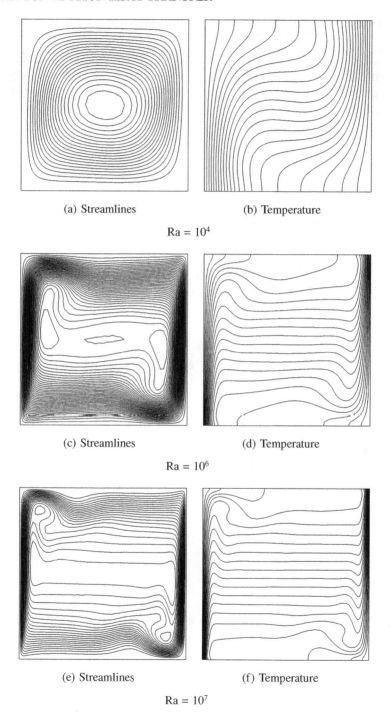

(a) Streamlines   (b) Temperature

Ra = 10^4

(c) Streamlines   (d) Temperature

Ra = 10^6

(e) Streamlines   (f) Temperature

Ra = 10^7

Figure 7.48   Natural convection in a square enclosure. Streamlines and temperature contours for different Rayleigh numbers, Pr = 0.71.

Table 7.3   Quantitative results for natural convection in a square cavity.

Ra	Nu_{av}	ψ_{max}	u_{2max}
10^3	1.116	1.175	3.692
10^4	2.243	5.075	19.63
10^5	4.521	9.153	68.85
10^6	8.806	16.49	221.6
10^7	16.40	30.33	702.3

this scale is that the nondimensional length of the channel can be considerably reduced. The analytical solution for a fully developed mixed convection profile is given (Aung and Worku 1986b) as

$$u_1 = \frac{Gr}{Re}(1 - r_T)\left(\frac{-x_1^3}{6} + \frac{x_1^2}{4} - \frac{x_1}{12}\right) - 6x_1^2 + 6x_1. \tag{7.204}$$

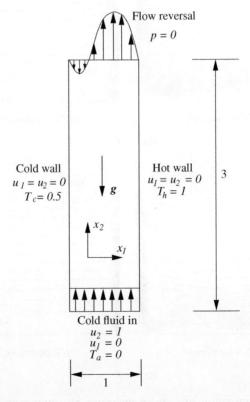

Figure 7.49   Mixed convection in a vertical channel. Geometry and boundary conditions.

where

$$r_T = \frac{T_c - T_a}{T_h - T_a}.$$ (7.205)

Two vertical plates serve as the channel walls, one of them being at a higher temperature ($T_h = 1$) than that of the other wall. The temperature T_c of the cold wall is 0.5 and the cold fluid entering the channel from the bottom is zero ($T_a = 0$). A uniform, nondimensional, vertical velocity of unity is imposed at the entrance ($u_2 = 1$). The direction of gravity is assumed to act in the negative x_2 direction. The inlet Reynolds number is 100 and the Grashof number is assumed to be 250 000, which results in a Gr/Re value of 250. At the exit, zero pressure values are imposed. The total length of the channel is three times the width of the channel. The Reynolds number is defined with respect to the width of the channel.

The problem defined here is the case of gravity opposing the forced flow from the bottom. This is an example of buoyancy aided convective heat transfer as the buoyancy is helping the flow to move quicker by creating a density driven upward flow close to the hot wall. However, at very high Richardson numbers, flow reversal is possible in this type of problem, as shown in Figure 7.49. It is quite possible in certain practical applications that the flow will be forced from the top of the channel (in the negative x_2 direction). Such a flow will be called "opposing" flow in which the buoyancy-driven flow is in the opposite direction of the forced flow.

The mesh used in the computations was fully unstructured and is shown in Figure 7.50. The mesh is fine close to the solid walls and a total number of 8956 elements and 4710 nodes were employed. Figure 7.51 shows the velocity profile distributions at various heights. As seen, the air flows upwards close to the inlet and flow reversal occurs somewhere between the vertical distances of 0.5 and 1.0 from the inlet. The flow is nearly fully developed at a vertical distance of 2 from the inlet. As mentioned previously, the ratio (GR/Re) is 250 and a further increase in this ratio will lead to a stronger flow reversal. Further details regarding this type of problem may be found in Aung and Worku (1986a). A comparison of the fully developed velocity profile with the analytical solution is given in Figure 7.52 (Aung and Worku 1986a) and as may be seen the agreement is excellent.

7.12   Extension to Axisymmetric Problems

The axisymmetric formulation of the heat conduction equations has been discussed in many of the earlier chapters. Here, an extension of the plane formulation to axisymmetric convection heat transfer problems will be discussed. The governing equations in cylindrical coordinates are given (Figure 7.53), as follows:

Conservation of mass:

$$\frac{1}{r}\frac{\partial(ru_r)}{\partial r} + \frac{\partial u_z}{\partial z} = 0.$$ (7.206)

r momentum component:

$$\frac{\partial u_r}{\partial t} + u_r\frac{\partial u_r}{\partial r} + u_z\frac{\partial u_z}{\partial z} = -\frac{1}{\rho}\frac{\partial p}{\partial r} + \nu\left[\frac{1}{r}\frac{\partial}{\partial r}\left(r\frac{\partial u_r}{\partial r}\right) + \frac{\partial^2 u_r}{\partial z^2} - \frac{u_r}{r^2}\right].$$ (7.207)

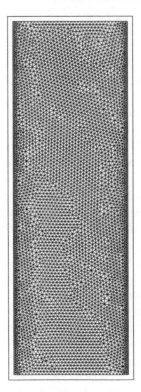

Figure 7.50   Mixed convection in a vertical channel. Unstructured finite element mesh.

z momentum component:

$$\frac{\partial u_z}{\partial t} + u_r \frac{\partial u_z}{\partial r} + u_z \frac{\partial u_r}{\partial z} = -\frac{1}{\rho} \frac{\partial p}{\partial z} + \nu \left[\frac{1}{r} \frac{\partial}{\partial r} \left(r \frac{\partial u_z}{\partial r} \right) + \frac{\partial^2 u_z}{\partial z^2} \right]. \tag{7.208}$$

Energy equation:

$$\rho c_p \left(\frac{\partial T}{\partial t} + u_r \frac{\partial T}{\partial r} + u_z \frac{\partial T}{\partial z} \right) = k \left[\frac{1}{r} \frac{\partial}{\partial r} \left(r \frac{\partial T}{\partial r} \right) + \frac{\partial^2 T}{\partial z^2} \right]. \tag{7.209}$$

The CBS procedure follows the same steps as for the plane problem. However, the integration of the matrices will be different as the area of the element will no longer be two-dimensional. For example, let us consider the diffusion matrix of the energy equation. The temperature diffusion matrix for the plane problem is given by Equation (7.156). We can rewrite this as

$$\begin{aligned}
[K_t]_e &= k \int_\Omega \left(\frac{\partial N^T}{\partial r} \frac{\partial N}{\partial r} + \frac{\partial N^T}{\partial z} \frac{\partial N}{\partial z} \right) d\Omega \\
&= k \int_\Omega \left(\frac{\partial N^T}{\partial r} \frac{\partial N}{\partial r} + \frac{\partial N^T}{\partial z} \frac{\partial N}{\partial z} \right) 2\pi r dA, \tag{7.210}
\end{aligned}$$

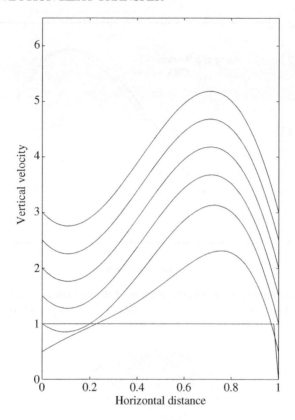

Figure 7.51 Mixed convection in a vertical channel. Developing velocity profiles are various vertical vertical sections.

where the radial coordinate r is expressed as

$$r = N_i r_i + N_j r_j + N_k r_k. \tag{7.211}$$

The formula used in the integration is the same as for any linear triangular element (Equation (7.117)). On applying Equation (7.117), then Equation (7.210) becomes

$$[K_m]_e = k\frac{2\pi}{12A}(r_i + r_j + r_k)\begin{bmatrix} b_i^2 & b_i b_j & b_i b_k \\ b_j b_i & b_j^2 & b_j b_k \\ b_k b_i & b_k b_j & b_k^2 \end{bmatrix}$$

$$+ k\frac{2\pi}{12A}(r_i + r_j + r_k)\begin{bmatrix} c_i^2 & c_i c_j & c_i c_k \\ c_j c_i & c_j^2 & c_j c_k \\ c_k c_i & c_k c_j & c_k^2 \end{bmatrix}. \tag{7.212}$$

All the other terms of the axisymmetric equations may be discretized in a similar fashion. In discretizing the r momentum diffusion terms, the term u_r/r^2 can be approximated by averaging r over an element.

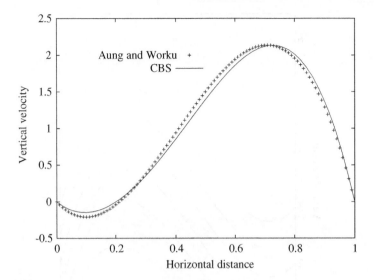

Figure 7.52   Mixed convection in a vertical channel. Comparison of velocity profile at exit with fully developed analytical solution. *Source*: Data from Aung and Worku (1986b).

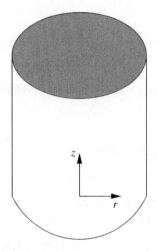

Figure 7.53   Coordinate system for axisymmetric geometries.

7.13   Summary

In this chapter, we have given a brief overview of convection heat transfer. However, the subject is vast in extent and it is difficult to cover all the aspects within a single chapter. Several details have been neglected on purpose in order to keep the discussion brief. For instance, higher order elements have not been discussed and few solution procedures have been touched upon.

However, the CBS scheme for convection heat transfer has been discussed in detail for linear triangular elements. A complete knowledge of such a single scheme will provide the reader with a strong starting point for understanding other relevant fluid dynamics and convection heat transfer solution procedures. In the following chapter the heat convection problems in the turbulent regime are discussed.

7.14 Exercises

Exercise 7.14.1 *Using a differential control volume approach, derive a three-dimensional convection-diffusion equation for pollution transport in a river.*

Exercise 7.14.2 *Following the derivation of the CG method for a convection-diffusion equation discussed in this chapter, derive the CG method for a convection-diffusion equation with a source term Q.*

Exercise 7.14.3 *Derive Navier-Stokes equations in cylindrical and spherical coordinates.*

Exercise 7.14.4 *Reduce the incompressible Navier-Stokes equations to solve a one-dimensional time-dependent convection heat transfer problem.*

Exercise 7.14.5 *For natural convection problems, if α is replaced by v in the nondimensional scaling, derive the new nondimensional form.*

Exercise 7.14.6 *Calculate laminar flow and heat transfer from a hot cylinder at $Re = 40$ placed inside a rectangular channel (assume the size) using the CBSflow code. Assume buoyancy effect is negligible. Determine the influence of the distance between the cylinder and the channel walls, inlet and exit.*

Exercise 7.14.7 *Write a program in any standard scientific language to calculate stream functions from a computed velocity field.*

Exercise 7.14.8 *In this example, you are asked to make appropriate assumptions and model flow past the heat exchanger tubes as shown in Figure 7.54.*

A schematic diagram of a typical cross flow heat exchanger arrangement is shown in Figure 7.54. As seen, the hot working fluid from the industry is passed through tubes and the coolant is pumped from the bottom and used to cool the working fluid. In this particular heat exchanger the tubes are arranged in a staggered style.

The flow and heat transfer analysis over these tubes is very important in determining an optimal tube arrangement. Neglecting the outer wall effects carry out a heat transfer analysis at a Reynolds number of 300. Assume that the flow is laminar and the buoyancy effects are negligibly small.

Assume that the vortex shedding effects can be neglected and simplify the three-dimensional problem to a two-dimensional problem. Set up the appropriate boundary conditions, generate the mesh and carry out the analysis either using the CBSflow code or any other available software.

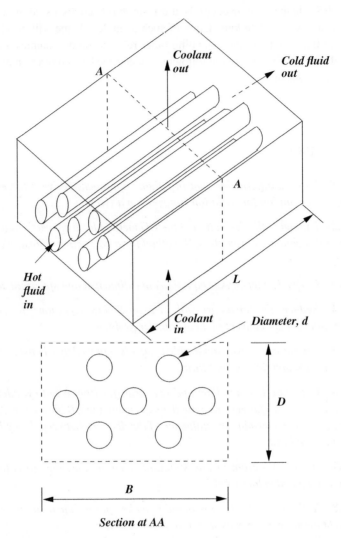

Figure 7.54   Schematic diagram of a cross flow heat exchanger, d = 1, B = 8, D = 6, pitch = 3, L = 42.

Exercise 7.14.9   *In this exercise, you are asked to simulate the liquid flow through a liquid processing plant as shown in Figure 7.55.*

In the liquid processing industry, liquid is passed through several tanks as shown in Figure 7.55. The diagram shows a simplified model of such a plant. With appropriate assumptions, simplify the problem further and determine the flow mechanism. The raw liquid is pumped into the plant from the left-hand side at a Reynolds number of 400 which is based on the width of the inlet channel and inlet velocity.

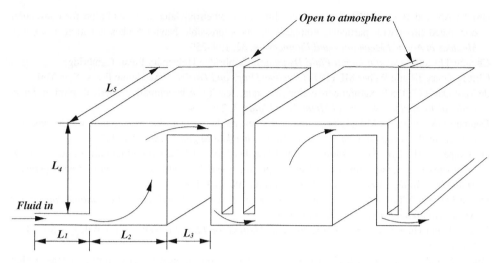

Figure 7.55 Schematic diagram of water processing plant, inlet/exit channel height = 1, $L_1 = 4, L_2 = 5, L_3 = 4, L_4 = 6, L_5 = 30$.

Include appropriate assumptions and formulate a simplified physical problem. The simplification should be in such a way that the model should not lose accuracy and at the same time should not be very expensive to solve. Discuss the project and design the boundary limits and conditions.

Once the problem has been simplified to two dimensions, generate a mesh and solve the problem using the CBSflow solver. Determine the temperature distribution if the bottom surface of the tank is hotter than the incoming fluid. Neglect the buoyancy effects and assume the liquid is water in the heat transport problem.

Exercise 7.14.10 *A two-dimensional square enclosure (all solid walls) filled with air is subjected to a linearly varying temperature on one of its vertical walls (say $T = (x_2/L)T_{max}$, where L is the characteristic dimension) and a constant temperature on the other vertical wall, which is less than that of T_{max}. If the horizontal walls are assumed to be adiabatic, obtain solutions for the flow and heat transfer inside the enclosure for different Rayleigh numbers. Refer to Chapter 7 for nondimensional scales.*

Exercise 7.14.11 *In the above problem if the linear variation of temperature is replaced with a constant heat flux, determine the temperature and flow patterns.*

References

Aung W and Worku G (1986a) Developing flow and flow reversal in a vertical channel with asymmetric wall temperatures. *ASME Journal of Heat Transfer*, **108**, 299–304.

Aung W and Worku G (1986b) Theory of fully developed, combined convection including flow reversal. *ASME Journal of Heat Trasnfer*, **108**, 485–488.

Brooks AN and Hughes TJR (1982) Streamline upwind/Petrov-Galerkin formulation for convection dominated flows with particular emphasis on incompressible Navier Stokes equation. *Computer Methods in Applied Mechanics and Engineering*, **32**, 199–259.

Cheung TJ (2002) *Computational Fluid Dynamics*. Cambridge University Press, Cambridge.

Clift R, Grace JR and Weber ME (1978) *Bubbles Drops and Particles*. Academic Press, New York.

de Vahl Davis G (1983) Natural convection in a square cavity: A benchmark numerical solution. *International Journal for Numerical Methods in Fluids*, **3**, 249–264.

Denham MK and Patrik MA (1974) Laminar flow over a downstream-facing step in a two-dimensional flow channel. *Transactions of the Institution of Chemical Engineers*, **52**, 361–367.

de Sampaio PAB, Lyra PRM, Morgan K and Weatherill NP (1993), Petrov-Galerkin solutions of the incompressible Navier-Stokes equations in primitive variables with adaptive remeshing. *Computer Methods in Applied Mechanics and Engineering*, **106**, 143–178.

Donea J (1984) A Taylor-Galerkin method for convective transport problems. *International Journal for Numerical Methods in Fluids*, **20**, 101–119.

Donea A and Huerta A (2003) *Finite Element Method for Flow Problems*. John Wiley & Sons Ltd, Chichester.

Feng ZG and Michaelides EE (2000) A numerical study on the heat transfer from a sphere at high Reynolds and Peclet numbers. *International Journal of Heat Mass Transfer*, **43**, 219–229.

Fletcher CAJ (1988) *Computational Techniques for Fluid Dynamics, Volume 1 Fundamentals and General Techniques*, Springer Verlag, London.

Ghia U, Ghia KN and Shin CT (1982) High Re solutions for incompressible flow using the Navier-Stokes equations and multigrid method. *Journal of Computational Physics*, **48**, 387–411.

Gowda YTK, Narayana PAA and Seetharamu KN (1998) Finite element analysis of mixed convection over in-line tube bundles. *International Journal of Heat and Mass Transfer*, **41**, 1613–1619.

Gresho PM and Sani RL (2000) *Incompressible Flow and the Finite Element Method, Vol. 2 Isothermal Laminar Flow*, John Wiley & Sons Ltd, Chichester.

Gülçat Ü and Aslan AR (1997) Accurate 3D viscous incompressible flow calculations with FEM. *International Journal for Numererical Methods in Fluids*, **25**: 985–1001.

Hirsch C (1989) *Numerical Computation of Internal and External Flows, Volume 1*, John Wiley & Sons, Inc., New York.

Jaluria Y (1986) *Natural Convection Heat and Mass Transfer*. Pergamon Press.

Jaluria Y and Torrancs KE (1986) *Computational Heat Transfer*, Hemisphere Publishing Corporation, New York.

Kondoh T, Nagano Y and Tsuji T (1993) Computational study of laminar heat transfer downstream of a backward-facing step. *International Journal of Heat and Mass Transfer*, **36**, 577–591.

Le Clair BP, Hamielec AE and Pruppacher HR (1970) A numerical study of the drag on a sphere at low and intermediate Reynolds numbers, *Journal of Atmospheric Sciences*, **27**, 308–315.

Laufer J (1951) *Investigation of Turbulent Flow in a Two Dimensioanl Channel*, Report 1053, NACA.

Launder BE and Spalding DB (1972) *Mathematical Models of Turbulence*. Academic Press, New York.

Lee S (2000) A numerical study of the unsteady wake behind a sphere in a uniform flow at moderate Reynolds numbers. *Computers and Fluids*, **29**, 639–667.

Lewis RW, Huang HC, Usmani AS and Cross J (1991) Finite element analysis of heat transfer and flow problems using adaptive remeshing. *International Journal for Numerical Methods in Engineeing*, **32** 767–782.

Lewis RW, Huang HC, Usmani AS and Cross J (1995) Efficient mould filling simulation in castings by an explicit finite element method. *International Journal for Numerical Methods in Fluids*, **20**, 493–506.

Lewis RW, Ravindran K and Usmani AS (1995) Finite element solution of incompressible flows using an explicit segregated approach, *Archives of Computational Methods in Engineering*, **2**, 69–93.

Lewis RW, Morgan K, Thomas HR and Seetharamu KN (1996) *The Finite Element Method for Heat Transfer Analysis*. John Wiley & Sons, Inc., New York.

Löhner R (2001) *Applied CFD Techniques*, John Wiley & Sons, Inc., New York.

Löhner R, Morgan K and Zienkiewicz OC (1984) The solution of non-linear hyperbolic equation systems by the finite element method. *International Journa of Numerical Methods in Fluids*, **4**, 1043–1063.

Magnaudet J, Rivero M and Fabre J (1995) Accelerated flows past a rigid sphere or a spherical bubble. Part I: Steady straining flow. *Journal of Fluid Mechanics*, **284**, 97–135.

Malan AG, Lewis RW and Nithiarasu P (2002a) An improved unsteady, unstructured, artificial compressibility, finite volume scheme for viscous incompressible flows: Part I. Theory and implementation. *International Journal for Numerical Methods in Engineering*, **54**, 695–714.

Malan AG, Lewis RW and Nithiarasu P (2002) An improved unsteady, unstructured, artificial compressibility, finite volume scheme for viscous incompressible flows: Part II Application. *International Journal for Numerical Methods in Engineering*, **54**, 715–729.

Massarotti N, Nithiarasu P and Zienkiewicz OC (1998) Characteristic–Based–Split (CBS) algorithm for incompressible flow problems with heat transfer. *International Journal of Numerical Methods for Heat and Fluid Flow*, **8**, 969–990.

Mohammadi B and Pironneau O (1994) *Analysis of k-ε Turbulence Model*, John Wiley & Sons, Inc., New York.

Moin P and Mahesh K (1998) Direct numerical simulation: A tool in turbulence research. *Annual Review of Fluid Mechanics*, **30**, 539–578.

Morgan K, Weatherhill NP, Hassan O, *et al.* (1999) A parallel framework for multidisciplinary aerospace engineering simulations using unstructured meshes. *International Journal for Numerical Methods in Fluids*, **31**, 159–173.

Nithiarasu P (2002) An adaptive remeshing scheme for laminar natural convection problems. *Heat and Mass Transfer*, **38**, 243–250.

Nithiarasu P (2003) An efficient artificial compressibility (AC) scheme based on the Characteristic Based Split (CBS) method for incompressible flows, *International Journal for Numerical Methods in Engineering*, **56**, 1815–1845.

Nithiarasu P, Codina R and Zienkiewicz OC (2006), Characteristic based split scheme (CBS), *International Journal for Numerical Methods in Engineering*, **66** 1514–1546.

Nithiarasu P, Massarotti N and Mathur JS (2004) Three dimensional convection heat transfer calculations using the Characteristic Based Split (CBS) scheme. *Proceedings of ASME/ISHMT Conference, Chennai, India, January.*

Nithiarasu P, Seetharamu KN and Sundararajan T (1998) Finite element analysis of transient natural convection in an odd-shaped enclosure, *International Journal of Numerical Methods for Heat and Fluid Flow*, **8**, 199–216.

Nithiarasu P and Zienkiewicz OC (2000) Adaptive mesh generation for fluid mechanics problems. *International Journal for Numerical Methods in Engineering*, **47**, 629–662.

Nithiarasu P and Zienkiewicz OC (2006) Analysis of an explicit and matrix free fractional step method for incompressible flows. *Computer Methods in Applied Mechanics and Engineering*, **195**, 5537–5551.

Patnaik BSV, Gowda YTK, Ravisankar MS, *et al.* (2001) Finite element simulation of internal flows with heat transfer using a velocity correction approach. *Sadhana – Academy Proceedings in Engineering Sciences*, **26**, 251–283.

Pironneau O (1989) *Finite Element Method for Fluids*. John Wiley & Sons, Ltd, Chichester.

Ravindran K and Lewis RW (1998) Finite element modelling of solidification effects in mould filling. *Finite Element Analysis and Design*, **31**, 99–116.

Rimon Y and Cheng SI (1969) Numerical solution of a uniform flow over a sphere at intermediate Reynolds numbers. *Physics of Fluids*, **12**: 949–959.

Sai BVKS, Seetharamu KN and Narayana PAA (1994) Solution of laminar natural convection in a square cavity by an explicit finite element scheme. *Numerical Heat Transfer Part A – Applications*, **25**, 593–609.

Sagaut P (1998) *Large Eddy Simulation for Incompressible Flows*. Springer, Berlin.

Shames IH (1982) *Mechanics of fluids*. McGraw-Hill, New York.

Schlichting H (1968) *Boundary layer theory*, 6th Edition. McGraw-Hill, New York.

Spalding DB (1972) A novel finite difference formulation for differential equations involving both first and second derivatives. *International Journal for Numerical Methods in Engineering*, **4**, 551–559.

Srinivas M, Ravisanker MS, Seetharamu KN and Aswathanarayana PA, (1994) Finite element analysis of internal flows with heat transfer. *Sadhana – Academy Proceedings in Engineering*, **19**, 785–816.

Tezduyar TE, Osawa Y, Stein K, *et al.* (2000) Computational methods for parachute aerodynamics. *Proceedings of Computational Fluid Dynamics for the 21st Century*, Kyoto, Japan.

Usmani AS, Lewis RW and Seetharamu KN (1992a) Finite element modelling of natural convection controlled change of phase. *International Journal for Numerical Methods in Fluids*, **14**, 1019–1036.

Usmani AS, Cross J and Lewis RW (1992b) A finite element model for the simulation of mould filling in metal casting and the associated heat transfer, *International Journal for Numerical Methods in Engineering*, **35**, 787–806.

Whitaker S (1983) *Fundamental Principles of Heat Transfer*. Krieger Publishing Company, Malabar.

Wilcox DC (1993) *Turbulence Modelling for CFD*. DCW Industries Inc., La Canada, CA.

Wolfstein M (1970) Some solutions of plane turbulent impinging jets, *ASME Journal of Basic Engineering*, **92**, 915–922.

Yuge T (1960) Experiments on heat transfer from spheres including combined natural and forced convection. *ASME Journal of Heat Transfer*, **82**, 214–220.

Zienkiewicz OC and Codina R (1995) A general algorithm for compressible and incompressible flow, Part I, The split characteristic based scheme. *International Journal for Numerical Methods in Fluids*, **20**, 869–885.

Zienkiewicz OC, Nithiarasu P, Codina R, *et al.* (1999) An efficient and accurate algorithm for fluid mechanics problems. The characteristic based split (CBS) algorithm. *International Journal for Numerical Methods in Fluids*, **31**, 359–392.

Zienkiewicz OC, Sai BVKS, Morgan K and Codina R (1996) Split characteristic based semi-implicit algorithm for laminar/turbulent incompressible flows. *International Journal for Numerical Methods in Fluids*, **23**, 1–23.

Zienkiewicz OC, Taylor RL and Nithiarasu P (2005) *The Finite Element Method, Vol. 3, Fluid Dynamics*. Elsevier, Amsterdam.

Zienkiewicz OC, Taylor RL and Nithiarasu P (2013) *The Finite Element Method. Vol. 3. Fluid Dynamics*, 7th Edition, Elsevier, Amsterdam.

8

Turbulent Flow and Heat Transfer

8.1 Introduction

Turbulent heat convection is defined as the flow and heat transfer, with random variation, of various convection heat transfer quantities such as velocity, pressure and temperature. It is important to note that turbulence is a property of the flow, not a property of the fluid. Despite considerable progress on the topic of turbulence modeling over the last century, one has to admit that molecular turbulence is still an unresolved problem and will remain so for several more years. In this chapter, we will provide an overview of the numerical solution of turbulent fluid dynamics and heat transfer equations, based upon existing turbulence models and the CBS scheme that was discussed in the previous chapter.

Before going into such details, we have summarized some important fundamental properties of turbulence in the following paragraphs. At this stage, it is worth noting that a truly turbulent flow is three-dimensional in nature and occurs at relatively high Reynolds numbers. Thus, any two-dimensional study should only be treated as an approximation. A turbulent flow is marked with random variation of quantities as shown in Figure 8.1.

The Navier-Stokes equations are sufficient to resolve all turbulent scales, if an adequate mesh resolution is used. However, this requires extremely large computer resources. With present day computers this is possible, within a reasonable amount of time, only for simple problems at low and moderate Reynolds numbers. Until sufficiently fast computing power is available, it is essential to employ Reynolds decomposition and turbulence models.

In a real turbulent flow scenarios, the kinetic energy is transferred from larger scales to smaller scales. At the smallest scale, the kinetic energy is transformed into internal energy and this process is called "dissipation" and the process of energy transfer between the scales is called "the cascade process." The smallest turbulent length scale is determined by the molecular

Fundamentals of the Finite Element Method for Heat and Mass Transfer, Second Edition.
P. Nithiarasu, R. W. Lewis, and K. N. Seetharamu.
© 2016 John Wiley & Sons, Ltd. Published 2016 by John Wiley & Sons, Ltd.

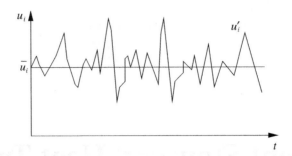

Figure 8.1   Random variation of velocity in a turbulent flow with respect to time.

viscosity and dissipation rate. Such a length scale is often referred to as the *Kolmogorov length scale* (Kolmogorov 1941) and is given as

$$\left(\frac{v^3}{\epsilon}\right)^{1/4}. \tag{8.1}$$

Similarly the Kolmogorov velocity and time scales are determined as

$$v = (v\epsilon)^{1/4} \tag{8.2}$$

and

$$\tau = \left(\frac{v}{\epsilon}\right)^{1/2} \tag{8.3}$$

respectively. The dissipation rate occurring at small scales can be linked to the energy of large eddies as

$$\epsilon = \frac{U^3}{l}, \tag{8.4}$$

where U is the large eddy velocity scale and l is the large eddy length scale. The turbulent kinetic energy of a flow is defined as

$$\kappa = \frac{1}{2}\overline{u_i' u_i'}, \tag{8.5}$$

where u' is the fluctuating component of the velocity as shown in Figure 8.1. The above relations are given in order to make the readers aware that the length scales, turbulent kinetic energy and dissipation are closely related (see also Section 8.1.2). The turbulence modeling procedures are developed based on these relationships.

8.1.1   Time Averaging

As mentioned previously, an extremely high mesh resolution is required to solve problems with the smallest of turbulence scales. This is very expensive and presently not possible for high Reynolds number flows and complex geometries within a reasonable period of time. It is, therefore, obvious that other alternatives are necessary to obtain an approximate solution.

The standard procedure is to employ time-averaged Navier-Stokes equations, along with a turbulence modeling approach, to determine the essential time averaged quantities, which reduces the excessive grid resolution otherwise needed. With reference to Figure 8.1, any turbulence quantity of interest may be expressed as

$$u = \bar{u} + u'. \tag{8.6}$$

The time averaged quantity may be obtained as

$$\bar{u} = \frac{1}{2T} \int_{-T}^{T} u(t)dt. \tag{8.7}$$

Let us consider the following one-dimensional steady-state incompressible momentum equation to demonstrate the concept of time averaging

$$\frac{du^2}{dx} + \frac{1}{\rho}\frac{dp}{dx} - \frac{d}{dx}\left(v\frac{du}{dx}\right) = 0 \tag{8.8}$$

Substituting a variation of the form of Equation (8.6) for the velocity u and pressure p into Equation (8.8), and time averaging, we obtain

$$\frac{d}{dx}\left[\overline{(\bar{u} + u')(\bar{u} + u')}\right] + \frac{1}{\rho}\frac{d}{dx}\overline{(\bar{p} + p')} - \frac{d}{dx}\left[v\frac{d}{dx}\overline{(\bar{u} + u')}\right] = 0, \tag{8.9}$$

where the overline indicates time averaging. In the above equation, the average of the fluctuating components u' and p' are equal to zero. Hence, the following simplified form of the above equation may be written as

$$\frac{d}{dx}\left[(\bar{u}^2 + \overline{u'^2})\right] + \frac{1}{\rho}\frac{d\bar{p}}{dx} - \frac{d}{dx}\left[v\frac{d}{dx}(\bar{u})\right] = 0. \tag{8.10}$$

Rearranging and rewriting the above momentum equation for incompressible flows in multi-dimensions and including the time term, we have

$$\rho\frac{\partial\bar{u}_i}{\partial t} + \rho\frac{\partial}{\partial x_j}(\bar{u}_j\bar{u}_i) = -\frac{\partial\bar{p}}{\partial x_i} + \frac{\partial\bar{\tau}_{ij}}{\partial x_j} - \frac{\partial}{\partial x_j}(\rho\overline{u_i'u_i'}), \tag{8.11}$$

where

$$\bar{\tau}_{ij} = \mu\left(\frac{\partial\bar{u}_i}{\partial x_j} + \frac{\partial\bar{u}_j}{\partial x_i}\right), \tag{8.12}$$

is the time-averaged deviatoric stress and $\rho\overline{u_iu_i}$ is a new unknown referred to as *Reynolds stress*. The *Boussinesq assumption* gives the Reynolds stress as

$$\bar{\tau}_{ij}^{R} = -\rho\overline{u_i'u_i'} = \mu_T\left(\frac{\partial\bar{u}_i}{\partial x_j} + \frac{\partial\bar{u}_j}{\partial x_i}\right). \tag{8.13}$$

From Equation (8.13), it is clear that the additional unknown quantity to be modeled is the turbulent kinematic viscosity μ_T (or $v_T = \mu_T/\rho$). In a similar fashion, the energy equation may also be time averaged by writing the temperature in the form given in Equation (8.6), i.e., $T = \bar{T} + T'$. Following the procedure described above, the time-averaged energy equation may be written as

$$\frac{\partial \bar{T}}{\partial t} + \frac{\partial}{\partial x_j}\left(\bar{u}_j \bar{T}\right) = \frac{\partial}{\partial x_j}\left(\alpha \frac{\partial \bar{T}}{\partial x_j} - \overline{u'_j T'}\right), \tag{8.14}$$

where α is the thermal diffusivity. The turbulent heat diffusion $\overline{u'_j T'}$ may be approximated as

$$\overline{u'_j T'} = -\alpha_T \frac{\partial \bar{T}}{\partial x_j}, \tag{8.15}$$

where α_T is the turbulent thermal diffusivity. For a two-dimensional problem, with coordinates x_1 and x_2, the equations may be summarized as

Continuity equation:

$$\frac{\partial \bar{u}_1}{\partial x_1} + \frac{\partial \bar{u}_2}{\partial x_2} = 0. \tag{8.16}$$

Momentum equation components:

$$\left(\frac{\partial \bar{u}_1}{\partial t} + \bar{u}_1 \frac{\partial \bar{u}_1}{\partial x_1} + \bar{u}_2 \frac{\partial \bar{u}_1}{\partial x_2}\right) = -\frac{1}{\rho}\frac{\partial \bar{p}}{\partial x_1} + (\nu + \nu_T)\left(\frac{\partial^2 \bar{u}_1}{\partial x_1^2} + \frac{\partial^2 \bar{u}_1}{\partial x_2^2}\right) \tag{8.17}$$

and

$$\left(\frac{\partial \bar{u}_2}{\partial t} + \bar{u}_1 \frac{\partial \bar{u}_2}{\partial x_1} + \bar{u}_2 \frac{\partial \bar{u}_2}{\partial x_2}\right) = -\frac{1}{\rho}\frac{\partial \bar{p}}{\partial x_2} + (\nu + \nu_T)\left(\frac{\partial^2 \bar{u}_2}{\partial x_1^2} + \frac{\partial^2 \bar{u}_2}{\partial x_2^2}\right). \tag{8.18}$$

Energy equation:

$$\frac{\partial \bar{T}}{\partial t} + \bar{u}_1 \frac{\partial \bar{T}}{\partial x_1} + \bar{u}_2 \frac{\partial \bar{T}}{\partial x_2} = \left[\alpha + \frac{\nu_T}{Pr_T}\right]\left(\frac{\partial^2 \bar{T}}{\partial x_1^2} + \frac{\partial^2 \bar{T}}{\partial x_2^2}\right). \tag{8.19}$$

To solve the above set of incompressible Navier-Stokes equations, we now need some additional information to compute the turbulent eddy viscosity.

8.1.2 Relationship between κ, ϵ, ν_T and α_T

The turbulent kinematic viscosity, or turbulent eddy viscosity ν_T, has the same dimensions as the physical kinematic viscosity. Thus, we can express the turbulent eddy viscosity in terms of the velocity and length scales of a large eddy, that is,

$$\nu_T = c_\mu U l, \tag{8.20}$$

where c_μ is a constant. The definitions of U and l are discussed in Section 8.1. In the above equation, U may be replaced with $\sqrt{\kappa}$. With such a substitution, the turbulent eddy viscosity may be determined by solving a scalar transport equation for κ and assuming an appropriate turbulent length scale l ($\kappa - l$ or "one-equation" models). However, a better expression for the turbulent eddy viscosity may be obtained by substituting Equation (8.4) into Equation (8.20) as

$$v_T = c_\mu \frac{\kappa^2}{\epsilon}. \tag{8.21}$$

To employ the above equation, we need to solve two transport equations, one for κ and another for ϵ ($\kappa - \epsilon$ or the "two-equation" models). Details of some one- and two-equation models are provided in the following sections (Cebeci and Smith 1974; Launder and Spalding 1972; Wilcox 1992). In addition to discussions on the Reynolds Averaged Navier-Stokes (RANS) models, we also provide a brief summary of Large Eddy Simulation (LES) and currently popular topics such as Detached Eddy Simulation (DES) and Monotonically Integrated LES (MILES) approaches. Before introducing the RANS and other approaches, the relationship between the eddy viscosity, v_T, and eddy diffusivity, α_T, are discussed here. The ratio between the physical kinematic viscosity and the thermal diffusivity is referred to as the Prandtl number ($Pr = v/\alpha$). In a similar fashion, the turbulent Prandtl number, Pr_T may be defined as

$$Pr_T = \frac{v_T}{\alpha_T} \tag{8.22}$$

or $\alpha_T = v_T/Pr_T$. The turbulent Prandtl number normally varies between values of 0.7 and 0.9.

8.2   Treatment of Turbulent Flows

8.2.1   Reynolds Averaged Navier-Stokes (RANS)

The equations derived in the previous section are summarized here for a general multi-dimensional problem. The Reynolds averaged Navier-Stokes equations (RANS) of motion for incompressible flows may be summarized as

Mean-continuity:

$$\frac{\partial \bar{u}_i}{\partial x_i} = 0. \tag{8.23}$$

Mean-momentum:

$$\rho \left(\frac{\partial \bar{u}_i}{\partial t} + \bar{u}_j \frac{\partial \bar{u}_i}{\partial x_j} \right) = -\frac{\partial \bar{p}}{\partial x_i} + \frac{\partial \bar{\tau}_{ij}}{\partial x_j} - \frac{\partial}{\partial x_j} (\rho \overline{u_i' u_i'}), \tag{8.24}$$

where $\bar{u}_i$ are the mean velocity components, p is the pressure, ρ is the density, and $\bar{\tau}_{ij}$ is the laminar shear stress tensor given by Equation (8.12). The Reynolds stress tensor, $\bar{\tau}_{ij}^R$, is given in Equation (8.13). As mentioned before, the extra variable to be modeled is the turbulent eddy viscosity μ_T or $\nu_T = \mu_T/\rho$. The energy equation, in terms of the turbulent Prandtl number and eddy viscosity, may be written as

Energy:

$$\frac{\partial \bar{T}}{\partial t} + \bar{u}_j \frac{\partial \bar{T}}{\partial x_j} = \frac{\partial}{\partial x_j}\left(\alpha \frac{\partial \bar{T}}{\partial x_j} + \frac{\nu_T}{Pr_T}\frac{\partial \bar{T}}{\partial x_j}\right). \tag{8.25}$$

If the turbulent Prandtl number, Pr_T, is known, the additional variable to be determined is again the turbulent eddy viscosity. In the following section we outline some very standard models available to calculate ν_T.

8.2.2   One-equation Models

8.2.2.1   Wolfstein $\kappa - l$ Model (Wolfstein 1970)

In this model the turbulent eddy viscosity is determined from a mixing length and turbulent kinetic energy as

$$\nu_T = c_\mu^{1/4}\kappa^{1/2}l_m, \tag{8.26}$$

where c_μ is a constant equal to 0.09, κ is the turbulent kinetic energy and l_m is a mixing length. The mixing length l_m is related to the length scale of the turbulence L as

$$l_m = \left(\frac{c_\mu'^3}{C_D}\right)^{1/4} L, \tag{8.27}$$

where C_D and c_μ' are constants. The transport equation for turbulent kinetic energy κ is

$$\frac{\partial \kappa}{\partial t} + \bar{u}_i \frac{\partial \kappa}{\partial x_i} = \left(\nu + \frac{\nu_T}{\sigma_\kappa}\right)\frac{\partial^2 \kappa}{\partial x_i^2} + \tau_{ij}^R\frac{\partial \bar{u}_i}{\partial x_j} - \varepsilon, \tag{8.28}$$

where σ_κ is the diffusion Prandtl number for turbulent kinetic energy. The dissipation, ε, is modeled as

$$\varepsilon = C_D\frac{\kappa^{3/2}}{L}. \tag{8.29}$$

Near solid walls, the Reynolds number tends to zero and the highest mean velocity gradient occurs at the solid boundary. Thus, the one-equation model has to be used in conjunction with empirical wall functions, that is, ν_T is multiplied by the damping function $f_\mu = 1 - e^{-0.160R_\kappa}$ and ε is divided by $f_b = 1 - e^{-0.263R_\kappa}$, where $R_\kappa = \sqrt{\kappa}y/\nu$, where y is the shortest distance to the nearest wall. The constants are $\sigma_k = 1$ and $C_D = 1.0$.

8.2.2.2   Spalart-Allmaras (SA) Model (Spalart and Allmaras 1992)

The Spalart-Allmaras (SA) model was first introduced for aerospace applications and is now adopted for incompressible flow calculations as well. This is another one equation model, which employs a single scalar equation and several constants to model turbulence. Here we provide the model without a trip point or curve. The model, which includes a trip, may be found in reference (Spalart and Allmaras 1992). The scalar transport equation used by this model is

$$\frac{\partial \hat{v}}{\partial t} + \bar{u}_j \frac{\partial \hat{v}}{\partial x_j} = c_{b1} \hat{S} \hat{v} + \frac{1}{\sigma} \left[(v + \hat{v}) \frac{\partial^2 \hat{v}}{\partial x_i^2} + c_{b2} \left(\frac{\partial \hat{v}}{\partial x_i} \right)^2 \right] - c_{w1} f_w \left[\frac{\hat{v}}{y} \right]^2, \tag{8.30}$$

where

$$\hat{S} = S + (\hat{v}/k^2 y^2) f_{v2} \tag{8.31}$$

and

$$f_{v2} = 1 - X/(1 + X f_{v1}). \tag{8.32}$$

In Equation (8.31), S is the magnitude of the vorticity and y is the shortest distance from a node to the nearest solid wall. The eddy viscosity is calculated as

$$v_T = \hat{v} f_{v1}, \tag{8.33}$$

where

$$f_{v1} = X^3 / \left(X^3 + c_{v1}^3 \right) \tag{8.34}$$
$$X = \hat{v}/v. \tag{8.35}$$

The parameter f_w is given as

$$f_w = g \left[\frac{1 + c_{w3}^6}{g^6 + c_{w3}^3} \right]^{1/6}, \tag{8.36}$$

where

$$g = r + c_{w2}(r^6 - r) \tag{8.37}$$
$$r = \frac{\hat{v}}{\hat{S} k^2 y^2}. \tag{8.38}$$

The constants are $c_{b1} = 0.1355, \sigma = 2/3, c_{b2} = 0.622, k = 0.41, c_{w1} = c_{b1}/k^2 + (1 + c_{b2})/\sigma, c_{w2} = 0.3, c_{w3} = 2$ and $c_{v1} = 7.1$.

8.2.3 Two-equation Models

8.2.3.1 The Standard $\kappa - \varepsilon$ Model

In this model, the transport equation for κ is the same as that in the one-equation model of Section 8.2.2. The second transport equation for calculating the turbulence energy dissipation rate ε is

$$\frac{\partial \varepsilon}{\partial t} + \bar{u}_i \frac{\partial \varepsilon}{\partial x_i} = \left(v + \frac{v_T}{\sigma_\varepsilon} \right) \frac{\partial^2 \varepsilon}{\partial x_i^2} + C_{\varepsilon 1} \frac{\varepsilon}{\kappa} \bar{\tau}_{ij}^R \frac{\partial \bar{u}_i}{\partial x_j} - C_{\varepsilon 2} \frac{\varepsilon^2}{\kappa}, \tag{8.39}$$

where $C_{\varepsilon 1} = 1.44$, $C_{\varepsilon 2} = 1.92$ and σ_ε is the diffusion Prandtl number for an isotropic turbulence energy dissipation rate and is equal to 1.3. These constants are proposed by Jones and Launder (1972).

In addition, ν_T is evaluated by

$$\nu_T = c_\mu \frac{\kappa^2}{\varepsilon}. \tag{8.40}$$

For near-wall treatments, modifications to the source terms of the ε equation are needed in the near-wall region. Multiplying the coefficients c_μ, $C_{\varepsilon 1}$ and $C_{\varepsilon 2}$, by the turbulence damping functions f_μ, $f_{\varepsilon 1}$ and $f_{\varepsilon 2}$ then an appropriate low Reynolds number status is achieved near the walls. Numerous wall damping functions have been proposed. The values suggested by Lam and Bremhorst (1981) for steady flows are:

$$f_\mu = (1 - e^{-0.0165R_\kappa})^2 \left(1 + \frac{20.5}{R_t} \right) \tag{8.41}$$

$$f_{\varepsilon 1} = 1 + \left(\frac{0.05}{f_\mu} \right)^3 \tag{8.42}$$

and

$$f_{\varepsilon 2} = 1 - e^{-R_t^2}, \tag{8.43}$$

where $R_t = \kappa^2 / \nu\varepsilon$. The damping functions of Fan *et al.* (1993) are

$$f_\mu = 0.4 \frac{f_w}{\sqrt{R_t}} \left(1 - 0.4 \frac{f_w}{\sqrt{R_t}} \right) \left[1 - exp\left(-\frac{R_y}{42.63} \right) \right]^3, \tag{8.44}$$

where

$$f_w = 1 - exp\left\{ -\frac{\sqrt{R_y}}{2.30} + \left(\frac{\sqrt{R_y}}{2.30} - \frac{R_y}{8.89} \right) \left[1 - exp\left(-\frac{R_y}{20} \right) \right]^3 \right\} \tag{8.45}$$

$$f_{\varepsilon 2} = \left\{ 1 - \frac{0.4}{0.8} exp\left[-\left(\frac{R_t}{6} \right)^2 \right] \right\} f_w^2 \tag{8.46}$$

and $f_{\varepsilon 1} = 1$. Note that $R_y = R_t$ (see Section 8.2.2.1). The constants are $c_\mu = 0.09$, $\sigma_k = 1.0$, $\sigma_\varepsilon = 1.3$, $C_{\varepsilon 1} = 1.4$ and $C_{\varepsilon 2} = 1.8$.

8.2.4   Nondimensional Form of the Governing Equations

A turbulent flow solution can be obtained by solving Equations (8.23) and (8.24) with appropriate boundary conditions along with one of the turbulence models. The following nondimensional scales may be used in the calculations

$$\bar{u}_i^* = \frac{\bar{u}_i}{u_\infty}; \quad x_i^* = \frac{x_i}{D}; \quad \bar{p}^* = \frac{\bar{p}}{\rho_\infty u_\infty^2}; \quad t^* = \frac{tu_\infty}{D}$$

$$\kappa^* = \frac{\kappa}{u_\infty^2}; \quad \varepsilon^* = \frac{\varepsilon D}{u_\infty^3}; \quad \nu_T^* = \frac{\nu_T}{\nu_\infty}; \quad \hat{\nu}^* = \frac{\hat{\nu}}{\nu_\infty}, \tag{8.47}$$

where D is a characteristic dimension and the subscript ∞ indicates a reference value. Substituting the nondimensional scales into Equations (8.24) to (8.25) and dropping the asterisks leads to

$$\frac{\partial \bar{u}_i}{\partial x_i} = 0 \tag{8.48}$$

and

$$\frac{\partial \bar{u}_i}{\partial t} + \bar{u}_j \frac{\partial \bar{u}_i}{\partial x_j} = -\frac{\partial \bar{p}}{\partial x_i} + \frac{\partial \left(\bar{\tau}_{ij} + \bar{\tau}_{ij}^R \right)}{\partial x_j} \tag{8.49}$$

with

$$\bar{\tau}_{ij} + \bar{\tau}_{ij}^R = \frac{(1.0 + \nu_T)}{Re} \left(\frac{\partial \bar{u}_i}{\partial x_j} + \frac{\partial \bar{u}_j}{\partial x_i} \right). \tag{8.50}$$

The nondimensional density in Equation (8.50) is unity for incompressible flow problems and the nondimensional form of the energy equation is

$$\frac{\partial \bar{T}}{\partial t} + \bar{u}_j \frac{\partial \bar{T}}{\partial x_j} = \frac{1}{Pe} \frac{\partial}{\partial x_j} \left(\frac{\partial \bar{T}}{\partial x_j} + \frac{\nu_T}{Pr_T} \frac{\partial \bar{T}}{\partial x_j} \right), \tag{8.51}$$

where Pe is the thermal Peclet number ($Pe = RePr$). The Reynolds number, Re, in the above equations is defined as

$$Re = \frac{\bar{u}_\infty D}{\nu_\infty} \tag{8.52}$$

and the Prandtl number is

$$Pr = \frac{\nu}{\alpha}. \tag{8.53}$$

Note that the viscosity, ν, is assumed to be constant and equal to ν_∞ in the above equations. The thermal diffusivity α is also assumed to be constant and equal to α_∞. The nondimensional form of the turbulence transport equations is given below.

8.2.4.1   One-equation Model

The nondimensional form of the κ equation is

$$\frac{\partial \kappa}{\partial t} + \bar{u}_i \frac{\partial \kappa}{\partial x_i} = \frac{1}{Re} \left(1 + \frac{\nu_T}{\sigma_\kappa} \right) \frac{\partial^2 \kappa}{\partial x_i^2} + \tau_{ij}^R \frac{\partial u_i}{\partial x_j} - \varepsilon. \tag{8.54}$$

The mixing length and the turbulence length scales are normalized using the characteristic dimension D. Here we assume $L = D$. The nondimensional form of R_k is $\sqrt{\kappa} y Re$.

8.2.4.2   Spalart-Allmaras Model

The nondimensional form of the transport equation is

$$\frac{\partial \hat{v}}{\partial t} + \bar{u}_j \frac{\partial \hat{v}}{\partial x_j} = c_{b1} \hat{S} \hat{v} + \frac{1}{Re\sigma} \left[(1 + \hat{v}) \frac{\partial^2 \hat{v}}{\partial x_i^2} + c_{b2} \left(\frac{\partial \hat{v}}{\partial x_i} \right)^2 \right] - \frac{c_{w1} f_w}{Re} \left[\frac{\hat{v}}{y} \right]^2, \tag{8.55}$$

where

$$\hat{S} = S + \frac{1}{Re}(\hat{v}/k^2 y^2) f_{v2} \tag{8.56}$$

and Eq. 8.38 becomes

$$r = \frac{1}{Re}\frac{\hat{v}}{\hat{S}k^2 y^2}. \tag{8.57}$$

The structures of all the remaining parameters are unchanged.

8.2.4.3 The $\kappa - \varepsilon$ Model

The κ equation is identical to that of the one-equation model and the dissipation equation is given as

$$\frac{\partial \varepsilon}{\partial t} + \bar{u}_i \frac{\partial \varepsilon}{\partial x_i} = \frac{1}{Re}\left(1 + \frac{v_T}{\sigma_\varepsilon}\right)\frac{\partial^2 \varepsilon}{\partial x_i^2} + C_{\varepsilon 1}\frac{\varepsilon}{\kappa}\tau_{ij}^R\frac{\partial \bar{u}_i}{\partial x_j} - C_{\varepsilon 2}\frac{\varepsilon^2}{\kappa}. \tag{8.58}$$

The parameter R_t in its nondimensional form is $\kappa^2 Re/\varepsilon$.

8.3 Solution Procedure

The solution procedure follows the steps of the CBS scheme as discussed in the previous chapter (Section 7.6, Equations (7.165)–(7.170)). If isothermal flow is of interest, then the temperature equation is ignored and a solution to the turbulence model equation becomes the fourth step. For nonisothermal problems, the temperature equation is solved at Step 4 (Equation (7.170)) and the turbulence model equation is solved at Step 5. At each and every time step, the turbulent eddy viscosity is calculated and substituted into the averaged momentum and energy equations. For the sake of completeness, the characteristic procedure for theSpalart-Allmaras model is presented here.

The temporal discretization of the Spalart-Allmaras (SA) model follows the characteristic Galerkin discretization of the convection-diffusion equation as discussed in the previous chapter (Section 7.4.1), that is,

$$\hat{v}^{n+1} = \hat{v}^n + \Delta t \left[-\bar{u}_j \frac{\partial \hat{v}}{\partial x_j} + \frac{1}{\sigma_{\hat{v}} Re}(1 + \hat{v})\frac{\partial^2 \hat{v}}{\partial x_j^2} + \frac{c_{b2}}{\sigma_{\hat{v}} Re}\left(\frac{\partial \hat{v}}{\partial x_j}\right)^2 - \frac{c_{w1}f_w}{Re}\left(\frac{\hat{v}}{y}\right)^2 \right]^n$$

$$+ \Delta t \left\{ c_{b1}\hat{S}\hat{v} + \frac{\Delta t}{2}\bar{u}_i \frac{\partial}{\partial x_i}\left[\bar{u}_j \frac{\partial \hat{v}}{\partial x_j} - \frac{1}{\sigma_{\hat{v}} Re}\frac{\partial}{\partial x_j}(1 + \hat{v})\frac{\partial \hat{v}}{\partial x_j} \right] \right\}^n \tag{8.59}$$

$$+ \frac{(\Delta t)^2}{2}\left\{ \bar{u}_i \frac{\partial}{\partial x_i}\left[-\frac{c_{b2}}{\sigma_{\hat{v}} Re}\left(\frac{\partial \hat{v}}{\partial x_j}\right)^2 + \frac{c_{w1}f_w}{Re}\left(\frac{\hat{v}}{y}\right)^2 - c_{b1}\hat{S}\hat{v} \right] \right\}^n.$$

Spatial discretization of the individual terms in the above equation is identical to that of the convection-diffusion equation as discussed in Chapter 7.

8.4   Forced Convective Flow and Heat Transfer

This section provides some examples of forced flow and heat transfer in passages and also flow over objects. The first example is the simple case of isothermal turbulent flow in a two-dimensional rectangular channel. This problem is then extended to include heat transfer in the second example. Other problems studied in this section include the backward facing step and external flow problem of flow past a circular cylinder.

Example 8.4.1   *Isothermal flow in a rectangular channel*
   Flow through a rectangular channel is often treated as a benchmark problem for turbulent flows due to its simplicity. A number of isothermal experiments have been conducted on channel flows and published by Laufer (1951). These experimental data may be used to check the turbulent flow models. We have considered here a simple, moderate Reynolds number flow problem through a rectangular channel. The channel is assumed to be 2 units wide and 40 units long. The Reynolds number is defined based on the half width of the channel. Several structured meshes were tried and the results presented here are generated on a nonuniform structured mesh with 6400 nodes and 12 446 elements. The first node from the wall is placed at a nondimensional distance of 0.005. The results may be further improved by increasing the resolution of the mesh.

   A nondimensional horizontal velocity component of unity and vertical velocity component of zero are assumed at the inlet. No slip conditions are applied on both walls of the channel. For the one equation $\kappa - l$ turbulence model, a fixed value of nondimensional $\kappa = 0.05$ is assumed at the inlet. On the walls, zero value is assumed for the turbulent kinetic energy. For the SA model, the nondimensional scalar variable $\hat{v}$ is prescribed equal to 0.05 at the inlet and zero on the walls. The boundary conditions for the two equation turbulence models are: inlet values of both κ and ε are prescribed ($\kappa = 0.05$ and $\varepsilon = 0.05$) based on the idea proposed in Johansson et al., (1993). On the walls, $\kappa = 0$ and $\varepsilon = (2/Re)(d\kappa^{1/2}/dy)^2$ are prescribed as proposed in Yang and Shih (1993).

   Figure 8.2 shows the comparison of fully developed velocity profiles obtained from all the three turbulence models with the experimental data of Laufer (1951). A rapid convergence to steady state was obtained using both one-equation and SA models. However, the two equation model has difficulties in to converging and took a longer time. All the methods seem to give solutions close to the experimental data. It may be possible to further improve the solution by fine tuning the boundary conditions of the turbulence variables and damping functions.

   Often, engineering problems are quite complex and thus domain discretization is easy to carry out if unstructured meshes are used. However, unstructured meshes in general are less accurate than structured meshes. In order to estimate the accuracy of using an unstructured mesh, we carried out a comparison between structured and unstructured mesh results of flow through a rectangular channel. In general, agreement with the experimental average velocity profile is close if the unstructured element size is comparable to the element size of a structured mesh near the solid walls (Nithiarasu et al., 2007). In order to further estimate the accuracy in terms of friction velocity or shear velocity, a comparison between the structured and unstructured mesh results for flow through a rectangular channel is shown in Figure 8.3.

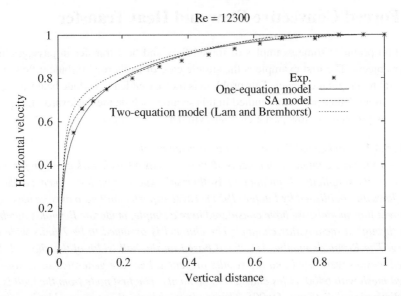

Figure 8.2 Turbulent flow in a two-dimensional rectangular channel. Comparison of fully developed velocity profiles at Re = 12 300.

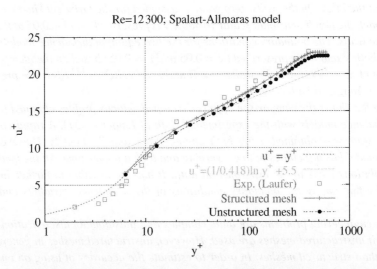

Figure 8.3 Turbulent incompressible flow in a rectangular channel using the matrix free CBS-AC scheme with the Spalart-Allmaras model at Re=12 300. Logarithmic representation of time-averaged velocity profile.

The friction velocity is defined in dimensional form as

$$u_\tau = \sqrt{\frac{\tau_w}{\rho}},$$
(8.60)

where τ_w is the local wall shear stress and ρ is the density. In nondimensional form, this may be written as

$$u_\tau^* = \sqrt{\frac{1}{Re}\frac{\partial \bar{u}_t^*}{\partial n^*}}$$
(8.61)

where the superscript $*$ indicates a nondimensional quantity, subscript t indicates the tangential direction and n is the normal direction. For more details on the calculation of wall shear stress, readers are referred to Chapter 7. The friction velocity is often normalized against the average velocity as follows:

$$u^+ = \frac{u_\tau}{\bar{u}},$$
(8.62)

where both u_τ and $\bar{u}$ can be used either in a dimensional or nondimensional form to obtain the above ratio, as long as the same scales are employed in nondimensionalising the velocities. Equation (8.62) may be plotted against a nondimensional normal distance from the wall. The non-dimensional distance is generally calculated as

$$y^+ = \frac{yu_\tau}{\nu}$$
(8.63)

or, in nondimensional form

$$y^+ = Rey^*u_\tau^*.$$
(8.64)

Figure 8.3 shows the variation of u^+ against the logarithmic nondimensional distance y^1. This figure shows the experimental data, structured and unstructured mesh results for the channel flow problem. As seen, only the SA model is used in the flow calculations. The first node of the structured mesh was placed at a distance of 0.005 and the unstructured mesh was placed roughly around 0.01. As seen, the logarithmic representation of the time averaged velocity variation is close to the experimental data of Laufer (1951). The differences between the structured and unstructured meshes are attributed to the larger element size of the unstructured mesh used close to the wall.

Example 8.4.2   *Nonisothermal flow in a rectangular channel*

An extension of the channel flow to study the effect of constant wall temperature conditions is straightforward. The solid walls of the channel may be assumed to be at a constant higher temperature than that of the inlet fluid temperature. We leave this study of developing temperature profile as an exercise for the readers. We provide here, a channel flow problem with wall heat flux boundary conditions.

The problem definition is similar to the isothermal problem discussed in the previous section. The differences here are that the Reynolds number is defined based on the full height of the channel and the bottom wall of the channel is subjected to a constant heat flux boundary condition entering the channel. All other flow boundary conditions are identical to the one discussed in the previous section, except that the fluid approaching the channel inlet is at a reference atmospheric temperature. The top wall is assumed to be adiabatic and zero heat flux is assumed at the exit. Since the boundary condition applied is a flux condition, the *nondimensional scaling used for temperature in this problem is redefined as*

$$T^* = \frac{T - T_\infty}{\frac{\bar{q}L}{k}}, \qquad (8.65)$$

where T_∞ is the reference temperature of the fluid approaching the inlet of the channel, $\bar{q}$ is a reference constant heat flux applied, L is the height of the channel and k is the thermal conductivity of the fluid. The nondimensional heat flux applied is $q^* = q/\bar{q}$ assumed to be unity. This nondimensional scale will result in a local Nusselt number relation of (refer to Chapter 7, Section 7.4.1)

$$Nu = \frac{1}{T^*}. \qquad (8.66)$$

Figure 8.4 shows the flat turbulent velocity profile and other variable distributions close the exit of the channel. To obtain a fully developed flow, the length of the channel needs to be much longer than the one used in this study. To avoid very long domains, periodic boundary conditions, in which the outlet solution is continuously fed into the inlet, are often preferred (Patankar 1980). Figure 8.5 shows the local Nusselt number distribution along the length of the channel. The Nusselt number is very large at the entrance of the channel due to very small wall temperature that is close to the reference temperature of approaching fluid, T_∞. As the distance increases, the Nusselt number value drops exponentially and reaches an almost constant valve, indicating that the temperature is approaching a fully developed value.

Example 8.4.3   *Isothermal flow over a backward-facing step*
    Another standard test case, commonly employed for testing turbulent incompressible flow models at moderate Reynolds numbers, is the recirculating flow past a backward facing

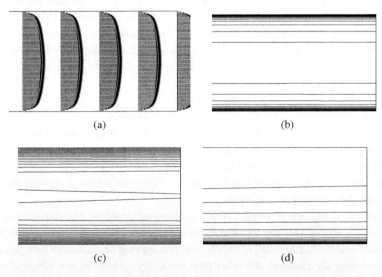

(a)   (b)

(c)   (d)

Figure 8.4   Forced convective turbulent flow and heat transfer in a rectangular channel. Horizontal velocity, SA model turbulent variable and temperature distribution, $Re = 5000$, $Pr = 0.71$. (a) Velocity vectors; (b) Horizontal velocity; (c) SA model variable; (d) Temperature.

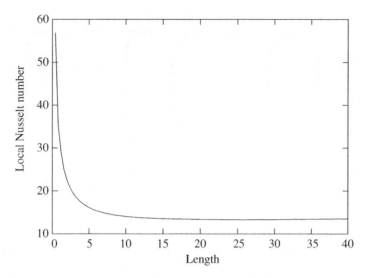

Figure 8.5   Forced convective turbulent flow and heat transfer in a rectangular channel. Local Nusselt number distribution along the hot wall, $Re = 5000$, $Pr = 0.71$.

step. Unlike the channel flow problem, the model here has to handle the recirculation region immediately downstream of the step. The definition of the problem is shown in Figure 8.6. The characteristic dimension of the problem is the step height. All other dimensions are defined with respect to the characteristic dimension. The inlet is located at a distance of 4 times the step height from the step. The inlet channel height is two times the step height. The total length of the channel is 40 times the step height.

The inlet velocity profile is obtained from experimental data reported by Denham et al. (1975). No slip conditions are applied at the solid walls. For the one-equation and two-equation models the inlet κ and ε profiles are obtained by solving a channel flow problem. For the SA model, a fixed value of 0.05 for the turbulent scalar variable at inlet was prescribed. On the walls, κ was assumed to be equal to zero. The wall conditions for ε are the same as in the previous problem. The scalar variable of the SA model was also assumed to be zero on the walls.

Both structured and unstructured meshes were employed in the calculation. Figure 8.7 shows the comparison of velocity profiles against the experimental data of Denham et al.

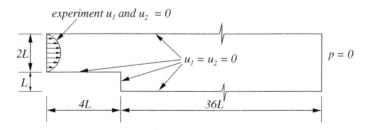

Figure 8.6   Turbulent flow past a two-dimensional backward facing step. Problem definition.

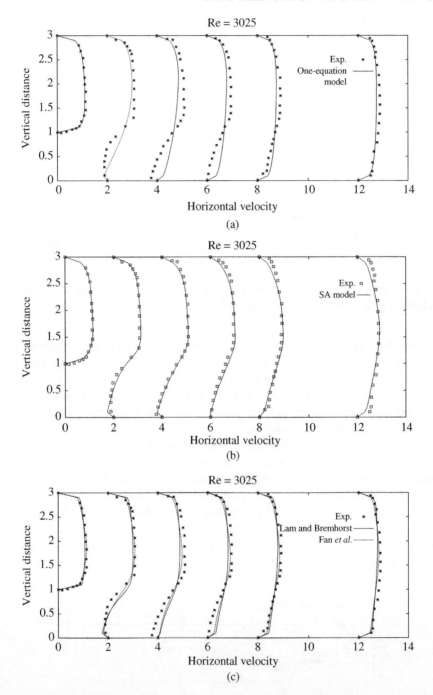

Figure 8.7   Incompressible turbulent flow past a backward-facing step. Velocity profiles at various downstream sections at Re = 3025 (a) One-equation model; (b) SA model; (c) Two-equation model.

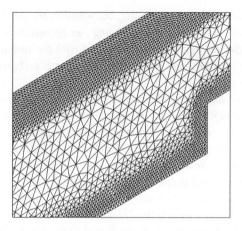

Figure 8.8 Turbulent incompressible flow past a backward-facing step. Unstructured mesh used. Number of elements: 297 054; Number of nodes: 65 372.

(1975). Among the different models, the SA-model seems to predict the recirculation more accurately. However, some differences between the experiment and the present SA model predictions are noticed along the top wall. Figures 8.8 and 8.9 show the backward facing problem demonstration in three dimensions. Figure 8.8 shows details of the unstructured mesh close to the recirculation zone. As seen, the mesh is refined along the walls to capture high gradients.

Example 8.4.4 *Unsteady RANS (URANS) flow calculation past a circular cylinder*

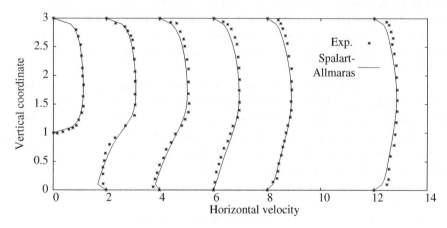

Figure 8.9 Turbulent incompressible flow past a backward facing step. Comparison of the velocity profile in the recirculation region, Re = 3025.

Although the Reynolds averaged equations, discussed in the previous sections, are time-averaged, these equations may be used in carrying out approximate transient calculations if the transient time scales are really large compared with the turbulence scales. One such problem is studied here. The problem considered is unsteady turbulent flow past a circular cylinder.

The Spalart-Allmaras model with the matrix free CBS-AC scheme is used here to solve three-dimensional turbulent flow past a stationary circular cylinder at a Reynolds number of 10 000. Both the top and bottom sides and inlet are placed at a distance of 4 times the diameter of the cylinder from the center of the cylinder. The exit of the domain is assumed at a distance of 10 times the diameter of the cylinder from the center of the cylinder. The length of the cylinder used is 0.5 times the diameter and the front and back surfaces of the domain are assumed to be symmetric (no flow). Uniform velocity conditions in the x_1 direction are assumed at the inlet. The size of the real time step was set at 0.05. The turbulent scalar variable (modified turbulent eddy kinematic viscosity) was assumed to be 10^{-8} at the inlet for the Spalart-Allmaras model. On the top and bottom sides, slip conditions were assumed and no turbulence quantity was prescribed. On the cylinder walls, no slip conditions were assumed and the turbulent scalar variable of the Spalart-Allmaras model was assumed to be zero.

The dual time-stepping method was employed (see Section 7.7) with the matrix free CBS-AC scheme. The local time step depends on each element size within every real time step. The two different meshes used to test the flow past a three-dimensional circular cylinder problem are shown in Figure 8.10. The fully unstructured mesh (mesh1) comprises 606 769 tetrahedral elements and 115 035 nodes. The hybrid mesh (mesh2) consists of three structured layers close to the cylinder surface and unstructured grid away from the wall. Figure 8.10(d) shows the mesh in the vicinity of the cylinder. Both meshes are refined close to the wall and in the wake region to predict the vortex shedding.

Figure 8.11 shows the time variation of the drag coefficient, lift coefficient and pressure coefficient using the unstructured and hybrid meshes. The average drag coefficient obtained is 1.311 from the unstructured mesh1. The Strouhal number is 0.152 ($St = fL/u_\infty$, f is the frequency of vortex shedding). The amplitude of lift coefficient is between 1 and −1. The averaged drag coefficient obtained by the hybrid mesh2 is 1.239, which is more accurate than the result of mesh1 in comparison with experimental data. The Strouhal number here is around 0.144.

In Figure 8.11(c), the pressure coefficient values at Re=10 000 are compared with two different turbulence procedures, one is the available LES modeling data (Lu et al., 1997) and another is the numerical data from non-linear eddy viscosity modeling (Lu et al., 2003a,b). As seen, the time-averaged pressure distribution on the hybrid mesh2 is in good agreement with LES and nonlinear models, except near the stagnation point. This may be attributed to turbulence modeling accuracy (Tutar and Hold, 2001).

In Figure 8.12 and Figure 8.13 the contours of horizontal velocity component, vertical velocity component, pressure and modified turbulent eddy kinematic viscosity obtained from mesh1 and mesh2 respectively are shown. Both results are almost identical. As seen, the origin of the vortex street shifts between the areas above and below the central axis. The behavior qualitatively confirms the periodic vortex shedding phenomenon.

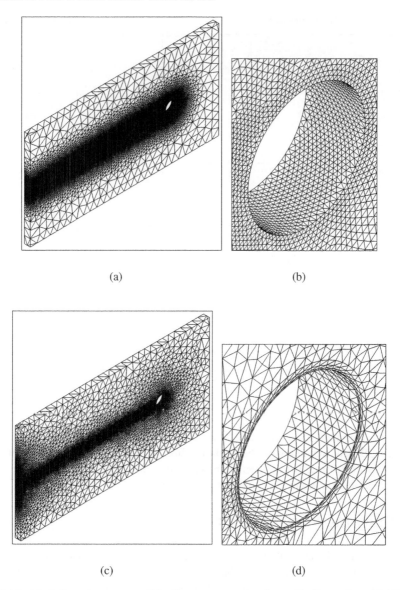

(a)   (b)

(c)   (d)

Figure 8.10 Turbulent incompressible flow over a circular cylinder at Re=10 000 using the matrix free CBS-AC scheme with the Spalart-Allmaras model: (a) unstructured mesh1 (Elements: 606 769, Nodes: 115 035); (b) unstructured mesh1 of close to solid wall (0.038 distance); (c) hybrid mesh2 (Elements: 489 463, Nodes: 88 964); (d) hybrid mesh2 of close to solid wall (0.01 distance).

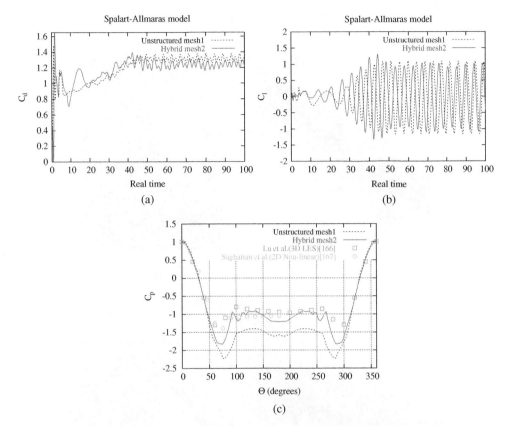

Figure 8.11   Turbulent incompressible flow over a circular cylinder at Re=10 000 using the matrix free CBS-AC scheme with the Spalart-Allmaras model: (a) drag coefficient variation with respect to real time; (b) lift coefficient variation with respect to real time; (c) pressure coefficient distribution along the cylinder surface at real time = 100.

8.5   Buoyancy-driven Flow

In this section, the Spalrat-Allmaras (SA) model is used to solve buoyancy-driven natural convection inside a square enclosure. The nondimensional form of the SA model for natural convection may be obtained by replacing Re in Equation (8.55) with $1/Pr$. The remaining equations are identical to those of laminar natural convection as discussed in Chapter 7, except that the momentum and energy equations should include additional turbulence terms to compute turbulent flows.

Example 8.5.1   *Natural convection in a square enclosure*

The problem definition is similar to the one discussed in Section 7.11.2. The difference here is that turbulence is also modeled. No slip conditions are enforced on all the four walls of the cavity. The left vertical wall of the cavity is assume to be at a higher temperature than that of the right vertical wall. Both the top and bottom walls are assumed to be insulated.

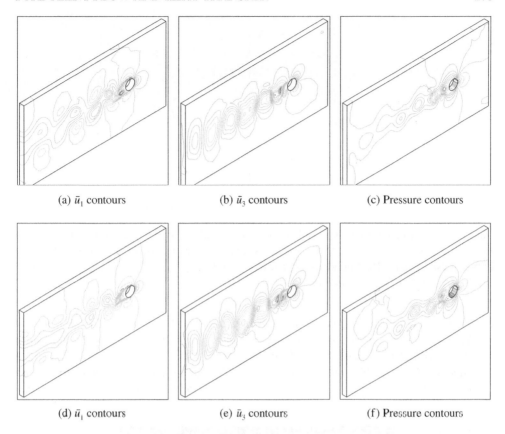

(a) $\bar{u}_1$ contours (b) $\bar{u}_3$ contours (c) Pressure contours

(d) $\bar{u}_1$ contours (e) $\bar{u}_3$ contours (f) Pressure contours

Figure 8.12 Turbulent incompressible flow over a circular cylinder at Re = 10 000 using the matrix free CBS-AC scheme with the Spalart-Allmaras model on unstructured mesh1 (up) and hybrid mesh2 (down). Real nondimensional time = 100. (a) $\bar{u}_{1_{min}} = -0.526$, $\bar{u}_{1_{max}} = 1.928$; (b) $\bar{u}_{3_{min}} = -1.223$, $\bar{u}_{3_{max}} = 1.437$; (c) $p_{min} = -1.090$, $p_{max} = 0.743$; (d) $\bar{u}_{1_{min}} = -1.135$, $\bar{u}_{1_{max}} = 1.973$; (e) $\bar{u}_{3_{min}} = -1.074$, $\bar{u}_{3_{max}} = 1.136$; (f) $p_{min} = -0.967$, $p_{max} = 0.704$.

The SA model used here needs at least one nonzero boundary to induce turbulence. For example, in forced convective flow problems a small nonzero value was always prescribed at the inlet to allow the variable to propagate into the domain and induce turbulence via vorticity. In the natural convection problem studied here, all the solid walls of the cavity can only be prescribed with a zero value for the SA turbulence parameter $\hat{\nu}$. Thus, we need to find a nonboundary location to prescribe a small nonzero value for this variable. Here, we follow a simple physics based procedure. At higher Rayleigh numbers, the flow at the center of the closed cavity is stratified and the flow is normally confined to the walls of the cavity. Due to this reason, the point at the center of the domain is used to prescribe a very small nonzero value of 0.001. This allows the turbulence variable to grow and distribute throughout the domain depending on the vorticity generation. There may be alternative ways of developing the SA model such as using a trip length, which needs to be investigated further.

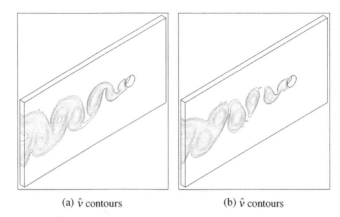

(a) $\hat{v}$ contours   (b) $\hat{v}$ contours

Figure 8.13   Turbulent incompressible flow over a circular cylinder at Re = 10 000 using the matrix free CBS-AC scheme with the Spalart-Allmaras model on unstructured mesh1 (left) and hybrid mesh2 (right). Real nondimensional time = 100. (a) $\hat{v}_{min}$(red) = 0.0, $\hat{v}_{max}$ = 368.329; (b) $\hat{v}_{min}$(red) = 0.0, $\hat{v}_{max}$ = 349.945.

The mesh used in the calculation is shown in Figure 8.14. This is a nonuniform structured mesh of 100x100 size. As seen the mesh is finely refined close to the walls of the cavity in order to capture the high temperature and velocity gradients.

Figures 8.15 and 8.16 show the results obtained for a buoyancy-driven flow in a square cavity using the CBS scheme and SA model. As seen in the vector plots of Figure 8.15 the

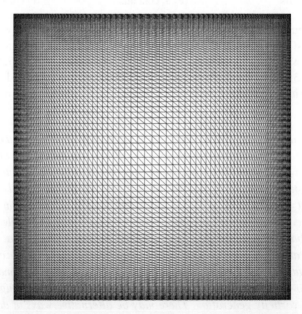

Figure 8.14   Turbulent natural convection in a square cavity. Structured finite element mesh.

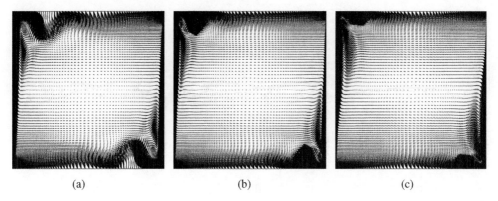

(a)   (b)   (c)

Figure 8.15   Turbulent natural convection in a square cavity. Velocity vectors. (a) $Ra = 10^8$, (b) $Ra = 10^9$; (c) $Ra = 10^{10}$.

flow is generally confined to the walls of the cavity. Figure 8.16 shows the stream traces, temperature and SA turbulent variable contours. The stream traces show very much a horizontal flow pattern except near the top left and bottom right corners and at the center line. The temperature contours clearly show the stratification with almost no temperature variation along a horizontal plane. This is an expected pattern. The SA turbulent variable distribution is symmetric and a strong variation in the variable is shown close to the top-left and bottom-right corners of the cavity.

8.6   Other Methods for Turbulence

8.6.1   Large Eddy Simulation(LES)

The idea of LES is developed based on splitting large-scale motions from small scales using a filtering operation such as

$$\overline{\phi}(x) = \int_{\Omega} f(x')G(x,x')dx'. \tag{8.67}$$

If the variables of the incompressible Newtonian equations are subjected to the above filtering operation, we get

$$\frac{\partial \overline{u}_i}{\partial x_i} = 0 \tag{8.68}$$

and

$$\frac{\partial \overline{u}_i}{\partial t} + \frac{\partial}{\partial x_j}(\overline{u_i u_j}) = -\frac{1}{\rho}\frac{\partial \overline{p}}{\partial x_i} + \frac{\partial \tau_{ij}}{\partial x_j} + \frac{\partial \tau_{ij}^{SGS}}{\partial x_j}, \tag{8.69}$$

where

$$\tau_{ij}^{SGS} = \overline{u}_i \overline{u}_j - \overline{u_i u_j}. \tag{8.70}$$

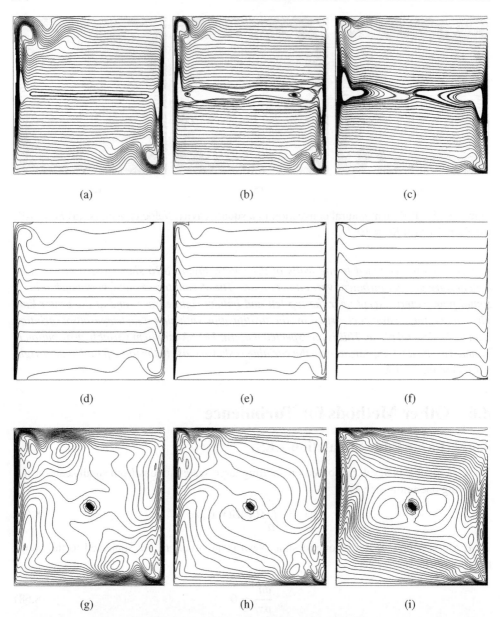

(a) (b) (c)

(d) (e) (f)

(g) (h) (i)

Figure 8.16   Turbulent natural convection in a square cavity. Stream traces (top), temperature contours (middle) and SA turbulence variable $\hat{v}$ at different Rayleigh numbers. Average Nusselt numbers at $Ra = 10^8$ is 30.67; at $Ra = 10^8$ is 55.21 and at $Ra = 10^{10}$ is 103.84. (a) $Ra = 10^8$, Stream traces; (b) $Ra = 10^9$, Stream traces; (c) $Ra = 10^{10}$, Stream traces; (d) $Ra = 10^8$, temperature; (e) $Ra = 10^9$, temperature; (f) $Ra = 10^{10}$, temperature; (g) $Ra = 10^8$, turbulence variable; (h) $Ra = 10^9$, turbulence variable; (i) $Ra = 10^{10}$, turbulence variable.

τ^{SGS} in the above equation is generally modeled using various subgrid scale (SGS) models. The standard SGS models (Smagorinsky, 1963), dynamic models (Germano, 1992) and nonlinear models are just a few that could be mentioned. This is a vast area of research and difficult to cover all the theory behind these models in a single chapter. For the sake of completeness, we provide the standard SGS model below. The SGS stress of Equation (8.70) is represented exactly as Equation (8.13). However, the eddy viscosity is modeled differently here.

8.6.1.1 Standard SGS Model

The eddy viscosity here is defined as

$$\nu_T = (C\Delta)^2 \bar{\omega}. \tag{8.71}$$

The most widely used eddy-viscosity model was proposed by the meteorologist Smagorinsky (1963). Smagorinsky was simulating a two-layer quasigeotrophic model in order to represent large (synoptic) scale atmospheric motions. He introduced an eddy viscosity that was supposed to model three-dimensional turbulence in the subgrid scales.

In Smagorinsky's model, a sort of mixing-length assumption is made, in which the eddy viscosity is assumed to be proportional to the subgrid scale characteristic length Δ and to a characteristic turbulent velocity based on the second invariant of the filtered field deformation tensor (i.e. strain-rate tensor). In other words, the well-known Smagorinsky's model, where the SGS time scaling, $\bar{\omega}$, in Equation (8.71) is set as the magnitude of the local resolved strain-rate tensor, namely

$$\bar{\omega} = |\bar{S}| = (2\bar{S}_{ij}\bar{S}_{ij})^{1/2} \tag{8.72}$$

$$C = C_s. \tag{8.73}$$

If one assumes that the cut off wave number in Fourier space, $k_c = \pi/\Delta$, lies within a $k^{-5/3}$ Kolmogorov cascade $E(k) = C_K \epsilon^{2/3} k^{-5/3}$ (where C_K is the Kolmogorov constant), one can adjust the constant C_s so that the ensemble-averaged subgrid kinetic energy dissipation is identical to ϵ. An approximate value for the constant is

$$C_s \approx \frac{1}{\pi}\left(\frac{3C_K}{2}\right)^{-3/4}. \tag{8.74}$$

For a Kolmogorov constant of 1.4, which is obtained by measurements in the atmosphere this yields $C_s \approx 0.18$. Most workers prefer $C_s = 0.1$ – a value for which Smagorinsky's model behaves reasonably well for free-shear flows and for channel flow. However, the Smagorinsky constant C_s is required to have a sensible value to avoid excessive damping of resolved structures. The grid size Δ, as an indication of characteristic length scale separates large and small scale eddies from each other and is considered to be an average cell size. It is calculated for two-dimensional elements as follows:

$$\Delta = f(\Delta x \Delta y)^{1/2}, \tag{8.75}$$

Despite increasing interest in developing more advanced subgrid scale stress models, Smagorinsky's model is still successfully used (Nakayama and Vengadesan 2002).

8.7   Detached Eddy Simulation (DES) and Monotonically Integrated LES (MILES)

The Large Eddy Simulation (LES), despite using fewer empirical relations, needs a higher computational overhead than the RANS models. Spalart *et al.* (1997) suggested an approach which attempts to combine the best features of RANS and LES. This approach is referred to as the Detached Eddy Simulation (DES). This hybrid method reduces to RANS near solid boundaries and LES away from the wall. A minor modification to the SA model presented previously achieves this. Such a model will take advantage of RANS in the thin shear layers close to the walls where RANS models are calibrated. Away from the wall in the separated regions large eddies are resolved.

The model used will be the same as the one in Section 8.2.2 except that the shortest distance to the wall, y, is modified in such a way that the model calculates a RANS eddy viscosity close to the walls and the SGS eddy viscosity away from the wall. The modification is simple and given as

$$\tilde{y} = min(y, C_{DES}\Delta). \tag{8.76}$$

Δ has the same meaning as discussed in the previous section. The constant C_{DES} was calibrated for homogeneous turbulence as 0.65. It is now possible to see that close to the walls Δ is larger than y and the model becomes a RANS model. However, away from the wall the model becomes one for calculating the SGS eddy viscosity (Constantinescu *et al.* 2003).

The MILES approach (Boris *et al.* 1992; Rider and Margolin 2003; Tucker 2004) is very similar to the DES approach but in the place of the LES eddy viscosity only the numerical diffusion viscosity is added. This alternative approach to DES has been the subject of very recent research and the people using this method believe that MILES eliminates some of the drawbacks of the DES approach. However, further investigations and results are necessary to further enhance the understanding of the DES and MILES approaches.

8.8   Direct Numerical Simulation (DNS)

The direct numerical simulation method can be used to solve all turbulence length scales including *Kolmogorov* length scale given by Equation (8.1). The standard Navier-Stokes equation without any modeling is adequate to compute all turbulence scales. The difficulty here is the prohibitively expensive computing cost required to carry out a calculation even at a very small Reynolds number. The enhanced computational over head is mainly due to the extremely fine mesh necessary to carry out the calculation and the need for higher order accurate numerical schemes. The number of nodes necessary to resolve all scales in a three dimensional flow problem may be written in terms of the Reynolds number as (based on the assumption that the Kolmogorov length scales are solved)

$$No.Nodes = Re^{9/4}. \tag{8.77}$$

Similarly the time step in a calculation is limited by the Kolmogorov time scale in Equation (8.3). The time-step size calculated from the time scale is (Kim *et al.* 1987)

$$\Delta t = \frac{0.003H}{u_T \sqrt{(Re_T)}},$$
(8.78)

where u_T is the shear velocity and $Re_T = u_T H / 2\nu$ is the turbulence Reynolds number (Kim *et al.* 1987).

Although DNS can resolve all turbulence length scales, the currently available computing facilities allow only very small Reynolds number turbulent flow calculations.

8.9 Summary

This chapter provided an introduction to using the finite element method to solve turbulent flow and heat transfer problems. The intention here was not to provide an in-depth analysis of turbulent heat transfer, rather to provide the basics of implementation. The problems provided include isothermal two- and three-dimensional flow problems and nonisothermal problems. Within nonisothermal problems, both forced and natural convection problems were treated. Although mainly the Spalart-Allmaras model was highlighted, other Reynolds Averaged Navier-Stokes (RANS) model implementation follows a similar route.

References

Boris JP, Grinstein FF, Oran ES and Kolbe RL (1992) New insights into large eddy simulation. *Fluid Dynamics Research*, **10**, 199–228.

Cebeci T and Smith AMO (1974) *Analysis of Turbulent Boundary Layers*, Academic Press, New York.

Constantinescu G, Matthieu C and Squires K (2003) Turbulence modelling applied to flow over a sphere. *AIAA Journal*, **41**, 1733–1742.

Denham MK, Briard P and Patrick MA (1975) A directionally-sensitive laser anemometer for velocity measurements in highly turbulent flows. *Journal of Physics E: Scientific Instruments*, **8**, 681–683.

Fan S, Lakshminarayana B and Barnett M (1993) Low-Reynolds-number $\kappa - \varepsilon$ model for unsteady turbulent boundary layer flows. *AIAA Journal*, **31**, 1777–1784.

Germano M (1992) Turbulence: the filtering approach. *J. Fluid Mech.*, **238**, 325.

Johansson SH, Davidson L and Olsson E (1993) Numerical simulation of vortex shedding past triangular cylinders at high Reynolds number using a $\kappa - \varepsilon$ model. *International Journal for Numerical Methods in Fluids*, **16**, 859–878.

Jones WP and Launder BE (1972) The prediction of laminarization with a two-equation model of turbulence. *International Journal of Heat and Mass Transfer*, **15**, 301–314.

Kim J, Moin P and Moser R (1987) Turbulence statistics in fully developed channel flow at low Reynolds numbers. *Journal of Fluid Mechanics*, **177**, 133–166.

Kolmogorov AN (1941) Local structure of turbulence in incompressible viscous fluid for very large Reynolds number. *Doklady AN. SSR*, **30**, 299–303.

Lam CKG, Bremhorst K (1981) A modified form of the $\kappa - \varepsilon$ model for predicting wall turbulence. *Transactions of the ASME Journal of Fluids Engineering*, **103**, 456–460.

Laufer J (1951) Investigation of turbulent flow in a two-dimensional channel. *NACA Report 1053*.

Launder BE and Spalding B (1972) *Mathematical Models of Turbulence*. Academic Press, New York.

Lu X, Dalton C, Zhang J (1997) Application of large eddy simulation to an oscillating flow past a circular cylinder. *Transactions of the ASME Journal of Fluids Engineering*, **119**, 519–525.

Lu X, Dalton C, Zhang J (2003) Application of large eddy simulation to an oscillating flow past a circular cylinder. *Journal of Fluids and Structures*, **17**,1213–1236.

Nakayama A and Vengadesan SN (2002) On the influence of numerical schemes and subgrid-stress models on large eddy simulation of turbulent flow past a square cylinder. *International Journal for Numerical Methods in Fluids*, **38**, 227–253.

Nithiarasu P and Liu CB and Massarotti N (2007) Laminar and turbulent flow through a model human upper airway. *Communications in Numerical Methods in Engineering*, **23**, 1057–1069.

Patankar SV (1980) *Numerical Heat Transfer and Fluid Flow*. Taylor & Francis, London.

Rider WJ and Margolin L (2003) From numerical analysis to implicit subgrid turbulence modelling. *AIAA Paper, AIAA 2003–4101*.

Smagorinsky J (1963) General circulation experiments with the primitive equations, I Basic experiment, *Monthly Weather Review*, **91**. 99.

Spalart PR and Allmaras SR (1992) A one-equation turbulence model for aerodynamic flows. *AIAA 30th Aerospace Sciences Meeting*, 92-0439.

Spalart PR, Jou WH, Strelets M and Allmaras SR (1997) Comments on the feasibility of LES for wings, and on a Hybrid RANS/LES approach, In C Liu and Z Liu (eds), *Advances in DNS/LES: First AFOSR International Conference on DNS/LES*. Greyden, Columbus, OH.

Tucker PG (2004) Novel MILES computations for jet flows and noise. *International Journal of Heat and Fluid Flow*, **25**, 625–635.

Tutar M and Hold AE (2001) Computational modelling of flow around a circular cylinder in sub-critical flow regime with various turbulence models. *International Journal for Numerical Methods in Fluids*, **35**, 763–784.

Wilcox DC (1992) *Turbulence modelling for CFD*. DCW Industries Inc., La Canada, CA.

Yang Z and Shih TH (1993) New time scale based $\kappa - \varepsilon$ model for near-wall turbulence, *AIAA Journal*, **31**, 1191–1198.

Wolfstein M (1970) Some solutions of plane turbulent impinging jets. *Transications of the ASME Journal of Basic Engineering*, **92**, 915–922.

9

Heat Exchangers

9.1 Introduction

A heat exchanger is a device built for heat transfer from one medium to another (generally fluids). The compact heat exchangers (CHE) are often defined as heat exchangers with large heat transfer surface area to volume ratio. In general, liquid heat exchangers with $>400\,\mathrm{m^2/m^3}$, and, gas heat exchangers with $>700\,\mathrm{m^2/m^3}$ are referred to as compact heat exchangers. The total heat transfer between two fluids in a heat exchanger may be written as

$$Q = UA\Delta T, \tag{9.1}$$

where Q is the total heat transfer, U is the overall heat transfer coefficient, A is the heat transfer area and ΔT is the average temperature difference between hot and cold fluids. Figure 9.1 shows the cross-sectional view along the length of a parallel flow tubular heat exchanger. As seen, the hot fluid is allowed to flow through the inner tube and the cooling fluid is allowed to pass through the outer tube. The heat transfer mechanism between the hot and cold fluid, in this simple heat exchanger, involves convection on the hot fluid side, conduction through the partitioning wall between the fluids and convection on the cold fluid side. Thus, the total heat transfer across the partitioning wall between the hot and cold fluids may be written as

$$Q = A_{h_f} h_h (T_{h_f} - T_{h_s}) = A_s \frac{k_s}{\delta} (T_{h_s} - T_{c_s}) = A_{c_f} h_c (T_{c_s} - T_{c_f}), \tag{9.2}$$

where h is the heat transfer coefficient, k is the thermal conductivity and δ is the thickness of the partitioning wall between the two fluids. The subscripts h, c, f and s represent hot, cold, fluid and solid respectively. As shown in Equation (9.2), the heat transfer takes place via three zones. The first is a convective heat transfer zone between the hot fluid and the hot side of the

Fundamentals of the Finite Element Method for Heat and Mass Transfer, Second Edition.
P. Nithiarasu, R. W. Lewis, and K. N. Seetharamu.
© 2016 John Wiley & Sons, Ltd. Published 2016 by John Wiley & Sons, Ltd.

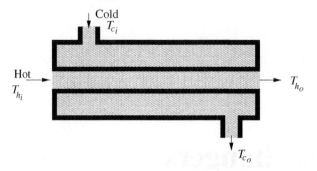

Figure 9.1   Schematic diagram of a parallel flow, tubular heat exchanger.

partitioning wall. From this zone, the average temperature difference between the hot fluid
and the wall may be written as (refer to Equation (9.2)),

$$(T_{h_f} - T_{h_s}) = \frac{Q}{A_{h_f} h_h}. \tag{9.3}$$

The average temperature difference in the second and third zones can also be determined
in a similar fashion to that of the Equation (9.3) as

$$(T_{h_s} - T_{c_s}) = \frac{\delta Q}{A_s k_s} \tag{9.4}$$

and

$$(T_{c_s} - T_{c_f}) = \frac{Q}{A_{c_f} h_c} \tag{9.5}$$

respectively. If we add all the average temperature differences in the three zones and simplify,
we obtain:

$$T_{h_f} - T_{c_f} = \frac{Q}{A_{h_f} h_h} + \frac{Q}{\frac{A_s k_s}{\delta}} + \frac{Q}{A_{c_f} h_c}. \tag{9.6}$$

Rearranging,

$$Q = \frac{1}{\left[\frac{1}{A_{h_f} h_h} + \frac{\delta}{A_s k_s} + \frac{1}{A_{c_f} h_c} \right]} (T_{h_f} - T_{c_f}). \tag{9.7}$$

Comparing with Equation (9.1), we have

$$UA = \frac{1}{\left[\frac{1}{A_{h_f} h_h} + \frac{\delta}{A_s k_s} + \frac{1}{A_{c_f} h_c} \right]}. \tag{9.8}$$

The above equation clearly is simple and easy to use to calculate the total heat transfer, if
the overall heat transfer coefficient and average temperature difference between the hot and
cold fluids can be determined. However, for a given heat transfer load on complex geometries,

the above relationship will be of little or no use and often more complex relationships are essential. Before discussing the numerical modeling options, the following section provides two other commonly employed heat exchanger design methods.

9.2   LMTD and Effectiveness-NTU Methods

Once again, we use the parallel flow heat exchanger shown in Figure 9.1 as an example for demonstrating the so-called logarithmic mean temperature and effectiveness-NTU methods. It is important to remark here that an exhaustive analysis of these methods is not within the scope of this chapter. The readers are referred to dedicated heat exchanger books for a detailed analysis (Kays and London 1998).

9.2.1   LMTD Method

Equation (9.7) makes no reference to inlet or exit temperature of either of the fluids. Often, heat exchanger design depends on the exit temperature requirement of the hot fluid. In order to incorporate the inlet and exit temperatures of the fluids in a design we should consider a part of the parallel flow heat exchanger, as shown in Figure 9.2. In this figure, an infinitesimal control volume is considered. The fraction of heat transfer taking place within this control volume between the hot and cold fluids is dQ. The hot fluid temperature is decreased by dT_h and the cold fluid temperature is increased by dT_c as the fluid passes through the control volume. The fractional heat transfer dQ may be expressed as (refer to Equation (9.1))

$$dQ = U\Delta T dA. \tag{9.9}$$

The fraction of heat transfer may also be expressed in terms of fluid properties as

$$dQ = -\dot{m}_h c_{p_h} dT_h = \dot{m}_c c_{p_c} dT_c, \tag{9.10}$$

where $\dot{m}$ is the mass flow rate in kg/s and c_p is the specific heat at constant pressure. Integration of the above equation between the inlet and exit of the heat exchanger gives

$$Q = -\dot{m}_h c_{p_h}(T_{h_o} - T_{h_i}) = \dot{m}_c c_{p_c}(T_{c_o} - T_{c_i}), \tag{9.11}$$

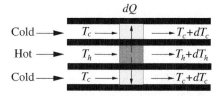

Figure 9.2   Local heat exchange in a parallel flow tubular heat exchanger.

where subscripts i and o refer to inlet and outlet (exit) respectively. From Equation (9.10), the change in incremental temperature difference between the hot and cold fluid may be established as

$$dT_h - dT_c = d(\Delta T) = -dQ \left(\frac{1}{\dot{m}_h c_{p_h}} + \frac{1}{\dot{m}_c c_{p_c}} \right). \tag{9.12}$$

Substituting dQ from Equation (9.9), rearranging and integrating between inlet and exit, that is,

$$\int_{in}^{out} \frac{d(\Delta T)}{\Delta T} = -U \left(\frac{1}{\dot{m}_h c_{p_h}} + \frac{1}{\dot{m}_c c_{p_c}} \right) \int_{in}^{out} dA. \tag{9.13}$$

Integration gives

$$\ln \left(\frac{\Delta T_{out}}{\Delta T_{in}} \right) = -UA \left(\frac{1}{\dot{m}_h c_{p_h}} + \frac{1}{\dot{m}_c c_{p_c}} \right), \tag{9.14}$$

where $\Delta T_{out} = T_{h_o} - T_{c_o}$ and $\Delta T_{in} = T_{h_i} - T_{c_i}$. Substituting Equation (9.11) into Equation (9.14) gives

$$\ln \left(\frac{\Delta T_{out}}{\Delta T_{in}} \right) = -UA \left(\frac{T_{h_i} - T_{h_o}}{Q} + \frac{T_{c_o} - T_{c_i}}{Q} \right). \tag{9.15}$$

This can be rearranged to obtain

$$\ln \left(\frac{\Delta T_{out}}{\Delta T_{in}} \right) = -\frac{UA}{Q}((T_{h_i} - T_{c_i}) - (T_{h_o} - T_{c_o})) = -\frac{UA}{Q}(\Delta T_{in} - \Delta T_{out}) \tag{9.16}$$

or

$$Q = UA\Delta T_m = UA \frac{\Delta T_{out} - \Delta T_{in}}{\ln \left(\Delta T_{out} / \Delta T_{in} \right)}, \tag{9.17}$$

where

$$\Delta T_m = \frac{\Delta T_{out} - \Delta T_{in}}{\ln \left(\Delta T_{out} / \Delta T_{in} \right)}. \tag{9.18}$$

ΔT_m is referred to as the Log Mean Temperature Difference (LMTD). The LMTD method is suitable for simple heat exchangers, such as the one shown in Figure 9.1, in which the inlet and exit temperatures are known. If the configuration of the solid structure, LMTD and overall heat transfer coefficient are known *a priori*, then Equation (9.17) can be used to determine the surface area. For complex heat exchanger structures, and, in cases where the mentioned parameters are not known, the LMTD method may not be a suitable method for design. This is especially true for modern day compact heat exchangers which have restrictions on the maximum space they can occupy.

9.2.2 Effectiveness – NTU Method

Even in a simple heat exchanger, if the exit temperatures are not known, the LMTD method is not applicable in computing the total heat transfer. For cases where only the inlet temperatures are known (this is often the case), the so-called Effectiveness Method is often used. The Effectiveness is defined as the ratio between the actual heat transfer and the maximum possible heat transfer. The maximum possible heat transfer is proportional to the maximum possible temperature difference, $(T_{h_i} - T_{c_i})$. In Equation (9.11), if $\dot{m}_c c_{p_c} < \dot{m}_h c_{p_h}$, then the cold fluid must experience a higher temperature difference than the hot fluid in order to satisfy the heat balance. If the opposite is true, the hot fluid experiences the higher temperature difference. Thus, the maximum possible heat transfer may be expressed as

$$Q_{max} = \left(\dot{m}c_p\right)_{min} (T_{h_i} - T_{c_i}). \tag{9.19}$$

Thus, the Effectiveness may be defined as

$$\epsilon = \frac{Q}{Q_{max}} = \frac{\dot{m}_h c_{p_h}(T_{h_i} - T_{h_o})}{\left(\dot{m}c_p\right)_{min}(T_{h_i} - T_{c_i})} = \frac{\dot{m}_c c_{p_c}(T_{c_i} - T_{c_o})}{\left(\dot{m}c_p\right)_{min}(T_{h_i} - T_{c_i})}. \tag{9.20}$$

If we assume $(\dot{m}c_p)_{min} = \dot{m}_h c_{p_h}$, then

$$\epsilon = \frac{T_{h_i} - T_{h_o}}{T_{h_i} - T_{c_i}}. \tag{9.21}$$

From Equation (9.14), we have (with $(\dot{m}c_p)_{min} = \dot{m}_h c_{p_h}$)

$$ln\left(\frac{\Delta T_{out}}{\Delta T_{in}}\right) = ln\left(\frac{T_{h_o} - T_{c_o}}{T_{h_i} - T_{c_i}}\right) = -\frac{UA}{\left(\dot{m}c_p\right)_{min}}\left(1 + \frac{\left(\dot{m}c_p\right)_{min}}{\left(\dot{m}c_p\right)_{max}}\right). \tag{9.22}$$

In the above equation, $\frac{UA}{\left(\dot{m}c_p\right)_{min}}$, is referred to as the number of transfer units or NTU. From Equation (9.11):

$$\frac{\dot{m}_h c_{p_h}(T_{h_i} - T_{h_o})}{\dot{m}_c c_{p_c}(T_{c_i} - T_{c_o})} = 1 \tag{9.23}$$

or

$$\frac{\dot{m}_h c_{p_h}}{\dot{m}_c c_{p_c}} = \frac{\left(\dot{m}c_p\right)_{min}}{\left(\dot{m}c_p\right)_{max}} = \frac{T_{c_i} - T_{c_o}}{T_{h_i} - T_{h_o}}. \tag{9.24}$$

Rearranging the above equation, we obtain

$$\frac{T_{h_o} - T_{c_o}}{T_{h_i} - T_{c_i}} = 1 - \epsilon\left(1 - \frac{\left(\dot{m}c_p\right)_{min}}{\left(\dot{m}c_p\right)_{max}}\right). \tag{9.25}$$

Substituting the above relation into Equation (9.22) and rearranging

$$\epsilon = \frac{1 - exp\left[-NTU\left(1 + \frac{(\dot{m}c_p)_{min}}{(\dot{m}c_p)_{max}}\right)\right]}{1 - \frac{(\dot{m}c_p)_{min}}{(\dot{m}c_p)_{max}}}. \tag{9.26}$$

This result is true for $(\dot{m}c_p)_{min} = \dot{m}_h c_{p_h}$. It is thus clear that the Effectiveness-NTU method can be used for cases in which the exit temperature distribution is not known *a priori*. However, the overall heat transfer coefficient needs to be determined. For simple configurations, determining the overall heat transfer coefficient is easy (see Equation (9.8)). For complex geometries, however, computational and experimental methods are essential. For compact heat exchanger configurations, the experimental results are available in terms of both the friction factor and the so-called Colburn j factor. The Colburn j factor is defined as

$$j = StPr^{2/3}, \tag{9.27}$$

where St is the Stanton number ($h/\rho v_{max} c_p$). Here, v_{max} is the maximum velocity. Once the distributions of friction and j factors against Reynolds number are available, the heat transfer coefficients can be obtained for use in the analysis using the LMTD or Effectiveness-NTU method. Such an analysis of compact heat exchangers is limited to the available experimental data. Thus, this method may not be completely useful in developing new compact designs. During the development of new compact designs, the computational approaches can help to reduce the number of prototypes necessary to carry out tests. In the following section, two different finite element approaches are explained.

9.3   Computational Approaches

The finite element method is a useful tool to refine a heat exchanger design. Although the accuracy of computational methods at high Reynolds numbers greatly depends on the turbulence model employed, computational methods can provide approximate qualitative and quantitative results of practical significance. Such results will be crucial in cutting down costs and will be of great use in designing heat exchangers with important constraints such as space occupied.

9.3.1   System Analysis

As discussed in the previous section, LMTD or Effectiveness-NTU method may be used either in simple heat exchanger configurations or when experimental friction and j factors are available (Holman 1989; Incropera and Dewitt 1990). The LMTD method requires the exit temperatures of the hot and cold fluids. These temperatures are often not available. In order to use the LMTD method for designing a heat exchanger, we have to calculate the outlet temperatures of both the hot fluid and the cooling fluid for the given inlet temperatures. The overall heat transfer coefficient may be a constant or could vary along the heat exchanger length, depending on the heat transfer area, material properties and heat transfer coefficient. In

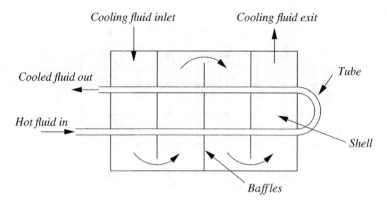

Figure 9.3   Schematic diagram of a shell and tube heat exchanger.

this section, we provide a simple system analysis procedure that may be employed to determine the temperature distribution in a heat exchanger.

For the purpose of illustration let us consider a shell and tube heat exchanger as shown in Figure 9.3 (Ravikumar *et al.* 1984). In this type of heat exchanger, the hot fluid flows through the tube and the tube is passed through the shell. The cooling fluid is pumped into the shell and thus the hot fluid in the tube is cooled.

Let us divide the given heat exchanger into eight cells as shown in Figure 9.4. It is assumed that both the hot and cold fluids will travel through the cell at least once. Let the overall heat transfer coefficient be U and the surface area of the tubes be A. These are assumed to be constant throughout the heat exchanger within each element. Let us assume that the hot and cold fluid temperatures vary linearly along the flow.

Now, the heat leaving node 1 and entering element 1 (Figure 9.4b) is

$$Q_1 = W_1 T_1, \tag{9.28}$$

where W_1 is ρc_p times the volume flow rate. The heat leaving element 1 and entering node 2 is as follows (the energy balance is considered with respect to the element; heat entering is taken as being positive and leaving the element is taken as negative):

$$Q_2 = W_1 T_1 - UA(T_{1,2} - T_{11,12}), \tag{9.29}$$

where

$$T_{1,2} = \frac{T_1 + T_2}{2} \quad \text{and} \quad T_{11,12} = \frac{T_{11} + T_{12}}{2}. \tag{9.30}$$

Similarly, the heat leaving node 11 and entering element 1 is

$$Q_{11} = W_2 T_{11} \tag{9.31}$$

and the heat leaving element 1 and entering node 12 is

$$Q_{12} = W_2 T_{11} - UA(T_{11,12} - T_{1,2}). \tag{9.32}$$

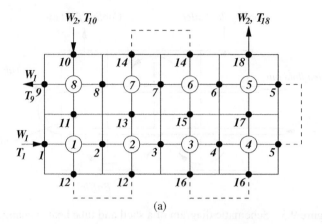

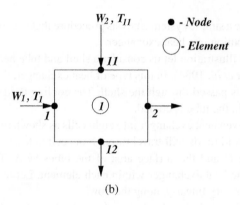

Figure 9.4   (a) Simplified model of a heat exchanger; (b) a single element within the mesh.

In this example the heat transfer between the fluids is given by $UA(T_{11,12} - T_{1,2})$ whereas some other models use $UA(T_{12} - T_2)$. The assumption in the present model is more logical in view of the continuous variation (linear in our case) of the temperature difference between the hot and cold fluids.

Equations (9.28), (9.29), (9.31) and (9.32) can be combined and recast in matrix form to give the element characteristics, that is,

$$
\begin{bmatrix}
W_1 & 0.0 & 0.0 & 0.0 \\
W_1 - C & -C & C & C \\
0.0 & 0.0 & W_2 & 0.0 \\
C & C & W_2 - C & -C
\end{bmatrix}
\begin{Bmatrix}
T_1 \\
T_2 \\
T_{11} \\
T_{12}
\end{Bmatrix}
=
\begin{Bmatrix}
Q_1 \\
Q_2 \\
Q_{11} \\
Q_{12}
\end{Bmatrix},
\tag{9.33}
$$

where $C = \dfrac{UA}{2}$.

Assembly of the element characteristics for elements 1 to 8 (see Figure 9.4(a)) will result in the global stiffness matrix in which Q_1, and Q_{10} are known (in other words T_1, and T_{10}

are known). The solution of the remaining equations will give the temperature distribution for both the fluids, that is, $T_2, T_3, T_4, T_5, T_6, T_7, T_8$ and T_9 for the incoming hot fluid and $T_{11}, T_{12}, T_{13}, T_{14}, T_{15}, T_{16}, T_{17}$ and T_{18} for the coolant. With the known exit temperatures T_9 and T_{18}, the LMTD method can be used as the design method.

9.3.2   Finite Element Solution to Differential Equations

The system analysis discussed in the previous section is simple but it has many restrictions. For simple geometries, where the assumptions hold, the system analysis is one of the best ways of obtaining the temperature distributions. The system analysis allows little manoeuvre to model complex heat exchanger passages. Thus, a full numerical simulation becomes essential for such complex geometries. A numerical simulation procedure may either be used in modeling a whole heat exchanger (Ismail *et al.* 2009, 2010; Nithiarasu 2008) or used to analyse a part of a heat exchanger (Atkinson *et al.* 1998; Ciofalo *et al.* 1996; Islamoglu and Parmaksizoglu 2004; Jang and Chen 1997; Saidi and Sunden 2001). The governing equations for a heat exchanger analysis are normally the incompressible Navier-Stokes equations (including the energy equation). Since the normal operating Reynolds numbers of heat exchangers are in the moderate range, modeling turbulence may be required. All the governing equations should be discretized using a numerical method to obtain an approximate solution. Such a finite element discretization of the equations, both spatial and temporal, are presented in Chapters 3 and 7. All the different options available for modeling turbulence in any heat transfer equipment are discussed in detail in Chapter 8. In the following sections, a few examples of single phase flow and heat transfer in heat exchangers are presented.

9.4   Analysis of Heat Exchanger Passages

As mentioned previously, heat exchanger parts or passages may be numerically analyzed to determine the influence of various parameters. For example, Figure 9.5 shows the contours of steady-state flow variables in a corrugated compact heat exchanger passage. With a forced flow situation, the drag force due to the changes in passage shape can be investigated without including the temperature. The figure shown is a result of a turbulent flow calculation using a one-equation $k - l$ model as discussed in Chapter 8. The inlet velocity in this case is assumed to be uniform and the exit pressure is constant. The contours of all the four variables are plotted in this figure. It may also be useful to plot stream lines to better understand the recirculation regions. It is often the quantitative results, such as friction and j factors, which are sought in a heat exchanger passage. Although the quantitative results are of ultimate interest, the qualitative results, such as contour plots, help a researcher to make sure that the boundary conditions are rightly applied and the variable is sensibly distributed. The following example of flow through another corrugated passage gives a more elaborate flow and thermal analysis.

Example 9.4.1   *Corrugated compact heat exchanger passage*
     Figure 9.6 shows a corrugated geometry considered for the finite element calculation. This is one of the common geometrical shapes employed in compact heat exchangers (Islamoglu

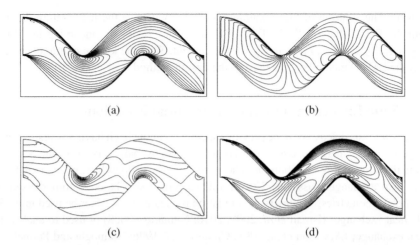

(a)   (b)

(c)   (d)

Figure 9.5   Flow through a corrugated compact heat exchanger passage. Contours of horizontal velocity component, vertical velocity component, pressure and turbulent kinetic energy at Re = 8280. (a) Horizontal velocity; (b) Vertical velocity; (c) Pressure; (d) Turbulent kinetic energy.

and Parmaksizoglu 2004). The results obtained can be easily generalized to other similar parts of the channel to understand the flow and heat transfer pattern. This geometry can be studied using both two and three-dimensional approximations depending on the accuracy requirement and conditions. The boundary conditions used include uniform velocity at inlet (parallel to the walls), no-slip conditions on the walls, constant pressure at the exit and the

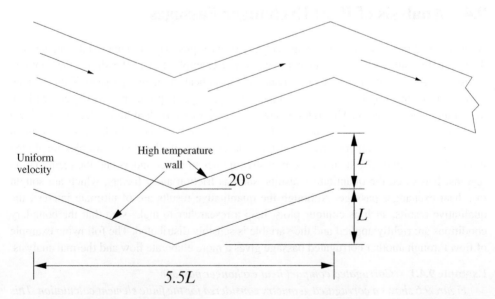

Figure 9.6   Flow through a corrugated compact heat exchanger passage. Actual passage and the simplified passage used in the analysis.

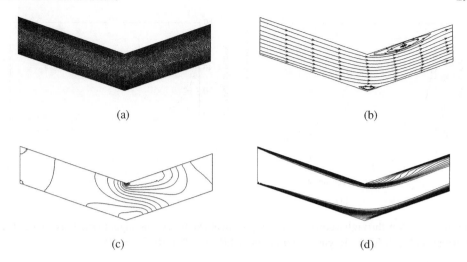

(a)
(b)

(c)
(d)

Figure 9.7   Flow through a corrugated compact heat exchanger passage. Mesh, stream lines, pressure and temperature at Re = 2000. (a) Mesh; (b) Stream lines; (c) Pressure; (d) Temperature.

wall temperature higher than that of the inlet fluid temperature. The Prandtl number used is 0.72 and the equations used are nondimensionalized using standard scales (see Chapter 7). Figure 9.7(a) shows the unstructured finite element mesh used. Although the mesh selected is not fully converged for all Reynolds numbers, this is sufficient to demonstrate the procedure. A reasonably converged mesh is essential for practical design and analysis calculations. The results are generated for different Reynolds numbers and are shown in Figures 9.7, 9.8 and 9.10 using one-equation, Spalart-Allmaras turbulence model (see Chapter 8).

The sample contour plots in Figure 9.7 and 9.8 are given here to make sure that the results show an anticipated pattern. The stream lines in Figures 9.7(b) and 9.8(b) clearly show a recirculation immediately after the turn along the top wall. A small recirculating pattern is also visible at the bottom turn. Since the flow separation takes place ahead of the turn along the bottom surface, the flow slows down towards the turn. Thus, the heat transfer rate is expected to

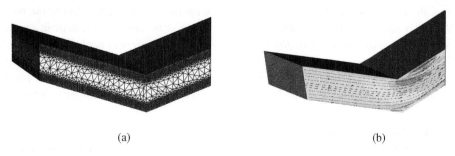

(a)
(b)

Figure 9.8   Flow through a corrugated compact heat exchanger passage. Mesh and stream lines at Re = 5000. (a) Mesh; (b) Stream lines.

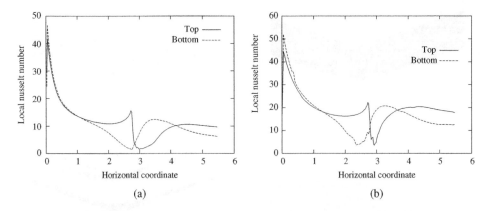

Figure 9.9   Flow through a corrugated compact heat exchanger passage. Local Nusselt number distribution for different Reynolds numbers. (a) $Re = 2000$; (b) $Re = 5000$.

drop significantly towards the turn along the bottom wall. However, the opposite is true along the top wall before the turn. This is due to the fact that no flow separation exists along the top wall ahead of the turn. Beyond the turn, presence of a strong recirculation and flow separation along the top wall are expected to reduce the heat transfer rate. The heat transfer will recover beyond the reattachment points along both walls. These flow trends clearly influence the temperature distribution as shown in Figure 9.7(d). As seen, the temperature contours are closely packed along the top wall all the way to the turn indicating a strong temperature gradient and rapid heat transfer. However, the temperature gradient is not consistently high all the way to the turn along the bottom wall. The temperature distribution along the bottom wall indicates a stronger heat transfer rate in the vicinity of the inlet and it reduces towards the turn. After the turn, the bottom wall recovers the high temperature gradient within a short distance from the turn while the top wall requires a longer distance to recover a high gradient in temperature. These trends are clearly shown by the local Nusselt number distribution in Figure 9.9. It is also important to note the adverse pressure gradients in Figure 9.7(c) at both the top and bottom recirculations.

The average Nusselt number and pressure drop variations with Reynolds number are shown in Figure 9.10. As expected the Nusselt number increases with Reynolds numbers on both walls. The value is almost identical on both walls to a Reynolds number of 1000. Beyond $Re = 1000$, the average Nusselt number on the top wall is higher than at the bottom wall. This is due to the fact that the recirculation zone at the top wall reduces in size with increase in Reynolds number and allows the Nusselt number to recover back to a higher value within a short distance from the turn as shown by the local Nusselt number distributions in Figure 9.9. The Nusselt number variation with Reynolds number shows a nonlinear pattern. The nondimensional pressure drop $(\Delta p / \rho u_\infty^2)$ reduces as the Reynolds number is increased.

Example 9.4.2   *Conjugate heat transfer in a model heat exchanger*

To demonstrate the incompressible fluid dynamics and conjugate heat transfer, forced convection flow and heat transfer in a model fin and tube heat exchanger are studied in this

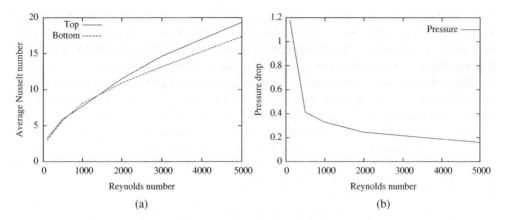

Figure 9.10   Flow through a corrugated compact heat exchanger passage. Average Nusselt number and pressure drop distribution for different Reynolds numbers. (a) Average Nusselt number; (b) Nondimensional pressure drop.

example (Nithiarasu 2008). Due to the low memory needs and easy implementation, the fully explicit solution algorithm is chosen here (see Chapter 7) (Nithiarasu 2003). The comparison of speed between the fully explicit method with local time stepping against other implicit methods shows that the explicit method is robust and in some cases outperforms the other methods (Codina et al. 2006; Massarotti et al. 2006).

Figure 9.11 shows a partial representation of a fin and tube heat exchanger model. The fins are attached to the solid wall of the heat exchanger as shown in the figure. The outside solid surface is assumed to be at a higher temperature than the air at the inlet to the heat exchanger. Thus, the heat is expected to transfer through the solid wall and then dissipated

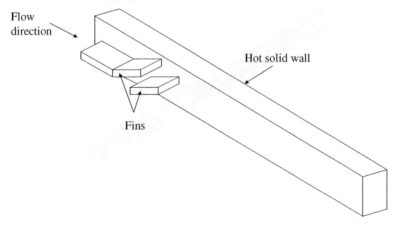

Figure 9.11   Conjugate heat transfer in a model fin and tube heat exchanger. Part of the geometry.

to the fluid via the inside solid wall surface and the fin surfaces. The fluid side, bottom and top boundaries are subjected to zero flux boundary conditions. The inlet velocity is assumed to be constant and uniform. To demonstrate the method, an inlet Reynolds number of 200 and thermal conductivity ratio between the solid and fluid of 10 are assumed.

The first step in the numerical modeling is generating a finite element mesh. We use the simplest of all type of elements, tetrahedral elements, to discretize the domain. Since unstructured meshing can be automated, we use a linear unstructured mesh. First, a sur-face mesh is generated by defining different curves and surfaces of the geometry. Once a surface mesh is available, a volume mesh is generated by filling the spaces between surfaces using tetrahedrons. In the present case, the solid fins should also be meshed. More details on the meshing methods is available in Chapter 14. Figure 9.12 shows the unstructured surface mesh used in the calculations. The mesh is generated using the meshing tools available within Swansea, Collage of Engineering (Morgan et al. 1999). As seen, the mesh is refined close to the

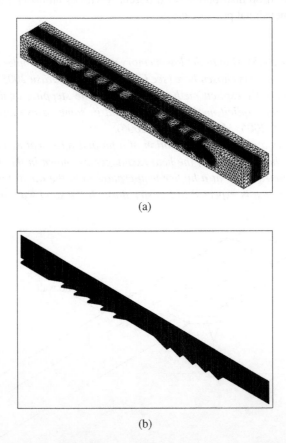

(a)

(b)

Figure 9.12   Conjugate heat transfer in a model fin and tube heat exchanger. Unstructured mesh. (a) Full view; (b) Fins and surface of the solid wall.

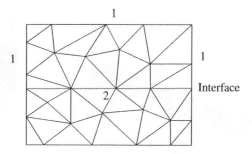

Figure 9.13   Conjugate heat transfer in a model fin and tube heat exchanger. Allocating material code.

solid–fluid interface to have smooth change in the temperature distribution. The total number of tetrahedron elements used is just over 1.7 million.

Once the mesh is generated, the solid and fluid elements should be labeled using appropriate material codes. In two-dimensional problems, the regions can be easily identified for allocating material codes. In three-dimensional problems, however, a special and automatic procedure may be used. This procedure is simple to implement. The procedure starts with zero material code for all the nodes in a domain, except the surfaces surrounding the domain(s) as shown in Figure 9.13. Once zero has been allocated to all the inside nodes, change the material code of the partitioning surface between two domains to a number, say 2, as shown in Figure 9.13. All the nodes on nonpartitioning surfaces should be labeled with the material code that would be given to the nodes inside the domain (1 in Figure 9.13). Now, check the nodes connected to the surface nodes labeled with 1 and if the nodes connected have a zero material code, replace it with 1. Continue checking until all the zero material codes are replaced with 1. Note that this process of checking will stop automatically as soon as the process reaches the surface nodes with label 2. This should be continued for all the different subdomains until all the nodes are labeled with appropriate material codes. Figure 9.14 shows contours of the material codes. As seen, the procedure outlined here easily identifies the solid and fluid domains with different material codes.

Often the interface conditions between the fluid and solid needs to be addressed before obtaining a solution. At the interface, both the temperature and flux should be continuous. Since a linear approximation for the temperature is assumed throughout the domain, temperature continuity across the interface is ensured. However, the heat flux is not continuous across the interface. Special conditions may be applied to ensure flux continuity. However, such treatment needs a solution to additional variables. The alternative is to approximate flux continuity by introducing finer mesh along the interface. For the heat exchanger problem solved here, the interface is treated using the latter approach.

Figures 9.15 to 9.17 show the results obtained. In Figure 9.15, the surface contours of pressure, velocity components and temperature are presented. As seen, the contours are generally smooth, including the pressure contours. In Figure 9.16 the temperature distribution at different sections, along the length (x_1 direction) of the heat exchanger, is presented. As

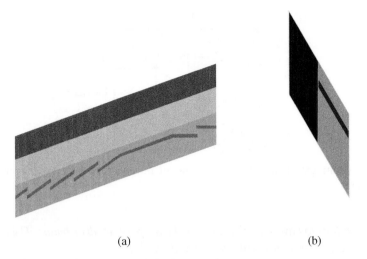

(a)   (b)

Figure 9.14   Conjugate heat transfer in a model fin and tube heat exchanger. Contours of material code. (a) Side view; (b) Cross-sectional view.

seen, the transition of the temperature from solid to fluid is smooth without any noticeable discontinuity.

Figure 9.17(a) shows the temperature distribution in the x_2 direction along the lines at the middle of the fins and Figure 9.17(b) shows the temperature distribution at $x_2 = 0.523$ along the x_3 direction. Figures 9.17 (c) and (d) show the temperature distributions along

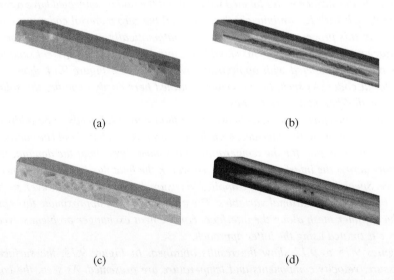

(a)   (b)

(c)   (d)

Figure 9.15   Conjugate heat transfer in a model fin and tube heat exchanger. Pressure, velocity and temperature contours. (a) Pressure; (b) u_1 velocity; (c) u_3 velocity; (d) Temperature contours.

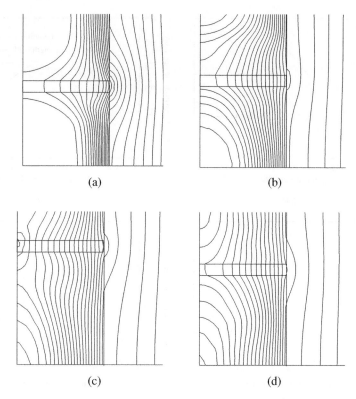

Figure 9.16 Conjugate heat transfer in a model fin and tube heat exchanger. Temperature contours at different sections. (a) $x_1 = 2$; (b) $x_1 = 6$; (c) $x_1 = 10$; (d) $x_1 = 14$.

x_2 direction, at sections $x_3 = 0.407$ and 1.064 respectively. As seen, a majority of the solid wall portions show a linear variation of temperature. In the fluid region, including fins, the temperature variation is nonlinear. It is also noticed that the no heat flux conditions at the fluid side boundary is effectively captured. In Figures 9.17 (c) and (d), a rapid change in temperature is clearly shown at the interface. In addition, the average temperature at section $x_1 = 2$ is much lower than at section $x_1 = 14$ as expected. All the results shown in Figure 9.17 are consistent with the qualitative solution expected.

9.5 Challenges

Despite the fact that tremendous progress has been made in heat exchanger design and development, computational methods have not been adopted for heat exchanger design as widely as in other fields. In general, confidence in modeling among practicing engineers is low. This is due to the expectation that the modeling should give precise answer to the design questions. This would never be possible. All the computational modeling methods are here to help the

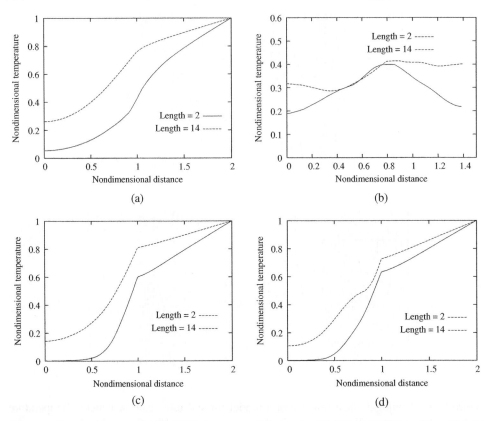

Figure 9.17   Conjugate heat transfer in a model fin and tube heat exchanger. Temperature distributions (a) Along the mid-horizontal line of the fin in the x_2 direction; (b) Along the x_3 direction, in the fluid section, at $x_2 = 0.523$; (c) Along $x_3 = 0.407$ below the fin in the x_2 direction; (d) Along $x_3 = 1.064$ above the fin in the x_2 direction.

engineers to cut down cost at the preliminary stages of design. It is important to remember that the prototype must be tested to make final adjustments to the design. The following are a few challenges facing the heat exchanger modeling community.

- Although turbulence modeling has seen enormous growth over the last thirty years, using state of the art turbulence modeling still remains expensive and time-consuming.

- Multi-phase flow modeling is also a challenge as the physical equation governing the multi-phase transport vary depending on the problem studied.

- Integration of various processes into modeling of heat exchangers is still lacking and needs further research. Integration of stress analysis and shape optimization modules with flow solvers is possible but the amount of work done in this area is negligible.

9.6 Summary

In this chapter, simple methods of analysis were presented before giving an overview of using modeling in heat exchanger design and analysis. The current state-of-the art in modeling clearly shows that computational methods are capable of guiding heat exchanger design procedures. Unlike other industries, such as the aerospace and car industries, the heat exchanger industry is not completely accustomed to employing computational modeling on a regular basis. This is slowly changing and hopefully, in the near future, computation will become an integral part of any heat exchanger design, especially compact heat exchanger design.

References

Atkinson KN, Drakulic R, Heikal MR and Cowell TA (1998) Two- and three-dimensional numerical models of flow and heat transfer over louvred fin arrays in compact heat exchangers. *International Journal of Heat and Mass Transfer*, **41**, 3952–3979.

Ciofalo M, Stasiek J and Collins MW (1996), Investigation of flow and heat transfer in corrugated passages II. Numerical simulations, *International Journal of Heat and Mass Transfer*, **39**, 165–192.

Codina R, Owen H-C, Nithiarasu P and Liu C-B (2006) Numerical comparison of CBS and SGS as stabilization techniques for the incompressible navier-stokes equations. *International Journal for Numerical Methods in Engineering*, **66**, 1672–1689.

Fabbri G (2000) Heat transfer optimization in corrugated wall channels. *International Journal of Heat and Mass Transfer*, **43**, 4299–4310.

Holman JP (1989) *Heat Transfer*, McGraw-Hill, New York.

Incropera FP and Dewitt DP (1990) *Fundamentals of Heat and Mass Transfer*. John Wiley & Sons, Inc., New York.

Islamoglu Y and Parmaksizoglu C (2004) Numerical investigation of convective heat transfer and pressure drop in a corrugated heat exchanger channel. *Applied Thermal Engineering*, **24**, 141–147.

Ismail LS, Ranganayakulu C and Shah RK (2009) Numerical study of flow patterns of compact plate-fin heat exchangers and generation of design data for offset wavy fins. *International Journal of Heat and Mass Transfer*, **52**, 3972–3983.

Ismail LS, Velraj R and Ranganayakulu C (2010) Studies on pumping power in terms of pressure drop and heat transfer characteristics of compact plate-fin heat exchangers – a review. *Renewable and Sustainable Energy Reviews*, **14**, 478–485.

Jang J-Y and Chen L-K (1997) Numerical analysis of heat transfer and fluid flow in a three-dimensional wavy-fin and tube heat exchanger. *International Journal of Heat and Mass Transfer*, **40**, 3981–3990.

Kays MW and London AL (1998) *Compact Heat Exchangers*. Krieger Publishing Company, Florida.

Lewis RW, Nithiarasu P and Seetharamu KN (2004) *Fundamentals of the Finite Element Method for Heat and Fluid Flow*. John Wiley & Son, Ltd, Chichester.

Massarotti N, Arpino F, Lewis RW, and Nithiarasu P (2006) Explicit and semi-implicit CBS procedures for incompressible viscous flows. *International Journal for Numerical Methods in Engineering*, **66**, 1618–1640.

Morgan K, Weatherhill NP, Hassan O, *et al.* (1999) A parallel framework for multidisciplinary aerospace engineering simulations using unstructured meshes. *International Journal for Numerical Methods in Fluids*, **31**, 159–173.

Nithiarasu P (2003) An efficient artificial compressibility (AC) scheme based on the characteristic based split (CBS) method for incompressible flows. *International Journal for Numerical Methods in Engineering*, **56**, 1815–1845.

Nithiarasu P (2008) A unified fractional step method for compressible and incompressible flows, heat transfer and incompressible solid mechanics. *International Journal of Numerical Methods for Heat & Fluid Flow*, **18**, 111–130.

Ravikumaur SG, Seetharamu KN and Aswatha Narayana PA (1984) Applications of finite elements in heat exchangers. *Communications in Applied Numerical Methods*, **2**(2), 229–234.

Saidi A and Sunden B (2001) A numerical investigation of heat transfer enhancement in offset strip fin heat exchangers in self-sustained oscillatory flows. *International Journal of Numerical Methods for Heat & Fluid Flow*, **11**, 699–716.

Spalart PR and Allmaras SR (1992) A one-equation turbulence model for aerodynamic flows. *AIAA paper 92-0439*.

Zienkiewicz OC, Taylor RL and Nithiarasu P (2005) *The Finite Element Method for Fluid Dynamics*. Elsevier Butterworth-Heinemann, Burlington, MA.

10

Mass Transfer

10.1 Introduction

Mass transfer is analogous to heat transport, as discussed in previous chapters. We know that heat transfer is a result of temperature difference and similarly, mass transport is a result of concentration difference. In other words, mass transfer is the transfer of mass from a high species concentration domain to low concentration domain. Some common examples of mass transfer processes are the evaporation of water into the atmosphere, the diffusion and transport of pollution from industrial exhausts into the atmosphere, transport of pollution in lakes, drying and processing of food, brick and other materials etc. (Bird *et al.* 1960; Comini and Lewis 1976; Holman 1989; Lewis and Malan 2005; Murugesan *et al.* 2001; Nithiarasu *et al.* 1996).

As mentioned previously, the driving force for mass transfer is the difference in concentration. The random motion of molecules causes a net transfer of mass from an area of high concentration to an area of low concentration. Mass transfer modes can also be classified under the categories of diffusion and convection as in the case of heat transport. The diffusion mass transport is described by Fick's law of diffusion, which is very similar to Fourier's law of heat conduction, that is,

$$\frac{M}{A} = j_x = -D\frac{\partial c}{\partial x},\tag{10.1}$$

where M is the total mass transfer per unit time, A is the area across which mass transfer takes place, D is the mass diffusion coefficient, c is the species concentration, x is the perpendicular direction to the area A and j_x is the mass transfer per unit time per unit area (mass flux). As seen, Equation (10.1) is analogous to Fourier's law of heat conduction, Equation (1.1).

Fundamentals of the Finite Element Method for Heat and Mass Transfer, Second Edition.
P. Nithiarasu, R. W. Lewis, and K. N. Seetharamu.
© 2016 John Wiley & Sons, Ltd. Published 2016 by John Wiley & Sons, Ltd.

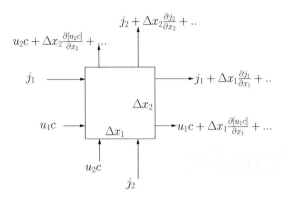

Figure 10.1   Infinitesimal control volume in a flow field. Derivation of conservation of species concentration.

The mass transfer coefficient, h_m, may be defined using an equivalent form of Newton's law of cooling (Equation (1.2)) as

$$j = h_m(c_w - c_a),$$ (10.2)

where subscripts w and a represent the wall and atmosphere or free stream respectively.

10.2   Conservation of Species

The scalar species conservation equation may be derived in a similar fashion to the equation of conservation of energy, as discussed in Section 7.2.3. However, the difference here is that the concentration of a species replaces the temperature. In order to derive this equation, let us consider a control volume as shown in Figure 10.1. The concentration of the species convected into the control volume, in the x_1 direction, is

$$u_1 c \Delta x_2.$$ (10.3)

Similarly, the concentration convected into the control volume, in the x_2 direction, is

$$u_2 c \Delta x_1.$$ (10.4)

As in Chapter 7, a Taylor series expansion may be used to express the concentration convected out of the control volume, in both the x_1 and x_2 directions, as

$$u_1 c \Delta x_2 + \frac{\partial(u_1 c)}{\partial x_1} \Delta x_1 \Delta x_2$$ (10.5)

and

$$u_2 c \Delta x_1 + \frac{\partial(u_2 c)}{\partial x_2} \Delta x_2 \Delta x_1$$ (10.6)

respectively. The mass diffusion into and out of the control volume is also derived using the above approach. The mass diffusing into the domain in the x_1 direction (Fick's law of mass diffusion) is

$$j_1 \Delta x_2 = -D_{x_1} \frac{\partial c}{\partial x_1} \Delta x_2 \tag{10.7}$$

and the diffusion entering the control volume in the x_2 direction is

$$j_2 \Delta x_1 = -D_{x_2} \frac{\partial c}{\partial x_2} \Delta x_1. \tag{10.8}$$

Using a Taylor series expansion , the mass diffusing out of the control volume may be written as

$$-D_{x_1} \frac{\partial c}{\partial x_1} \Delta x_2 + \frac{\partial}{\partial x_1} \left(-D_{x_1} \frac{\partial c}{\partial x_1} \right) \Delta x_2 \Delta x_1 \tag{10.9}$$

in the x_1 direction and

$$-D_{x_2} \frac{\partial c}{\partial x_2} \Delta x_1 + \frac{\partial}{\partial x_2} \left(-D_{x_2} \frac{\partial c}{\partial x_2} \right) \Delta x_1 \Delta x_2 \tag{10.10}$$

in the x_2 direction.

Finally, the rate of change of concentration within the control volume is

$$\Delta x_1 \Delta x_2 \frac{\partial c}{\partial t}. \tag{10.11}$$

Now, it is a simple matter of balancing the mass entering and exiting the control volume. The mass balance may be obtained as

```
mass entering the control volume by convection +
mass entering the control volume by diffusion =
mass exiting the control volume by convection +
mass exiting the control volume by diffusion +
rate of change of mass within the control volume
```

Following the above mass balance approach and rearranging, we obtain

$$\frac{\partial c}{\partial t} + \frac{\partial (u_1 c)}{\partial x_1} + \frac{\partial (u_2 c)}{\partial x_2} = \left[\frac{\partial}{\partial x_1} \left(D_{x_1} \frac{\partial c}{\partial x_1} \right) + \frac{\partial}{\partial x_2} \left(D_{x_2} \frac{\partial c}{\partial x_2} \right) \right]. \tag{10.12}$$

On differentiating by parts the convection terms and substituting Equation (7.11) (continuity) into Equation (10.12), we obtain the simplified concentration equation in two dimensions as

$$\frac{\partial c}{\partial t} + u_1 \frac{\partial c}{\partial x_1} + u_2 \frac{\partial c}{\partial x_2} = \left[\frac{\partial}{\partial x_1} \left(D_{x_1} \frac{\partial c}{\partial x_1} \right) + \frac{\partial}{\partial x_2} \left(D_{x_2} \frac{\partial c}{\partial x_2} \right) \right]. \tag{10.13}$$

If the mass diffusivity D is assumed to be constant and $D = D_{x_1} = D_{x_2}$, the concentration equation is reduced to

$$\frac{\partial c}{\partial t} + u_1 \frac{\partial c}{\partial x_1} + u_2 \frac{\partial c}{\partial x_2} = D \left(\frac{\partial^2 c}{\partial x_1^2} + \frac{\partial^2 c}{\partial x_2^2} \right). \tag{10.14}$$

The concentration equation in vector form is

$$\frac{\partial c}{\partial t} + \mathbf{u}.\nabla c = D\nabla^2 c \tag{10.15}$$

and in indicial form

$$\frac{\partial c}{\partial t} + u_i \frac{\partial c}{\partial x_i} = D \frac{\partial^2 c}{\partial x_i^2} \tag{10.16}$$

and finally in conservation form

$$\frac{\partial c}{\partial t} + \frac{\partial}{\partial x_i}(u_i c) = D \frac{\partial^2 c}{\partial x_i^2}. \tag{10.17}$$

The above equation is applicable in any space dimension. If the time-dependent term $\partial/\partial t$ is neglected, from Equation (10.17) we obtain a steady-state equation for the species transport. A pure mass diffusion equation may be obtained by substituting the velocity components $u_i = 0$ in Equation (10.17). In problems with multiple species transport, each species should be represented by a convection-diffusion equation of the type shown in Equation (10.17). In some cases, the interaction between the species may be very complex. For incompressible flow problems, either a conservative (Equation (10.17)) or nonconservative (Equation (10.16)) form of the concentration equation may be employed.

10.2.1 Nondimensional form

The nondimensional form of Equation (10.16) may be useful in carrying out some general mass transfer calculations. The nondimensional scales may be different, depending on the type of problem studied. For forced convective mass transfer with given Dirichlet conditions for concentration, we recommend the following nondimensional scales.

$$c^* = \frac{c - c_\infty}{c_w - c_\infty}; t^* = \frac{tL}{u_\infty}; x_i^* = \frac{x_i}{L}; u_i^* = \frac{u_i}{u_\infty}, \tag{10.18}$$

where L is a characteristic dimension, the subscript ∞ indicates a reference quantity and subscript w indicates a wall. The above scales result in the following nondimensional equation.

$$\frac{\partial c^*}{\partial t^*} + u_i^* \frac{\partial c^*}{\partial x_i^*} = \left(\frac{D}{u_\infty L}\right) \frac{\partial^2 c^*}{\partial x_i^{*2}}. \tag{10.19}$$

In the above equation

$$\frac{D}{u_\infty L} = \left(\frac{D}{v}\right)\left(\frac{v}{u_\infty L}\right) = \frac{1}{ScRe}, \tag{10.20}$$

where $Sc = v/D$ is the Schmidt number and $Re = u_\infty L/v$ is the Reynolds number. A typical example where the above formulation can be used is the case of a channel flow with a higher concentration of species along the walls and a lower concentration in the fluid approaching the channel. If the mass flux conditions are given, then the scales need to be redefined using the given mass flux.

10.2.2 Buoyancy-driven Mass Transfer

So far the mass transport has been discussed from a forced convection perspective. The mass transport can also take place under the influence of pure buoyancy, or, in a mixed convective form. This is very similar in nature to free and mixed convection heat transfer as discussed in Chapter 7. Buoyancy-driven mass convection is generated by the density difference induced by the concentration differences within a fluid system. Because of the small density variations present in these type of flows, a general incompressible flow approximation is normally adopted. To represent the variation in density, a body force term needs to be added to the momentum equations to include the effect of local density differences, that is,

$$\frac{g(\rho - \rho_a)}{\rho_a} = g\beta_c(c - c_\infty),$$ (10.21)

where g is the acceleration due to gravity (9.81 m/s^2) and β_c is the coefficient of solutal expansion. The above body force term is added to the momentum equations in the gravity direction. In a normal situation, the body force is added to the x_2 momentum in 2D flows (if the gravity direction x_2 is negative), that is,

$$\frac{\partial u_2}{\partial t} + u_1\frac{\partial u_2}{\partial x_1} + u_2\frac{\partial u_2}{\partial x_2} = -\frac{1}{\rho}\frac{\partial p}{\partial x_2} + v\left(\frac{\partial^2 u_2}{\partial x_1^2} + \frac{\partial^2 u_2}{\partial x_2^2}\right) + g\beta_c(c - c_\infty).$$ (10.22)

In practice, the following nondimensional scales are adopted for buoyancy-driven mass convection in the absence of a reference velocity value.

$$x_1^* = \frac{x_1}{L}; \quad x_2^* = \frac{x_2}{L}; \quad t^* = \frac{tD}{L^2}; \quad u_1^* = \frac{u_1L}{D};$$

$$u_2^* = \frac{u_2L}{D}; \quad p^* = \frac{pL^2}{\rho D^2}; \quad c^* = \frac{c - c_\infty}{c_w - c_\infty}.$$ (10.23)

On introducing the above nondimensional scales into the governing equations, we obtain nondimensional form of the equations as follows:

Continuity equation:

$$\frac{\partial u_1^*}{\partial x_1^*} + \frac{\partial u_2^*}{\partial x_2^*} = 0.$$ (10.24)

x_1 momentum equation:

$$\frac{\partial u_1^*}{\partial t^*} + u_1^*\frac{\partial u_1^*}{\partial x_1^*} + u_2^*\frac{\partial u_1^*}{\partial x_2^*} = -\frac{\partial p^*}{\partial x_1^*} + Sc\left(\frac{\partial^2 u_1^*}{\partial x_1^{*2}} + \frac{\partial^2 u_1^*}{\partial x_2^{*2}}\right).$$ (10.25)

x_2 momentum equation:

$$\frac{\partial u_2^*}{\partial t^*} + u_1^*\frac{\partial u_2^*}{\partial x_1^*} + u_2^*\frac{\partial u_2^*}{\partial x_2^*} = -\frac{\partial p^*}{\partial x_2^*} + Sc\left(\frac{\partial^2 u_2^*}{\partial x_1^{*2}} + \frac{\partial^2 u_2^*}{\partial x_2^{*2}}\right) + Gr_cSc^2c^*.$$ (10.26)

Concentration equation:

$$\frac{\partial c^*}{\partial t^*} + u_1^*\frac{\partial c^*}{\partial x_1^*} + u_2^*\frac{\partial c^*}{\partial x_2^*} = \left(\frac{\partial^2 c^*}{\partial x_1^{*2}} + \frac{\partial^2 c^*}{\partial x_2^{*2}}\right).$$ (10.27)

where Gr_c is the solutal Grashof number given as

$$Gr_c = \frac{g\beta_c \Delta c L^3}{\nu^2}. \tag{10.28}$$

where $\Delta c = c_w - c_\infty$. Often, another nondimensional number called the solutal Rayleigh number   is used in the calculations. This is given as

$$Ra_c = Gr_c Sc = \frac{g\beta_c \Delta c L^3}{\nu D}. \tag{10.29}$$

On comparing the nondimensional equations of natural and forced convection, it is easy to identify the differences. If we substitute $1/Sc$ in place of the Reynolds number for the forced convection equations, we revert to a natural convection scaling. Obviously, the extra buoyancy term needs to be added to appropriate component(s) of the momentum equation for natural convection flows.

10.2.3 Double-diffusive Natural Convection

As mentioned in the previous section, buoyancy-driven mass convection is possible and has been widely studied. In many application problems, mass transfer is very much influenced also by the presence of energy transport. One such example is the so called double-diffusive natural convection. This has been widely studied in many application areas, including saturated porous media (see Chapter 11). In this section we explain this phenomenon briefly. Here, in addition to buoyancy-driven mass convection, the buoyancy-driven heat convection is also important. Thus, two body forces, one each for mass and energy, appear in the momentum equation, that is, in addition to Equation (10.21),

$$\frac{g(\rho - \rho_a)}{\rho_a} = g\beta(T - T_\infty) \tag{10.30}$$

will also appear in the momentum equation in the gravity direction. Assuming, the gravity is in a negative x_2 direction, the dimensional form of the momentum equation may be written in two dimension as

$$\frac{\partial u_2}{\partial t} + u_1 \frac{\partial u_2}{\partial x_1} + u_2 \frac{\partial u_2}{\partial x_2} = -\frac{1}{\rho}\frac{\partial p}{\partial x_2} + \nu\left(\frac{\partial^2 u_2}{\partial x_1^2} + \frac{\partial^2 u_2}{\partial x_2^2}\right) + g\beta(T - T_\infty) + g\beta_c(c - c_\infty). \tag{10.31}$$

Using the standard nondimensional scales for buoyancy-driven heat convection (Chapter 7), we obtain

$$\frac{\partial u_2^*}{\partial t^*} + u_1^* \frac{\partial u_2^*}{\partial x_1^*} + u_2^* \frac{\partial u_2^*}{\partial x_2^*} = -\frac{\partial p^*}{\partial x_2^*} + Pr\left(\frac{\partial^2 u_2^*}{\partial x_1^{*2}} + \frac{\partial^2 u_2^*}{\partial x_2^{*2}}\right) + RaPrT^* + BRaPrc^*, \tag{10.32}$$

where B is the buoyancy ratio, given as

$$B = \frac{\beta_c \Delta c}{\beta \Delta T}. \tag{10.33}$$

In addition to the continuity and momentum equations, double-diffusive convection requires a solution to the natural convective heat and mass transport equations, that is,

$$\frac{\partial T^*}{\partial t^*} + u_1^* \frac{\partial T^*}{\partial x_1^*} + u_2^* \frac{\partial T^*}{\partial x_2^*} = \left(\frac{\partial^2 T^*}{\partial x_1^{*2}} + \frac{\partial^2 T^*}{\partial x_2^{*2}} \right) \tag{10.34}$$

and

$$\frac{\partial c^*}{\partial t^*} + u_1^* \frac{\partial c^*}{\partial x_1^*} + u_2^* \frac{\partial c^*}{\partial x_2^*} = \frac{1}{Le} \left(\frac{\partial^2 c^*}{\partial x_1^{*2}} + \frac{\partial^2 c^*}{\partial x_2^{*2}} \right), \tag{10.35}$$

where Le is the Lewis number defined as $Le = \alpha/D$.

10.3 Numerical Solution

The temporal and spatial discretizations of the conservation of species equation follows the discretizations of the energy equation as given in Section 7.6. In the case of pure mass diffusion, discretization of the diffusion equation follows the discretization used for heat conduction as in Chapters 4, 5 and 6 depending on the type of diffusion required. For pure convective mass transport, without the influence of temperature, step four of the algorithm described in Section 7.6 will be replaced with a concentration calculation using the species concentration Equation (10.19). If the calculation involves the influence of temperature, the energy equation is solved at step four as shown in Section 7.6 and the species concentration equation forms the fifth step of the algorithm. This is the case for double-diffusive convection.

Example 10.3.1   *Mass diffusion at steady state*
Figure 10.2 shows the problem definition used here to study mass diffusion. The domain is one unit high and ten units long. The bottom side is assumed to be at a higher concentration than the top side. Both the vertical sides are assumed to have no flux exiting or entering the domain to mimic a one-dimensional mass flow in the vertical direction. The mass diffusion here is assumed to be time independent and thus the steady-state mass diffusion equation,

$$\frac{\partial^2 c}{\partial x_1^2} + \frac{\partial^2 c}{\partial x_2^2} = 0 \tag{10.36}$$

is sufficient to determine the distribution of the species concentration. This equation can be solved in a variety of ways as discussed in Chapter 5. The above equation can also be solved using the time-dependent equation,

$$\frac{\partial c}{\partial t} = D \left(\frac{\partial^2 c}{\partial x_1^2} + \frac{\partial^2 c}{\partial x_2^2} \right), \tag{10.37}$$

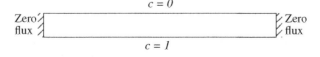

Figure 10.2   Steady-state mass diffusion in a rectangular domain.

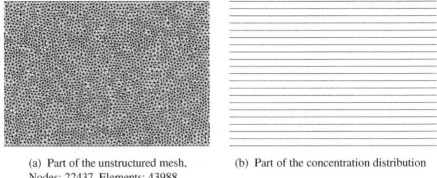

(a) Part of the unstructured mesh,   (b) Part of the concentration distribution
Nodes: 22437, Elements: 43988

Figure 10.3   Steady-state diffusion of a species from the bottom side of a channel. Part of the
mesh and concentration distributions.

*by iterating in time to steady state. Note that the mass diffusion in the above equation is
assumed to be isotropic, that is, the mass diffusion coefficient D is the same in both x_1 and x_2
directions. Although any method described in Chapters 5 and 6 can be used to solve the above
equations, an explicit scheme (Chapter 5) with mass lumping is employed here to obtain the
solution for the problem shown in Figure 10.2.*

*A uniform unstructured mesh as shown in Figure 10.3(a) is used in the calculations. Fig-
ure 10.3(b) shows the concentration distribution at steady state. As seen, the isoconcentration
lines are equally distanced and the difference between two consecutive isoconcentration lines
are the same throughout the domain. This indicates a linear decrease in concentration from a
nondimensional value of unity at the bottom wall to a nondimensional value of zero at the top
wall. This is not surprising as the equation solved is a linear Laplace equation and in one
dimension, it can easily be shown that the analytical solution is linear (Chapter 4). It should
be noted here that the solution for steady state is independent of the diffusion coefficient.
Irrespective of the media involved, the solution will be identical in steady state.*

Example 10.3.2   *Transient mass diffusion in one dimension*
*In the following problem, we demonstrate a model problem of time-dependent species
diffusion. For time-dependent diffusion, the solution depends heavily on all aspects of the
problem, including the diffusion coefficient D. To compare the finite element results with
analytical solution in one dimension, the following one-dimensional equation is considered.*

$$\frac{\partial c}{\partial t} = D\left(\frac{\partial^2 c}{\partial x^2}\right). \tag{10.38}$$

*An analytical solution to this one-dimensional equation can be obtained for a simplified
problem, in a semi-infinite domain (Holman 1989; Schneider 1955). This solution is originally
derived for heat conduction and adopted here for mass diffusion. Imagine an infinitely long
rod maintained at a concentration of c_i (e.g. moisture) and suddenly the concentration at one*

end of the rod is dropped down to a smaller value c_w. The analytical solution to this problem is given as

$$\frac{c(x,t) - c_w}{c_i - c_w} = erf\left(\frac{x}{2\sqrt{Dt}}\right),$$ (10.39)

where t is the time, x is the distance from the end where concentration is c_w and the Gauss error function is defined as

$$erf\left(\frac{x}{2\sqrt{Dt}}\right) = \frac{2}{\sqrt{\pi}} \int_0^{\frac{x}{2\sqrt{Dt}}} e^{-\zeta^2} d\zeta,$$ (10.40)

where ζ here is a dummy variable. Thus, the space- and time-dependent species concentration can be determined at any point at any time using Equation (10.39).

To compare the numerical solution with the analytical solution of Equation (10.39), a two-dimensional domain of 1 cm hight and 100 cm length is considered. The concentration through out the domain is assumed to be unity at $t = 0$ and the concentration on the left vertical side of the domain is suddenly reduced to zero. All the other three sides of the domain are assumed to have zero mass flux condition. A mass diffusivity of $D = 0.726 \, cm^2/s$ is used (diffusion of H_2 into CH_4 with a diffusivity of $0.726 \, cm^2/s$). The assumption here is that the domain is static and isothermal. This is not often the case in real problems.

Figure 10.4 shows the concentration distribution with respect to time at a point 24.6686 cm from the left vertical wall of the domain. As seen, a non-linear decay of concentration is observed with respect to time. It is also shown that the analytical and numerical solutions agree excellently.

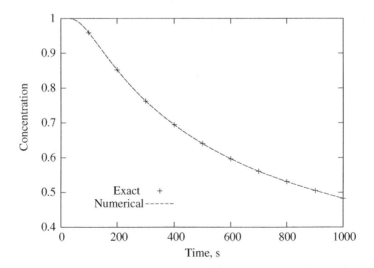

Figure 10.4 Unsteady-state mass diffusion in a rectangular domain.

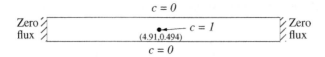

Figure 10.5   Unsteady-state mass diffusion in a rectangular domain.

Example 10.3.3   *Transient heat diffusion in two-dimension*
    *The second problem of transient diffusion considered is mass diffusion from a point source.
Figure 10.5 shows the model problem used for the transient solution of the mass diffusion
equation in two dimension. As seen the domain is unchanged from the previous steady-
state problem but the boundary conditions are changed. As seen, a point source with a
nondimensional species concentration of unity is introduced near the center of the domain
in addition to zero concentration boundary conditions on both top and bottom horizontal
walls. The vertical walls are assumed to be imposed with zero mass flux conditions. The
mass diffusion coefficient is same as the previous problem and is equal to 0.726 cm²/s. The
initial concentration throughout the domain is assumed to be zero. The concentration at point
(4.91,0.494) is then suddenly increased to unity at t > 0. The distribution of the concentration
with respect to time is then monitored.*
    *Figure 10.6 shows the concentration distribution with respect to time. The large value of
diffusion coefficient used here, and close vicinity of both the top and bottom walls (with zero
concentration) to the source, force the concentration to reach a steady-state very fast. From the*

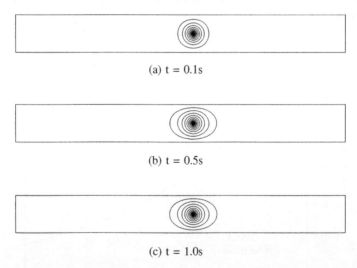

Figure 10.6   Unsteady-state diffusion of gas H_2 into CH_4 with a diffusivity of 0.726 cm²/s
at 298 K. Top and bottom sides with zero concentration and both vertical sides insulated.
Concentration distribution at different time periods.

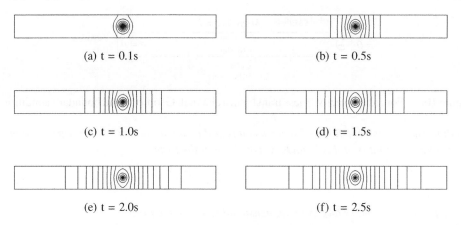

(a) t = 0.1s   (b) t = 0.5s

(c) t = 1.0s   (d) t = 1.5s

(e) t = 2.0s   (f) t = 2.5s

Figure 10.7   Unsteady-state diffusion of gas H_2 into CH_4 with a diffusivity of $0.726\,\text{cm}^2/\text{s}$ at 298 K. Top and bottom sides insulated and both vertical sides with zero concentration. Concentration distribution at different time periods.

figure, it is extremely difficult to distinguish between the results at 0.5s and 1s. This shows that an equilibrium state has been reached between the source and the boundary conditions within 0.5 seconds. Since the vertical walls are far away from the source, the concentration reaches a steady state equilibrium before any trace of the concentration from the source reaching the vertical walls on either side.

In order to examine the effect of outer boundary conditions, the boundary conditions between the vertical and horizontal walls in the previous problem are interchanged in the next model problem, that is, the vertical walls are assumed to have zero concentration and the horizontal walls are assumed to be prescribed with zero mass flux. All the other conditions remain the same.

The results for the problem with rearranged boundary conditions are shown in Figure 10.7. Since the zero concentration boundary condition now is not in the vicinity of the source, the problem has not reached a steady-state equilibrium even at 2.5 s. This time is much higher compared to the previous problem in which the steady state was reached before 0.5 s.

Example 10.3.4   *Forced mass convection*

All the problems discussed in the previous examples are on the topic of pure mass diffusion. The medium into which the concentration of a species is allowed to diffuse was assumed to be static, although this is not always true in real situations. In many industrial situations, such as food, brick and certain kind of mold drying, controlling the drying process is essential to obtain the required quality. In such situations, air is forced over the material. In this section, one such model problem in two dimensions is considered.

Figure 10.8 shows the geometry and boundary conditions used in the study. As seen the problem domain is unchanged from the previous example with one unit height and ten units length. A fully developed parabolic velocity profile is assumed at the inlet and the

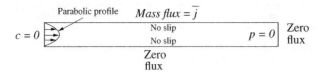

Figure 10.8   Forced convective mass transfer in a channel. Geometry and boundary conditions.

nondimensional concentration of species entering the domain is zero. At the top wall a mass flux is assumed, that is, a flux boundary condition of the form

$$\bar{j} = -D\frac{\partial c}{\partial y}, \tag{10.41}$$

where $\bar{j}$ is the given mass flux. A nondimensional scale of the form

$$c^* = \frac{c - c_a}{\dfrac{\bar{j}L}{D}} \tag{10.42}$$

is used here. Thus, the nondimensional form of Equation (10.41) becomes

$$-\frac{\partial c^*}{\partial y^*} = 1. \tag{10.43}$$

This relationship is easy to implement as discussed in Chapter 7. Apart from the flux boundary condition on the top, all other boundaries are assumed to have a zero concentration flux condition. A zero pressure boundary condition is prescribed at the exit.

Let us assume that the fluid flowing into the domain be air and the water vapour is the species that is transported into the domain through the top wall via the constant mass flux condition discussed above. Both the air and water vapor are assumed to be at the same temperature. The diffusivity of the water vapor diffusing into the air is approximately 2.5×10^{-5} m^2/s and the kinematic viscosity of air is 1.0×10^{-5} m^2/s at room temperature. This gives a Schmidt number of approximately 0.4. This value is used here.

Figure 10.9 shows the isoconcentration lines at different inlet Reynolds numbers. The Reynolds number here is defined based on the average inlet velocity and channel height. As seen, the solutal boundary layer is becoming thinner as the Reynolds number is increased. Figure 10.10 shows the local Sherwood number $\left(sh = \frac{h_t L}{D}\right)$ distribution along the top side of the channel. As seen, the Sherwood number increases as the Reynolds number is increased. It is also important to note that the Sherwood number is very high near the inlet, indicating a large mass transport. This is due to the fact that the air with zero concentration approaching the channel will remove large quantities of mass at the entrance and the mass transfer reduces as we move along the length of the channel.

Example 10.3.5   *Double-diffusive convection in a square cavity*

The double-diffusive convection is the convection that has both thermal and mass convection. The double-diffusive convection can be of forced and natural type. In this section, we consider only the natural type to demonstrate the double-diffusive convection. As discussed earlier, there are many nondimensional parameters involved in a double-diffusive natural

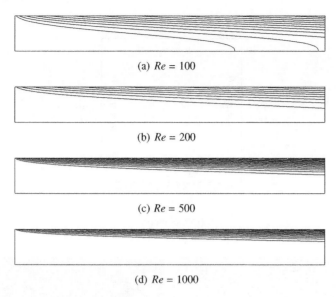

(a) $Re = 100$

(b) $Re = 200$

(c) $Re = 500$

(d) $Re = 1000$

Figure 10.9   Forced convective mass transfer in a channel. Concentration distribution at different Reynolds numbers, $Sc = 0.4$.

convection problem. They are Rayleigh number, Prandtl number, Lewis number and Buoyancy ratio. Only small number of parameters are considered here to demonstrate the finite element solver. The problem solved here is very similar to the natural convection solved in Chapter 7. The difference here is that the additional variable of concentration is also solved. The problem domain consists of a closed rectangular cavity with rigid walls. All the walls are subjected to

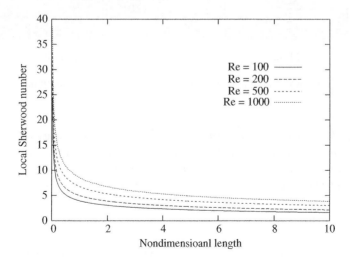

Figure 10.10   Forced convective mass transfer in a channel. Local Sherwood number distribution at different Reynolds numbers, $Sc = 0.4$.

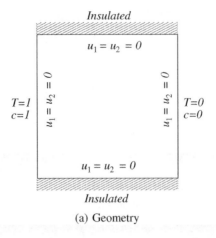

(a) Geometry

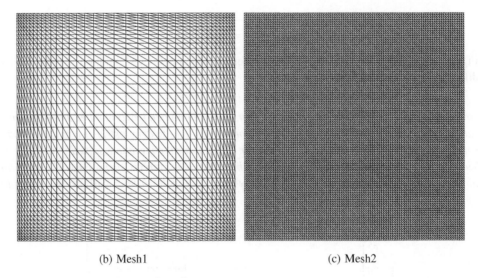

(b) Mesh1   (c) Mesh2

Figure 10.11   Buoyancy-driven convection in a cavity. Domain and boundary conditions and meshes used.

no slip velocity boundary conditions. The top and bottom walls are assumed to have zero heat and mass flux entering or exiting the domain. The left-side vertical wall is assumed to be at a higher temperature and concentration than that of the right-side vertical wall. Figure 10.11(a) shows the complete problem definition. For the sake of simplicity, the Prandtl number is fixed as 0.72 and Lewis number is assumed to be unity in the first case studied. The aspect ratio of the first case studied is unity (square cavity). The finite element meshes used are shown in Figure 10.11(b) and (c).

Figure 10.12 shows the velocity vectors and contours of concentration distribution at different buoyancy ratios. The temperature contours are identical to that of the concentration contours and thus they are not presented here. Only one Rayleigh number solution is presented here to demonstrate the problem. The Rayleigh number selected is 10^5. Figure 10.12(a) shows

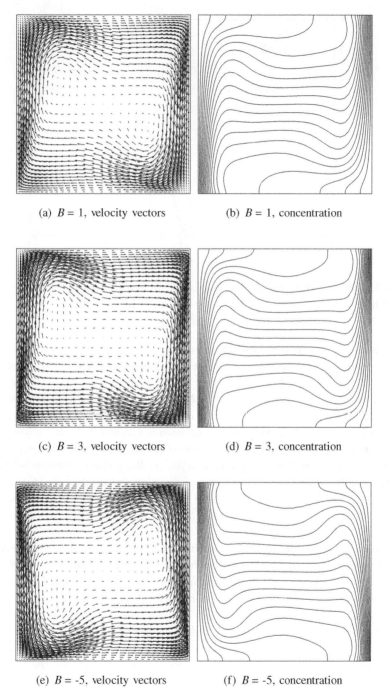

(a) $B = 1$, velocity vectors   (b) $B = 1$, concentration

(c) $B = 3$, velocity vectors   (d) $B = 3$, concentration

(e) $B = -5$, velocity vectors   (f) $B = -5$, concentration

Figure 10.12 Double diffusive natural convection in a square cavity. Concentration and velocity distributions at difference buoyancy rations, $Ra = 10^5$, $Pr = 0.72$, $Le = 1.00$.

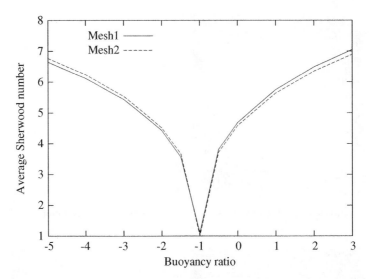

Figure 10.13 Double diffusive natural convection in a square cavity. Average Sherwood number distribution at different buoyancy ratios, $Ra = 10^5$, $Pr = 0.72$, $Le = 1.00$.

the results at a buoyancy ratio of unity. This means the effect of Rayleigh number doubles and results in a combined Rayleigh number of 2×10^5. As seen, the flow pattern is close to the natural convective flow at a Rayleigh number of 10^5, given in Chapter 7. As the buoyancy ratio is increased to 3 as shown in Figure 10.12(b), the combined Rayleigh number increases to 4×10^5 and the effect on flow, concentration and temperature distributions are very similar to the effect of increase in Rayleigh number discussed in Chapter 7. As the buoyancy ratio is reduced to −1, the thermal and solutal buoyancies cancel each other and the temperature and concentration transport will be through pure diffusion. Reduction of B below −1 will result in a net negative buoyancy value in the upward direction and the flow direction reverses as shown in Figure 10.12(c) at B = −5. At this B value, a combined Rayleigh number value of -4×10^5 is obtained and results in a solution that is a mirror image of the solution at B = 3 as shown in Figures 10.12(b) and (c).

Figure 10.13 shows the average Sherwood number distribution at a Rayleigh number of 10^5 at different buoyancy ratios. The Nusselt number distribution is identical to the distribution shown and thus not plotted separately. The Sherwood numbers calculated on both meshes are plotted to show the effect of number of points. The mesh1 has a small number of points compared to mesh2. Due to the small number of nodes and the orientation of triangles, the average Sherwood number distribution is slightly unsymmetric with respect B = −1. When the number of points are increased, as in mesh2, the birdseye view distribution of the average Sherwood number as shown in Figure 10.13 is more or less symmetric. The birdseye view distribution is typical in such applications.

Example 10.3.6 *Double-diffusive convection in a rectangular cavity*

The second double-diffusive natural convection problem studied here again is flow in a rectangular cavity but now with an aspect ratio of 7. The number of nodes and elements used

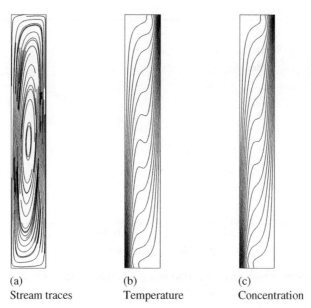

| (a) | (b) | (c) |
| Stream traces | Temperature | Concentration |

Figure 10.14   Double diffusive natural convection in a rectangular cavity with aspect ratio 7. Stream traces, temperature and concentration distributions at, $Ra = 10^4$, $Pr = 0.72$, $Le = 0.857$, $B = 1$.

are same as the problem studied previously but the meshes in Figure 10.11 are stretched here to increase the aspect ratio to 7. The boundary conditions remain the same and to match the conditions of Wee et al. *(1989), a Lewis number of 0.857 is adopted here inline with the Prandtl and Schmidt numbers of 0.7 and 0.6 used in the experiments of Wee* et al. *(1989). A thermal diffusivity of air 2.216×10^{-5} is used to compute the Lewis number.*

Figure 10.14 and 10.15 show the results for this problem. As seen, at $Ra = 10^4$ the flow pattern is dominated by one large vortex. The important aspect here is that the temperature and concentration patterns are not identical due to Lewis number being not unity. As seen, the concentration distribution is convectively less dominated than that of isotherms. This is indicated by isotherms with rapid change in structure at the center of the cavity. Figure 10.15 shows the average Sherwood number distribution with respect the combined Rayleigh number (thermal plus solutal Rayleigh numbers). As seen the Sherwood number increases with the Rayleigh number and the mesh with larger number of points give a better and more smoother Sherwood number distribution. Although not shown, the Sherwood number distribution is in close agreement with the experimental data provided by Wee et al. *(1989).*

10.4   Turbulent Mass Transport

We herein provide a brief summary of turbulent mass transport. Turbulent mass transport is very similar to the turbulent heat transport discussed in Chapter 8. The time averaging, used

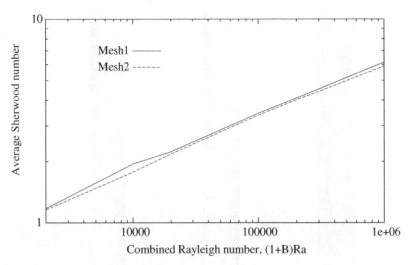

Figure 10.15   Buoyancy ndriven convection in a cavity with aspect ratio 7. Average Sherwood number for different Rayleigh numbers, $Le = 0.857$.

in Chapter 8, can also be used here for averaging the concentration equation. For deriving a time averaged equation, the concentration may be expressed as

$$c = \bar{c} + c'$$ (10.44)

where $\bar{c}$ is an averaged field and c' represents fluctuating part (refer to Figure 8.1). Substituting Equation (10.44) into Equation (10.19) and time averaging (see Section 8.1) we obtain

$$\frac{\partial \bar{c}}{\partial t} + \frac{\partial}{\partial x_i}(\overline{u_i c}) = D\frac{\partial^2 \bar{c}}{\partial x_i^2} - \frac{\partial}{\partial x_i}(\overline{u_i' c'})$$ (10.45)

or

$$\frac{\partial \bar{c}}{\partial t} + \frac{\partial}{\partial x_i}(\overline{u_i c}) = -\frac{\partial j_i}{\partial x_i} - \frac{\partial j_i^t}{\partial x_i}.$$ (10.46)

where

$$j_i = -D\frac{\partial c}{\partial x_i}$$ (10.47)

is the laminar mass flux and

$$j_i^t = -\overline{u_i' c'}$$ (10.48)

is the turbulent mass flux. In order to close the equation, the turbulent flux needs to be related to the time-averaged concentration, that is,

$$j_i^t = D_t\frac{\partial \bar{c}}{\partial x_i},$$ (10.49)

where D_t is the turbulent eddy mass diffusivity. If we invoke the analogy with heat transfer, we may be able write $D_t/\alpha_t = 1$. This ratio may be referred to as the turbulent Lewis number, Le_t. Recollect that α_t is calculated in Chapter 8.

10.5   Summary

In this chapter, we have provided brief, but sufficient information, on the finite element solution of mass transport equations. The chapter explained both diffusion and convection mass transport in addition to double-diffusive mass transport, in which both the mass and heat transport influence each other. The numerical solution of the mass transport equation followed the identical procedure followed in Chapters 4 to 7. Several sample examples are provided to demonstrate the application of the finite element method to solving mass transport equations. The chapter was concluded with a short description of how to incorporate turbulence into the mass transport equation.

References

Bird RB, Stewart WE and Lightfoot EN (1960) *Transport Phenomena*. John Wiley & Sons, Inc., New York.

Comini G and Lewis RW (1976) Numerical solution of 2-dimensional problems involving heat and mass transfer. *International Journal of Heat and Mass Transfer*, **19**, 1387–1392.

Holman JP (1989) *Heat Transfer*, McGraw-Hill, New York.

Lewis RW and Malan AG (2005) Continuum thermodynamic modeling of drying capillary particulate materials via an edge-based algorithm. *Computer Methods in Applied Mechanics and Engineering*, **194**, 2043–2057.

Murugesan K, Suresh HN, Seetharamu KN, *et al.* (2001) A theoretical model of brick drying as a conjugate problem. *International Journal of Heat and Mass Transfer*, **44**, 4075–4086.

Nithiarasu P, Seetharamu KN and Sundararajan T (1996) Double-diffusive natural convection in an enclosure filled with fluid-saturated porous medium: a generalized non-Darcy approach, *Numerical Heat Transfer Part A Applications*, **30**, 413–426.

Schneider PJ, 1955 *Conduction Heat Transfer*. Addison-Wesley Publishing Company, Inc., Reading, MA.

Wee HK, Kee RB and Cunningham MJ (1989) Heat and moisture transfer by natural convection in a rectangular cavity. *International Journal of Heat and Mass Transfer*, **32**, 1765–1778.

11

Convection Heat and Mass Transfer in Porous Media

11.1 Introduction

The phenomenon of fluid flow, heat and mass transfer in porous media has been recognized as a separate engineering topic for the last few decades. Several books have been published on this topic (Kaviany 1991; Lewis and Schrefler 1998; Nield and Bejan 1992; Zienkiewicz *et al.* 1999). Convection in porous media occurs in many engineering applications including packed beds, thermal insulation, metal solidification and geothermal problems. Advanced applications such as petroleum reservoirs, multi-phase flows and drying have also been studied using finite elements (Lewis and Ferguson 1990, Lewis *et al.* 1983, 1989; Lewis and Sukirman 1993; Murugesan *et al.* 2001; Pao *et al.* 2001). A wide variety of solution methodologies, both analytical and numerical, are available for solving porous media flow and heat transfer. Analytical methods are limited by many factors and the solution of realistic field problems is normally intractable by such techniques. With the advent of computing power in the last three decades, solutions to many practical porous medium problems are feasible using numerical methods (Lewis and Schrefler 1998; Zienkiewicz *et al.* 1999). Such numerical solution procedures have their own limitations, such as accuracy, implementation difficulties etc. However, with a proper combination of algorithms and discretization techniques, it is possible to obtain reasonably accurate solutions for complex problems where analytical approaches would not be feasible. In this chapter, the finite element modeling of incompressible flow, heat and transfer through porous media will be outlined in some detail.

The flow of fluid in a saturated porous media was obtained by a simple, phenomenological, linear relation by Darcy in the 19th century (Darcy 1856). Darcy's law relates the pressure

Fundamentals of the Finite Element Method for Heat and Mass Transfer, Second Edition.
P. Nithiarasu, R. W. Lewis, and K. N. Seetharamu.
© 2016 John Wiley & Sons, Ltd. Published 2016 by John Wiley & Sons, Ltd.

drop (head) to the flow rate across a porous column. The following relation can be written from such observations, that is,

$$u_i = -\frac{\kappa}{\mu}\frac{\partial p}{\partial x_i}. \tag{11.1}$$

Where u_i are the seepage velocity components, κ (m^2) is the permeability of the medium, μ is the dynamic viscosity of the fluid, p is the pressure and x_i are the coordinate axes. For two-dimensional flow, we can rewrite the velocity components as

$$u_1 = -\frac{\kappa}{\mu}\frac{\partial p}{\partial x_1}$$

$$u_2 = -\frac{\kappa}{\mu}\frac{\partial p}{\partial x_2}. \tag{11.2}$$

It is interesting to note that the above equation is very similar to Ohm's law for the flow of electricity, Fourier's law of heat conduction and Fick's law for mass diffusion. Substitution of the Darcy's law into the conservation of mass (divergence free velocity field) equation for incompressible flow yields (without body forces)

$$\frac{\partial}{\partial x_i}\left(-\frac{\kappa}{\mu}\frac{\partial p}{\partial x_i}\right) = 0. \tag{11.3}$$

This equation is often employed in problems with very low porosity (example: petroleum reservoirs). However, simple relations such as Darcy's law are not always applicable and further modifications, or extensions, are necessary in order to accurately predict the flow field in porous media.

Several years after the introduction of Darcy's law, two major extensions to the model have extended its use in many engineering disciplines including chemical, mechanical and civil engineering. The first extension was due to Forchheimer in 1901 (Forchheimer 1901) and this modification accounted for moderate and high Reynolds number effects with the addition of a nonlinear term in the Darcy equation. A relation of the form

$$D_p = au_i + bu_i^2 \tag{11.4}$$

for drag force was introduced by Forchheimer (Figure 11.1), which is balanced by the pressure force as follows:

$$au_i + bu_i^2 = -\frac{\partial p}{\partial x_i}. \tag{11.5}$$

In the above equation, the first term on the left-hand side is, in essence, similar to the linear drag term introduced by Darcy and the second term is the nonlinear drag term. The parameters a and b are determined by empirical relations and one such correlation was given by Ergun (Ergun 1952), that is,

$$a = 150\frac{(1-\epsilon^2)}{\epsilon^3}\frac{\mu_f}{d_p^2} \tag{11.6}$$

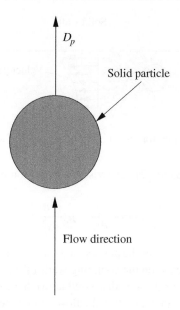

D_p

Solid particle

Flow direction

Figure 11.1   Drag force on a porous medium grain.

and

$$b = 1.75\frac{(1 - \epsilon)}{c^3}\frac{\rho_f}{d_p}.$$ (11.7)

It should be noted, however, that other suitable correlations may also be employed in different ranges of the bed porosity, ϵ, to obtain the non-Darcian flow behavior inside a porous medium. In the above equations, d_p is the solid particle size in a porous medium and ρ_f is the fluid density. The above solid matrix drag relation can also be expressed in terms of the medium permeability κ by defining

$$\kappa = \frac{\epsilon^3 d_p^2}{150(1 - \epsilon)^2}.$$ (11.8)

The flow relationship given by Equation (11.5) can be rewritten in terms of permeability as

$$\frac{\mu_f u_i}{\kappa} + \frac{1.75}{\sqrt{150}}\frac{\rho_f}{\sqrt{\kappa}}\frac{|\mathbf{V}|}{\epsilon^{3/2}}u_i = -\frac{\partial p}{\partial x_i}$$ (11.9)

Although the above equation gives an accurate solution at higher Reynolds numbers, it is not accurate enough to solve flow in highly porous and confined media. In order to deal with the viscous and higher porosity effects, *Brinkman* introduced an extension to the Darcy model in 1947, which included a second-order viscous term with an equivalent viscosity for

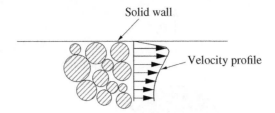

Figure 11.2　Viscous forces on a bounding wall of a porous medium.

the porous medium (Brinkman 1947). The viscous extension as given by Brinkman can be written as (Figure 11.2)

$$au_i = -\frac{\partial p}{\partial x_i} + \mu_e \frac{\partial^2 u_i}{\partial x_i^2}, \tag{11.10}$$

where μ_e is the equivalent viscosity of the porous medium. This modification takes into account the no-slip conditions, which exist on the confining walls (Tong and Subramanian 1985).

The Darcy model and the extensions discussed above have been widely used in the past. However, a generalized model, incorporating the flow regimes covered by both Darcy's model and its extension, will have several advantages (Hsu and Cheng 1990; Massarotti *et al.* 2003; Nithiarasu *et al.* 1997, 2002; Vafai and Tien 1981; Whitaker 1961). One of these is that the generalized flow model approaches the standard incompressible Navier-Stokes equations when porosity approaches a value of unity. The discussion on convection in porous media in this chapter will be brief and based on the generalized porous medium approach. Readers should be aware of the CBS scheme and the notations used in Chapter 7 before reading this chapter.

11.2　Generalized Porous Medium Flow Approach

In this section, a generalized model for solving porous medium flows will be presented. Let us consider the balance of mass, momentum, energy and species for two dimensional flow in a fluid saturated porous medium of variable porosity. The derivations are very similar to the one discussed in Chapter 7. We shall assume the medium to be isotropic with constant physical properties, except for the medium porosity. Let a_f be the fraction of area available for flow per unit cross-sectional area (Figure 11.3), at a location in a given direction. In fact, a_f is an averaged quantity, the average being taken over the length scale of the voids (or the length scale of the particles, if the porous bed is made up of particles), in the flow direction. For an isotropic porous bed, a_f will be identical in all directions and can also be equal to the local bed porosity, ϵ. In spite of averaging over the void length scale, the fractional area a_f may vary from location to location on the macro-length scale 'L' of the physical problem, due to the variation of the bed porosity.

The porosity, ϵ, of the medium is defined as

$$\epsilon = \frac{void \quad volume}{total \quad volume} = \frac{a_f \Delta x_1 \Delta x_2}{\Delta x_1 \Delta x_2} = a_f. \tag{11.11}$$

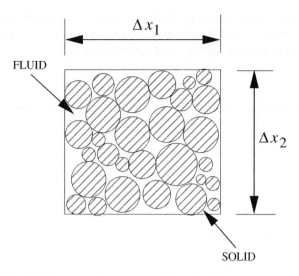

Figure 11.3 Fluid saturated porous medium. Infinitesimal control volume.

Now, the mass balance of an arbitrary control volume, as shown in Figure 11.3, gives (refer to Chapter 7)

$$\frac{\partial \rho_f}{\partial t} + \frac{\partial(\rho_f u_{1f})}{\partial x_1} + \frac{\partial(\rho_f u_{2f})}{\partial x_2} = 0, \tag{11.12}$$

where the subscript f stands for fluid, ρ is the density and u_1 and u_2 are the velocity components in the x_1 and x_2 directions respectively. The volume averaged velocity components may be defined as (Nield and Bejan 1992)

$$u_1 = \epsilon u_{1f} \quad \text{and} \quad u_2 = \epsilon u_{2f}. \tag{11.13}$$

Equation (11.12) can be simplified for an incompressible flow (constant density) as follows,

$$\frac{\partial u_1}{\partial x_1} + \frac{\partial u_2}{\partial x_2} = 0. \tag{11.14}$$

Similarly, the equation for momentum balance can be derived. For instance, in the x_2 direction, the momentum balance gives

$$\frac{\rho_f}{\epsilon} \left[\frac{\partial u_2}{\partial t} + \frac{\partial}{\partial x_1} \left(\frac{u_1 u_2}{\epsilon} \right) + \frac{\partial}{\partial x_2} \left(\frac{u_2^2}{\epsilon} \right) \right]$$
$$= -\frac{1}{\epsilon} \frac{\partial}{\partial x_2} (p_f \epsilon) + \frac{\mu_e}{\epsilon} \left(\frac{\partial^2 u_2}{\partial x_1^2} + \frac{\partial^2 u_2}{\partial x_2^2} \right) + (\rho_{ref} - \rho_f) g - D_{x_2}, \tag{11.15}$$

where μ_e is the equivalent viscosity; p_f the fluid pressure; g the acceleration due to gravity and D_{x_2} is the matrix drag per unit volume of the porous medium. The particle drag can be expressed in the following form, as discussed in Section 11.1,

$$D_p = aV + bV^2 \tag{11.16}$$

for a one-dimensional flow with velocity V. For two-dimensional flow the drag in the x_2 direction is given as

$$D_{x_2} = au_2 + b(u_1^2 + u_2^2)^{1/2}u_2 \tag{11.17}$$

by resolving the vertical drag expression along the x_2 direction. In the present formulation Ergun's correlation for the constants a and b, given in Equations (11.6) and (11.7), will be used.

Now, the solid matrix drag component D_{x_2} can be written as

$$D_{x_2} = \frac{\mu_f u_2}{\kappa} + \frac{1.75}{\sqrt{150}} \frac{\rho_f}{\sqrt{\kappa}} \frac{|V|}{\epsilon^{3/2}} u_2, \tag{11.18}$$

where V is the velocity vector in the field. By substituting Equation (11.18) into Equation (11.15) we obtain

$$\frac{\rho_f}{\epsilon} \left[\frac{\partial u_2}{\partial t} + \frac{\partial}{\partial x_1} \left(\frac{u_1 u_2}{\epsilon} \right) + \frac{\partial}{\partial x_2} \left(\frac{u_2^2}{\epsilon} \right) \right] = -\frac{1}{\epsilon} \frac{\partial}{\partial x_2} (p_f \epsilon) + \frac{\mu_e}{\epsilon} \left(\frac{\partial^2 u_2}{\partial x_1^2} + \frac{\partial^2 u_2}{\partial x_2^2} \right)$$
$$+ (\rho_{ref} - \rho_f)g - \frac{\mu_f u_2}{\kappa} - \frac{1.75}{\sqrt{150}} \frac{\rho_f}{\sqrt{\kappa}} \frac{|V|}{\epsilon^{3/2}} u_2. \tag{11.19}$$

Similarly, other momentum components can also be derived and the final form of the governing equations for incompressible flow through a porous medium in dimensional form can be given in indicial notation as

Continuity:

$$\frac{\partial u_i}{\partial x_i} = 0. \tag{11.20}$$

Momentum:

$$\frac{\rho_f}{\epsilon} \left[\frac{\partial u_i}{\partial t} + \frac{\partial}{\partial x_j} \left(\frac{u_i u_j}{\epsilon} \right) \right] = -\frac{1}{\epsilon} \frac{\partial}{\partial x_i} (p_f \epsilon) + \frac{\mu_e}{\epsilon} \frac{\partial^2 u_i}{\partial x_i^2}$$
$$+ (\rho_{ref} - \rho_f)g\gamma_i - \frac{\mu_f u_i}{\kappa} - \frac{1.75}{\sqrt{150}} \frac{\rho_f}{\sqrt{\kappa}} \frac{|V|}{\epsilon^{3/2}} u_i. \tag{11.21}$$

The previous equation can be simplified by substituting Equation (11.20) into Equation (11.21). The energy conservation equation is also derived in a similar manner. The final form of the energy equation is:

Energy:

$$\left[\epsilon(\rho c_p)_f + (1 - \epsilon)(\rho c_p)_s\right] \frac{\partial T}{\partial t} + (\rho c_p)_f u_i \frac{\partial T}{\partial x_i} = k\left(\frac{\partial^2 T}{\partial x_i^2}\right). \tag{11.22}$$

In the above equations, t is the time; c_p is the specific heat; γ_i is an unit vector in the buoyancy direction; T is the temperature and k is the equivalent thermal conductivity. The subscripts f and s stand for the fluid and solid phases respectively.

Finally, the conservation of mass (species conservation) equation may be written as

$$\epsilon \frac{\partial c}{\partial t} + u_i \frac{\partial c}{\partial x_i} = D\left(\frac{\partial^2 c}{\partial x_i^2}\right). \tag{11.23}$$

It should be noted that the permeability and thermal conductivity values can be directional, in which case they are tensors.

11.2.1 Nondimensional Scales

The nondimensional form of the equations simplify most of the calculations. The following final form of the nondimensional equations may be obtained by suitable scaling.

Continuity equation:

$$\frac{\partial u_i^*}{\partial x_i^*} = 0. \tag{11.24}$$

Momentum equations:

$$\frac{1}{\epsilon} \frac{\partial u_i^*}{\partial t^*} + \frac{1}{\epsilon} u_j^* \frac{\partial}{\partial x_j^*}\left(\frac{u_i^*}{\epsilon}\right) = -\frac{1}{\epsilon} \frac{\partial}{\partial x_i^*}\left(\epsilon p_f^*\right) - \frac{u_i^*}{ReDa}$$

$$- \frac{1.75}{\sqrt{150}} \frac{|\mathbf{V}^*|}{\sqrt{Da}} \frac{u_i^*}{\epsilon^{3/2}} + \frac{J}{Re\epsilon}\left(\frac{\partial^2 u_i^*}{\partial x_i^{*2}}\right) + \gamma_i \frac{Gr}{Re^2} T^*. \tag{11.25}$$

Energy equation:

$$\sigma \frac{\partial T^*}{\partial t^*} + u_i^* \frac{\partial T^*}{\partial x_i^*} = \frac{k^*}{RePr}\left(\frac{\partial^2 T^*}{\partial x_i^{*2}}\right). \tag{11.26}$$

Species equation:

$$\epsilon \frac{\partial c^*}{\partial t^*} + u_i^* \frac{\partial c^*}{\partial x_i^*} = \frac{1}{ReSc}\left(\frac{\partial^2 c^*}{\partial x_i^{*2}}\right). \tag{11.27}$$

In the previous equations, the parameters governing the flow and heat transfer are the Darcy number (Da), Reynolds number (Re), Prandtl number (Pr), Grashoff number (Gr), the ratio of heat capacities (σ), porosity of the medium (ϵ), conductivity ratio (k^*), viscosity ratio (J), Schmidt number (Sc) and the anisotropic property ratios for the case of an anisotropic medium. The definitions for the scales and nondimensional parameters are:

$$x_i^* = \frac{x_i}{L}; u_i^* = \frac{u_i}{u_a}; t^* = \frac{tu_a}{L}; p_f^* = \frac{p_f}{\rho_f u_a^2}; T^* = \frac{T - T_a}{T_w - T_a};$$

$$J = \frac{\mu_e}{\mu_f}; \sigma = \frac{\epsilon(\rho c_p)_f + (1 - \epsilon)(\rho c_p)_s}{(\rho c_p)_f}; k^* = \frac{k}{k_f} \tag{11.28}$$

and the nondimensional numbers are given as

$$Re = \frac{\rho_f u_a L}{\mu_f}; Pr = \frac{\nu_f}{\alpha_f}; Da = \frac{\kappa}{L^2}; Gr = \frac{g\beta\Delta T L^3}{\nu_f^2}; Sc = \frac{\mu_f}{\rho_f D}, \tag{11.29}$$

where the subscript a in the above equations refers to free stream or atmosphere, f is for fluid and s for solid. The above scales are suitable for most forced and mixed convection problems. However, for buoyancy-driven flows, it is convenient to handle the equations using the following definition of the Rayleigh number (Ra), that is,

$$Ra = \frac{g\beta\Delta T L^3}{\nu\alpha}, \tag{11.30}$$

where the following different scales need to be employed in solving natural convection problems.

$$u_i^* = \frac{u_i L}{\alpha_f}; t^* = \frac{t\alpha_f}{L^2}; p^* = \frac{pL^2}{\rho_f \alpha_f^2}. \tag{11.31}$$

The nondimensional governing equations for natural convection are:

Continuity equation:

$$\frac{\partial u_i^*}{\partial x_i^*} = 0. \tag{11.32}$$

Momentum equations:

$$\frac{1}{\epsilon}\frac{\partial u_i^*}{\partial t^*} + \frac{1}{\epsilon}u_j^*\frac{\partial}{\partial x_j^*}\left(\frac{u_i^*}{\epsilon}\right) = -\frac{1}{\epsilon}\frac{\partial}{\partial x_i^*}(\epsilon p_f^*) - \frac{Pr u_i^*}{Da}$$

$$-\frac{1.75}{\sqrt{150}}\frac{|V^*|}{\sqrt{Da}}\frac{u_i^*}{\epsilon^{3/2}} + \frac{JPr}{\epsilon}\left(\frac{\partial^2 u_i^*}{\partial x_i^{*2}}\right) + \gamma_i Ra Pr T^*. \tag{11.33}$$

Energy equation:

$$\sigma\frac{\partial T^*}{\partial t^*} + u_i^*\frac{\partial T^*}{\partial x_i^*} = k^*\left(\frac{\partial^2 T^*}{\partial x_i^{*2}}\right). \tag{11.34}$$

Species equation:

$$\epsilon \frac{\partial c^*}{\partial t^*} + u_i^* \frac{\partial c^*}{\partial x_i^*} = \frac{1}{Le} \left(\frac{\partial^2 c^*}{\partial x_i^{*2}} \right). \tag{11.35}$$

where $Le = \alpha_f/D$ is the Lewis number. Other alternative scales are possible depending of the problem studied. In general any constant free stream quantity may be used as a scale. In the above formulation, the buoyancy effects are incorporated by invoking the Boussinesq approximation as discussed in Chapter 7. The kinematic viscosity v, used in the above scales, is defined as

$$v = \frac{\mu}{\rho} \tag{11.36}$$

and α is the thermal diffusivity, given as

$$\alpha_f = \frac{k_f}{(\rho c_p)_f}. \tag{11.37}$$

It may be observed that the scales and nondimensional parameters are defined by using the fluid properties. Often, a quantity called the Darcy-Rayleigh number is used in the literature as a governing nondimensional parameter for Darcy flow. This is the product of the Darcy (Da) and fluid Rayleigh (Ra) numbers as defined previously.

11.2.2 Limiting Cases

The equations discussed above represent a porous medium which tends to a solid as porosity, $\epsilon \to 0$. Thus, a conjugate problem, where part of the domain is completely solid, can be dealt with by using the above equations.

Another limiting case of these equations is that they approach the incompressible Navier-Stokes equations as $\epsilon \to 1$. Again, a very general problem where a porous medium and a single phase fluid are part of a domain (porous-fluid interface (Massarotti *et al.* 2001)) can be solved by using the above equations. Thus, many applications such as alloy solidification (Sinha *et al.* 1992) and heat exchanger design can be analyzed via these equations.

11.3 Discretization Procedure

The CBS scheme will be employed to solve the porous medium flow equations. In this context the same four steps, with minor modifications, will be utilized as discussed in Chapter 7.

In the following subsections, the temporal and spatial discretization scheme are given, which will then be employed to solve the porous medium equations. Use will be made only of simple, linear triangular elements to study porous medium flow problems.

11.3.1 Temporal Discretization

Before going into the details of the CBS split, let us first consider the temporal discretization of the governing equations. The momentum equation is subjected to the characteristic Galerkin procedure, as discussed in the Chapter 7, viz.

$$\frac{u_i^{n+1} - u_i^n}{\epsilon \Delta t} =$$

$$-\frac{1}{\epsilon}\frac{\partial(p\epsilon)}{\partial x_i}^{n+\theta} - \left[\frac{u_j}{\epsilon}\frac{\partial}{\partial x_j}\left(\frac{u_i}{\epsilon}\right)\right]^{n+\theta_1} + \left[\frac{1}{\epsilon Re}\frac{\partial^2 u_i}{\partial x_i^2}\right]^{n+\theta_2} - \left[\frac{u_i}{ReDa} + C\frac{|V|}{\sqrt{Da}}\frac{u_i}{\epsilon^{3/2}}\right]^{n+\theta_3}$$

$$+ \frac{\Delta t}{2}u_k\frac{\partial}{\partial x_k}\left(\frac{1}{\epsilon}\frac{\partial(p\epsilon)}{\partial x_i} + \left[\frac{u_j}{\epsilon}\frac{\partial}{\partial x_j}\left(\frac{u_i}{\epsilon}\right)\right] + \left[\frac{u_i}{ReDa} + C\frac{|V|}{\sqrt{Da}}\frac{u_i}{\epsilon^{3/2}}\right]\right)^n. \tag{11.38}$$

The body force terms are neglected in the above equation in order to simplify the presentation. Equation (11.38), the parameter C is a constant equal to $1.75/\sqrt{150}$ (see Equation (11.18)). The parameters θ, θ_1, θ_2 and θ_3 all vary between zero and unity and with appropriate values, different schemes of interest can be established. The superscript θ should be interpreted as

$$f^{n+\theta} = \theta f^{n+1} + (1 - \theta)f^n, \tag{11.39}$$

where the superscript n indicates the nth time iteration.

In the CBS scheme the velocities are calculated by splitting Equation (11.38) into two parts as below. In order to simplify the presentation, θ_1, θ_2 and θ_3 are assumed to be equal to zero. It is important to note, however, that such an assumption severely restricts the time step which can be employed in the calculations. The semi- and quasi- implicit schemes, as discussed in Section 11.3.3, are widely employed for porous medium flow calculations.

In Step 1, the pressure term is completely removed from Equation (11.38) and the intermediate velocity components $\tilde{u}_i$ are calculated (similar to Step 1 of the CBS scheme discussed in Chapter 7), as

$$\frac{\Delta \tilde{u}_i}{\epsilon \Delta t} = \frac{\tilde{u}_i - u_i^n}{\epsilon \Delta t} = -\left[\frac{u_j}{\epsilon}\frac{\partial}{\partial x_j}\left(\frac{u_i}{\epsilon}\right)\right]^n + \left[\frac{1}{\epsilon Re}\frac{\partial^2 u_i}{\partial x_i^2}\right]^n - \left[\frac{1}{ReDa}u_i + C\frac{|V|}{\sqrt{Da}}\frac{u_i}{\epsilon^{3/2}}\right]^n$$

$$+ \frac{\Delta t}{2}u_k\frac{\partial}{\partial x_k}\left(\frac{1}{\epsilon}\frac{\partial(p\epsilon)}{\partial x_i} + \left[\frac{u_j}{\epsilon}\frac{\partial}{\partial x_j}\left(\frac{u_i}{\epsilon}\right)\right] + \left[\frac{u_i}{ReDa} + C\frac{|V|}{\sqrt{Da}}\frac{u_i}{\epsilon^{3/2}}\right]\right)^n. \tag{11.40}$$

The velocities can be corrected using the following equation which has been derived by subtracting Equation (11.40) from Equation (11.38), that is,

$$\frac{\Delta u_i}{\epsilon \Delta t} = \frac{u_i^{n+1} - u_i^n}{\epsilon \Delta t} = \frac{\Delta \tilde{u}_i}{\epsilon \Delta t} - \frac{1}{\epsilon}\frac{\partial(p\epsilon)}{\partial x_i}^{n+\theta}. \tag{11.41}$$

However, the value of the pressure in the above equation is not known. In order to establish the pressure field, a pressure Poisson equation can be derived from the above equation and may be written as (see Chapter 7, Section 7.6)

$$\frac{1}{\epsilon}\frac{\partial^2}{\partial x_i^2}(p\epsilon)^{n+\theta} = \frac{\partial \tilde{u}_i}{\partial x_i}. \tag{11.42}$$

The above simplified equation has been derived by substituting the equation of continuity. Thus, the conservation of mass is satisfied indirectly without explicitly solving for the mass conservation Equation (11.24).

We have a total of three steps to obtain a solution for the momentum and continuity equations. As discussed in Chapter 7, Equation (11.40) is solved at the first step followed by Equation (11.42) in the second step and Equation (11.41) in the third step. Additional steps, such as temperature, or concentration calculations, can be added as an addition to the above three steps.

In problems where nonisothermal and mass transfer effects are involved, then after velocity correction, additional equations will be solved. If no coupling exists between the velocities and the other variables, such as temperature and concentration and the steady-state solution is only of interest, then the steady velocity and pressure fields can be established first and the rest of the variables can be calculated using the steady-state velocity and pressure values.

11.3.2 Spatial Discretization

Once a temporal discretization of the equations has been achieved, then spatial discretization may be carried out. In this text, the finite element discretization will be carried out using linear triangular elements. Assuming a Galerkin approximation, the variables can be expressed as

$$u_i = [\mathbf{N}]\{\mathbf{u_i}\}; \Delta u_i = [\mathbf{N}]\{\Delta \mathbf{u_i}\}; \Delta \tilde{u}_i = [\mathbf{N}]\{\Delta \tilde{\mathbf{u}}_i\}; p = [\mathbf{N}]\{\mathbf{p}\}; \epsilon = [\mathbf{N}]\{\epsilon\}, \quad (11.43)$$

where [$\mathbf{N}$] are the shape functions. We assume that the equations are solved in the order mentioned before i.e. first, the intermediate velocity components, then the pressure field and, finally, the velocity correction. On considering the intermediate velocity calculation, we have the following weak form where porosity is assumed to be an averaged quantity over an element and body forces are neglected for the sake of simplicity,

$$\int_\Omega \frac{1}{\epsilon}[\mathbf{N}]^T \Delta \tilde{u}_i d\Omega = \frac{\Delta t}{\epsilon}\left[-\int_\Omega [\mathbf{N}]^T u_j \frac{\partial}{\partial x_j}\left(\frac{u_i}{\epsilon}\right) d\Omega \right]^n$$

$$- \left[\frac{\Delta t}{Re}\int_\Omega \frac{1}{\epsilon}\frac{\partial[\mathbf{N}]^T}{\partial x_i}\frac{\partial u_i}{\partial x_i} d\Omega\right]^n - \int_\Omega [\mathbf{N}]^T \left[\frac{\Delta t}{ReDa}u_i + \frac{C}{\sqrt{Da}}\frac{\Delta t|V|}{\epsilon^{3/2}}u_i\right]^n d\Omega$$

$$+ \frac{\Delta t^2}{2}\int_\Omega [\mathbf{N}]^T u_k \frac{\partial}{\partial x_k}\left(\frac{1}{\epsilon}\frac{\partial(p\epsilon)}{\partial x_i} + \left[\frac{u_j}{\epsilon}\frac{\partial}{\partial x_j}\left(\frac{u_i}{\epsilon}\right)\right] + \left[\frac{u_i}{ReDa} + C\frac{|V|}{\sqrt{Da}}\frac{u_i}{\epsilon^{3/2}}\right]\right)^n d\Omega.$$

$$(11.44)$$

The weak form of the Step 2, calculation for the pressure field can be written as (assuming $\theta = 1$)

$$-\frac{1}{\epsilon} \int_\Omega \frac{\partial [\mathbf{N}]^T}{\partial x_i} \frac{\partial (\epsilon p)^{n+1}}{\partial x_i} d\Omega = \frac{1}{\Delta t} \int_\Omega [\mathbf{N}]^T \frac{\partial \tilde{u}_i}{\partial x_i} d\Omega. \tag{11.45}$$

Finally Step 3, can be written in a weak form as

$$\int_\Omega [\mathbf{N}]^T \Delta u_i d\Omega = \int_\Omega [\mathbf{N}]^T \Delta \tilde{u}_i d\Omega - \Delta t \int_\Omega [\mathbf{N}]^T \frac{\partial p}{\partial x_i}^{n+1} d\Omega. \tag{11.46}$$

Other field variables, such as temperature and concentration, can be established in a similar fashion via Step 1 and will be discussed later. For the integration and relevant matrices, refer to Chapter 7.

11.3.3   Semi- and Quasi- Implicit Forms

Single phase incompressible fluid flow problems can be solved in a fully explicit form, which is quite popular in fluid dynamics calculations (Malan *et al.* 2002; Nithiarasu 2003). However, a solution for the generalized porous medium equations using a fully explicit form has been less successful. This is mainly due to the large values of the solid matrix drag terms, especially at smaller Darcy numbers. In order to eliminate some of the time-step restrictions imposed by these terms, schemes other than the fully explicit forms are discussed below.

In the semi-implicit (SI) form (Nithiarasu and Ravindran 1998), the porous medium source terms and pressure equation are treated implicitly. In other words, $\theta = \theta_3 = 1$ and $\theta_1 = \theta_2 = 0$. The split in the momentum equation (Equation (11.40)) will be different, that is,

$$\frac{\tilde{u}_i - u_i^n}{\epsilon \Delta t} + \frac{1}{ReDa} \tilde{u}_i + C \frac{|\mathbf{V}|}{\sqrt{Da}} \frac{\tilde{u}_i}{\epsilon^{3/2}} = -\left[\frac{u_j}{\epsilon} \frac{\partial}{\partial x_j} \left(\frac{u_i}{\epsilon} \right) \right]^n + \left[\frac{1}{\epsilon Re} \frac{\partial^2 u_i}{\partial x_i^2} \right]^n \tag{11.47}$$

or

$$\tilde{u}_i \left(\frac{1}{\epsilon \Delta t} + \frac{1}{ReDa} + C \frac{|\mathbf{V}|}{\sqrt{Da}} \frac{1}{\epsilon^{3/2}} \right) = \frac{u_i^n}{\epsilon \Delta t} - \left[\frac{u_j}{\epsilon} \frac{\partial}{\partial x_j} \left(\frac{u_i}{\epsilon} \right) \right]^n + \left[\frac{1}{\epsilon Re} \frac{\partial^2 u_i}{\partial x_i^2} \right]^n. \tag{11.48}$$

Step 2, the pressure calculation becomes

$$\frac{1}{\epsilon} \frac{\partial^2}{\partial x_i^2} (p\epsilon)^{n+\theta} = \left(\frac{1}{\Delta t \epsilon} + \frac{1}{ReDa} + \frac{C}{\sqrt{Da}} \frac{|\mathbf{V}|}{\epsilon^{3/2}} \right) \frac{\partial \tilde{u}_i}{\partial x_i}. \tag{11.49}$$

Step 3 is also different and is given as

$$\left(\frac{1}{\Delta t \epsilon} + \frac{1}{ReDa} + \frac{C}{\sqrt{Da}} \frac{|\mathbf{V}|}{\epsilon^{3/2}} \right) u_i^{n+1}$$

$$= \left(\frac{1}{\Delta t \epsilon} + \frac{1}{ReDa} + \frac{C}{\sqrt{Da}} \frac{|\mathbf{V}|}{\epsilon^{3/2}} \right) \tilde{u}_i - \frac{1}{\epsilon} \frac{(\partial p\epsilon)^{n+\theta}}{\partial x_i}. \tag{11.50}$$

Although extra complications were introduced in the semi-implicit form at Step 1, for steady-state solutions, we can avoid simultaneous solution of the algebraic equations by taking the coefficient

$$CO = \left(\frac{1}{\Delta t \epsilon} + \frac{1}{ReDa} + \frac{C}{\sqrt{Da}} \frac{|\mathbf{V}|}{\epsilon^{3/2}} \right) \tag{11.51}$$

on to the RHS. Thus, the system can be enabled for the mass lumping procedure (Nithiarasu and Ravindran 1998) when discretized in space.

The quasi-implicit (QI) form is very similar to that of the above scheme but now the viscous, second-order terms are also treated implicitly ($\theta_2 = 1$) (Nithiarasu *et al.* 1997c). The important difference, however, is that the quasi-implicit scheme does not benefit from mass lumping when solving for the intermediate velocity values. A simultaneous solution of the LHS matrices is essential here. It has been proven that both the QI and SI schemes generally perform well (Nithiarasu 2001).

11.4 Nonisothermal flows

Several examples of porous medium flow problems are nonisothermal in nature. The main focus in this case will be to demonstrate nonisothermal flow through a porous medium. As mentioned previously, an energy equation needs to be solved, in addition to the momentum and pressure equations if the flow is nonisothermal. For steady-state problems, if no coupling exists between the momentum and energy equation, the temperature field can be established after calculation of the velocity fields. The temporal discretization of the energy equation can be written in a similar form to the momentum equation and is given (since element Peclet number is expected to be below unity, the stabilization terms are neglected) as

$$\sigma \frac{T^{n+1} - T^n}{\Delta t} = - \left[u_i \frac{\partial T}{\partial x_i} \right]^{n+\theta_1} + \frac{k^*}{RePr} \left[\frac{\partial^2 T}{\partial x_i^2} \right]^{n+\theta_2}, \tag{11.52}$$

where θ_1 and θ_2 have the same meaning as previously discussed in Section 11.3. The variable involved in this case is temperature and can be spatially approximated as

$$T = [\mathbf{N}]\{\mathbf{T}\}. \tag{11.53}$$

The weak form of the energy equation can be written (assuming θ_1 and θ_2 are both equal to zero) as

$$\int_{\Omega} \sigma [\mathbf{N}]^T \Delta T d\Omega = -\Delta t \int_{\Omega} \left[[\mathbf{N}]^T u_i \frac{\partial T}{\partial x_i} \right]^n d\Omega - \frac{k^* \Delta t}{RePr} \int_{\Omega} \left[\frac{\partial [\mathbf{N}]^T}{\partial x_i} \frac{\partial T}{\partial x_i} \right]^n d\Omega + b.t., \tag{11.54}$$

where

$$\Delta T = T^{n+1} - T^n. \tag{11.55}$$

The substitution of Equation (11.53) into Equation (11.54) yields the final global matrix form of the energy equation, that is,

$$\sigma[M_p]\{\Delta T\} = -\Delta t \left[[C_p]\{T\} + [K_T]\{T\} - \{f_4\}\right]^n, \tag{11.56}$$

where the elemental matrices are

$$[K_{Te}] = \frac{1}{4A} \frac{k^*}{RePr} \begin{bmatrix} b_i^2 & b_i b_j & b_i b_k \\ b_j b_i & b_j^2 & b_j b_k \\ b_k b_i & b_k b_j & b_k^2 \end{bmatrix} + \frac{1}{4A} \frac{k^*}{RePr} \begin{bmatrix} c_i^2 & c_i c_j & c_i c_k \\ c_j c_i & c_j^2 & c_j c_k \\ c_k c_i & c_k c_j & c_k^2 \end{bmatrix} \tag{11.57}$$

and the forcing vector is

$$\{f_4\} = \frac{\Gamma}{4A} \frac{1}{RePr} \begin{bmatrix} b_i T_i + b_j T_j + b_k T_k \\ b_i T_i + b_j T_j + b_k T_k \\ 0 \end{bmatrix}^n n_1$$

$$+ \frac{\Gamma}{4A} \frac{1}{RePr} \begin{bmatrix} c_i T_i + c_j T_j + c_k T_k \\ c_i T_i + c_j T_j + c_k T_k \\ 0 \end{bmatrix}^n n_2.$$

It should be noted that both the flux and convective heat transfer boundary conditions are treated by using the boundary integral as discussed in the Chapter 7. In a similar fashion, the species equation may also be discretized in time, that is,

$$\epsilon \frac{c^{n+1} - c^n}{\Delta t} = - \left[u_i \frac{\partial c}{\partial x_i} \right]^{n+\theta_1} + \frac{1}{ReSc} \left[\frac{\partial^2 c}{\partial x_i^2} \right]^{n+\theta_2}. \tag{11.58}$$

The spatial discretization procedure is identical to that of the energy equation starting with spatial discretization of concentration similar to Equation (11.53). The rest of the procedure also follows similar lines to that of the energy equation. For further details on discretization and matrices, refer to Chapter 7.

Example 11.4.1   *Forced convection*

Flow through packed beds are important in many chemical engineering applications. Generally, the grain size in the packed beds will vary depending on the application. As the particle size increases, the packing close to the walls will become nonuniform thereby creating a channeling effect close to the solid walls. In such cases, the porosity value can be close to unity near the walls but will decrease to a free stream value away from the walls.

In such situations, the ability to vary the porosity within the domain itself is essential in order to obtain a correct solution. Although, the theoretical determination of the near wall porosity variation is difficult, there are some experimental correlations available to tackle this issue. One such widely employed correlation, given by Benenati and Brosilow (1962) will be used, that is,

$$\epsilon = \epsilon_e \left[1 + exp\left(-\frac{cx}{d_p} \right) \right], \tag{11.59}$$

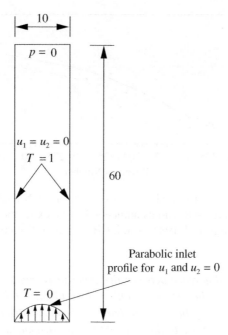

Figure 11.4   Forced convection in a channel filled with a variable porosity medium. Geometry and boundary conditions.

where ϵ_e is the free stream bed porosity and taken to be equal to 0.39, c is an empirical constant ($c = 2$ for $d_p = 5$ mm). In general, the problem in this case is formulated based on the particle size d_p, that is, the Reynolds number is based on the particle size.

Figure 11.4 shows the problem definition of forced flow through a packed bed. The inlet channel width is 10 times the size of the grain. The length of the channel is 6 times that of the inlet width. Zero pressure conditions are assumed at the exit. The inlet velocity profile is parabolic and no-slip boundary conditions apply on the solid side walls. Both the walls are assumed to be at a higher, uniform temperature than that of the inlet fluid temperature. The analysis is carried out for different particle Reynolds numbers ranging from 150 to 350. The quasi-implicit (QI) scheme with $\theta = 1$, $\theta_1 = 0$ and $\theta_2 = \theta_3 = 1$ has been employed to solve this problem. A nonuniform mesh with triangular elements was also used in the analysis. The mesh is fine close to the walls and coarse towards the center. The total number of nodes and elements used in the calculation are 3003 and 5776 respectively.

Figure 11.5 shows a comparison of the calculated steady-state average Nusselt number distribution on a hot wall with the available experimental and numerical data. The Nusselt number is calculated as

$$Nu = \frac{hL}{k} = \int_0^L \frac{\partial T}{\partial x_1} dx. \tag{11.60}$$

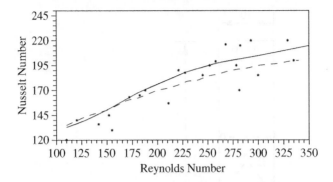

Figure 11.5   Forced convection in a channel. Comparison of Nusselt number with experimental data for different particle Reynolds numbers. Points: experimental (Vafai *et al.* 1984); dashed line: numerical (Vafai *et al.* 1984); solid: CBS. Source: Data from Vafai *et al.* 1984.

Figure 11.6 shows the difference between the generalized model and the Brinkman and Forcheimmer extensions for the velocity profiles close to the wall in a variable porosity medium at steady state. As may be seen the Forcheimmer and Brinkman extensions fail to predict the channeling effect close to the wall. Whilst the Brinkman extension is insensitive to porosity values, the Forcheimmer model does not predict the viscous effect close to the channel walls.

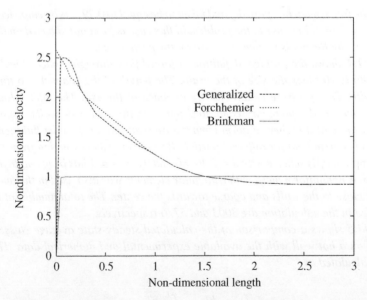

Figure 11.6   Forced convection in a channel. Comparison between the generalized model, Forcheimmer and Brinkman extensions to Darcy's law.

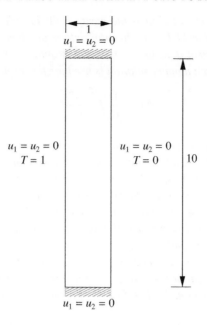

Figure 11.7 Natural convection in a fluid saturated variable porosity medium. Problem boundary conditions.

Example 11.4.2 *Natural convection*

The fluid flow in a variable porosity medium within an enclosed cavity under the influence of buoyancy is another interesting and difficult problem to analyse. In order to study such a problem, an enclosure packed with a fluid saturated porous medium is considered. The aspect ratio of the enclosure is 10 (ratio between height and width). All the enclosure walls are subjected to "no slip" boundary conditions. The left vertical wall is assumed to be at a higher, uniform temperature than that of the right-side wall. Both the horizontal walls are assumed to be insulated (Figure 11.7). The properties of the saturating fluid are assumed to be constant other than that of the density. The density variation is invoked by the Boussenesq approximation.

Table 11.1 shows the steady-state quantitative results and a comparison with the available numerical and experimental data. These data were obtained on a nonuniform structured 61 × 61 mesh. The accuracy of the prediction can be improved by further refinement of the mesh.

Table 11.1 Average hot wall Nusselt number distribution for natural convection in a variable porosity medium, aspect ratio = 10

Fluid	d_p	ϵ_e	Pr	k^*	Ra	Experimental	Numerical	CBS
Water	5.7	0.39	7.1	1.929	1.830×10^7	2.595	2.405	2.684
					3.519×10^7	3.707	3.496	3.892
Ethyl alcohol	5.7	0.39	2.335	15.4	2.270×10^8	12.56	13.08	12.17
					3.121×10^8	15.13	15.57	14.28

An extremely fine mesh is essential near the cavity walls in order to predict the channeling effect in this region. In Table 11.1, experimental data is obtained from reference (Inaba and Seki 1981) and the numerical data for comparison is obtained from reference (David et al. 1991). The following Nusselt number relation is used for this problem.

$$Nu = \frac{1}{L} \int_0^L \frac{\partial T}{\partial x} dx. \tag{11.61}$$

Example 11.4.3 *Natural convection – constant porosity medium*

The problems when the variation in porosity is of less significance are normally porous media with small solid particle size. For instance, thermal insulation is one such example where the variation in porosity near the solid walls is not important but the uniform free stream porosity value can be very high. In order to investigate such media, a benchmark problem involving buoyancy-driven convection in a square cavity has been solved.

The problem definition is similar to the one shown in Figure 11.7, the difference being that the aspect ratio is unity. The square enclosure is filled with a fluid saturated porous medium with constant and uniform properties except for the density which is again incorporated via the Boussenessq approximation. A 51 × 51 nonuniform mesh (Figure 11.8) is employed for this problem.

The Darcy and non-Darcy flow regime classifications and the Darcy number limits have been discussed by many researchers. One important suggestion was given in the paper by Tong and Subramanian (1985). In Figure 11.9, we show the velocity and temperature

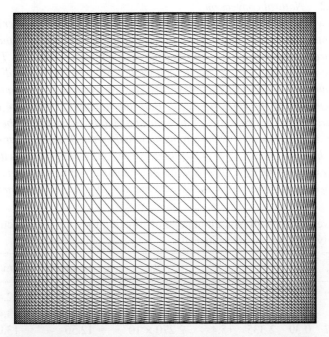

Figure 11.8 Buoyancy-driven flow in a fluid saturated porous medium. Finite element mesh. Nodes: 2601, Elements: 5000.

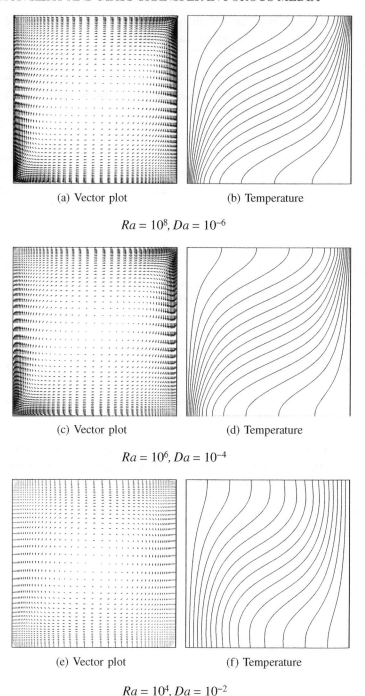

(a) Vector plot   (b) Temperature

$Ra = 10^8, Da = 10^{-6}$

(c) Vector plot   (d) Temperature

$Ra = 10^6, Da = 10^{-4}$

(e) Vector plot   (f) Temperature

$Ra = 10^4, Da = 10^{-2}$

Figure 11.9   Natural convection in a fluid saturated porous, square enclosure. Vector plots and temperature contours for different Rayleigh and Darcy numbers, $Pr = 0.71$.

Table 11.2   Average Nusselt number comparison with analytical and numerical results

$Ra^* = RaDa$	Nu			
	Analytical	Numerical1	Numerical2	CBS
10	–	1.07	–	1.08
50	1.98	–	2.02	1.96
100	3.09	3.09	3.27	3.02
500	8.40	–	–	8.38
1000	12.49	13.41	18.38	12.52

distribution at different Darcy and Rayleigh numbers. In this case the product of the Darcy and Rayleigh numbers is kept at a constant value in order to bring out the non-Darcy effects. It is clearly obvious that the maximum velocity in the Darcy flow regime, at a Darcy number of 10^{-6}, is located very close to the solid walls. The non-Darcy velocity profile, at a Darcy number of 10^{-2}, on the other hand looks very similar to that of a single phase fluid and the maximum velocity is located away from the solid walls. At a Darcy number of 10^{-4} the flow undergoes a transition from a Darcy flow regime to a non-Darcy flow regime. The temperature contours also undergo noticeable changes as the Darcy number increases from 10^{-6} to 10^{-2}.

Both the scheme and the model implementation have been designed in such a way that as the Darcy number increases, the flow approaches a single phase fluid flow, which is evident from Figure 11.9.

In Table 11.2, the quantitative results obtained from the above analysis (only for the Darcy flow regime, $Da < 10^{-5}$) are compared with other available analytical and numerical results. As seen the results are in excellent agreement with the reported results. In Table 11.2, analytical solution has been obtained from reference (Walker and Homsy 1978), "Numerical1" and "Numerical2" have been obtained from references (Lauriat and Prasad 1989) and (Trevisan and Bejan 1985) respectively.

Example 11.4.4   *Natural convection – axisymmetric problems*

In order to compare the numerical results with experimental data, an axisymmetric model was developed and a buoyancy-driven flow problem was studied. The boundary and initial conditions are the same as for the previous problem. The main difference being in the definition of the geometry. In this case geometry is an annulus with a radius ratio (ratio between outer and inner radii) of 5.338 (see Figure 11.10). The fluid used to saturate the medium is water with a Prandtl number of 5. The results are generated for different Grashof numbers (Ra/Pr) and compared with the experimental Nusselt number predictions as shown in Figure 11.11. In general the comparison is excellent for the Grashof number range considered.

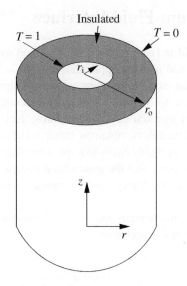

Figure 11.10 Natural convection in a fluid saturated constant porosity medium. Problem definition.

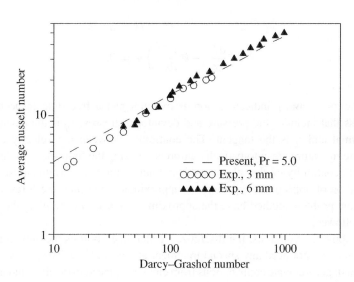

Figure 11.11 Natural convection in a fluid saturated constant porosity medium within an annular enclosure. Comparison of hot wall steady-state Nusselt number with the experimental and numerical data. Source: Data from Prasad *et al.* (1985).

11.5   Porous Medium-Fluid Interface

The interface between free fluid and a porous medium saturated with the same free fluid is very important in many industrial and real-life applications. Alloy solidification, heat exchanger pipes, petroleum recovery, heat recovery systems, as well as thermal insulation and ground water pollution, are just a few to mention. In these problems, the domain can be partly filled with a saturated porous medium and the remaining part is left to a free fluid. The solution of this type of problem can be obtained either using a single or multiple domain approach. The former, which is probably easier to implement than the latter, is the one adopted here and it is based on the property that the generalized porous medium model approaches the Navier-Stokes equations as the Darcy number increases and the porosity approaches unity.

When continuity of mass, momentum and energy is assumed, the following matching conditions need to be satisfied at the interface:

$$\left(u_i^p - u_i^f\right)n_i = 0$$

$$\left(\sigma_{ij}^p - \sigma_{ij}^f\right)n_j = 0$$

$$\left(\tau_{ij}^p - \tau_{ij}^f\right)n_j = 0$$

$$T^p - T^f = 0$$

$$\left(k^{*p}\frac{\partial T^p}{\partial x_i} - k^f\frac{\partial T^f}{\partial x_i}\right)n_i = 0. \qquad (11.62)$$

where superscripts p and f indicate a porous medium and a free fluid respectively, σ_{ij} is the total stress that includes the pressure and deviatoric stresses, τ_{ij} is the deviatoric stress, n_i is the normal and t_i is the tangent. The continuity between the velocity components, pressure and temperature are automatically enforced in using the finite element discretization. Stress and flux continuity on the other hand is not automatic and thus this must be enforced via discretization of appropriate equations or approximated via finer mesh resolution at the interface. In the problem studied here, the approximate approach of refining the mesh at the interface is followed.

With the above assumptions, the discretized generalized porous medium equations presented previously do not need any special treatment at the interface. The nodes placed along the interface will get adequate contributions from elements placed in the fluid and in the porous medium regions. In the free fluid domain, porosity is allowed to approach unity. Thus, the Darcy number approaches infinity.

Example 11.5.1   *Natural convection – vertically divided enclosure*

In the first problem studied, the domain is divided vertically into two equal parts and one part is filled with a free fluid and the other with a porous medium saturated by the same fluid

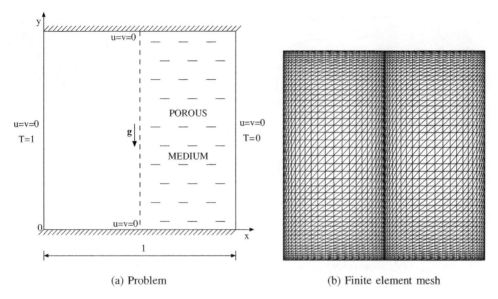

(a) Problem   (b) Finite element mesh

Figure 11.12   Natural convective heat transfer in a square cavity equally and vertically divided into saturated porous medium and free fluid. Problem domain, boundary conditions and the finite element mesh used.

as shown in Figure 11.12(a). The vertical wall adjacent to the free fluid, on the left-hand side of the cavity, is considered to be hot, while the opposite wall, adjacent to the porous matrix, is cold.

Figure 11.12 shows the problem definition and the finite element mesh used. The mesh is finely refined near all walls and along the interface of the problem which is at the middle of the cavity. The mesh contains 4608 elements and 2401 nodes (Massarotti et al. 2001).

The stream lines and isotherm patterns obtained for the vertical interface are shown in Figures 11.13(a) and (b) for the Darcy regime ($Ra = 3.028 \times 10^7$, $Da = 7.354 \times 10^{-7}$, $Pr = 6.97$, $\epsilon = 0.36$, $k^* = 1.397$) (Massarotti et al. 2001). As seen, the CBS procedure predicts the interface transition smoothly without any strong discontinueties. Since this figure corresponds to the Darcy regime, the porous medium part is dominated mainly by the conduction mode of heat transfer. The flow in the porous medium part is weaker than that in the free fluid region. A maximum stream function value of $|\psi|_{max} = 15.38$ is observed in the fluid region, while the value obtained by Beckermann et al. (1987) is $|\psi|_{max} = 15.98$.

Figure 11.13(c) shows the comparison of temperature distribution at different horizontal sections across the cavity with available experimental and numerical data (Beckerman et al. 1987). The agreement is excellent, in particular it can be noticed that the change in the slope of the isotherms at the interface, is well predicted. The results can be improved using the proper value of the effective viscosity (Massarotti et al. 2001).

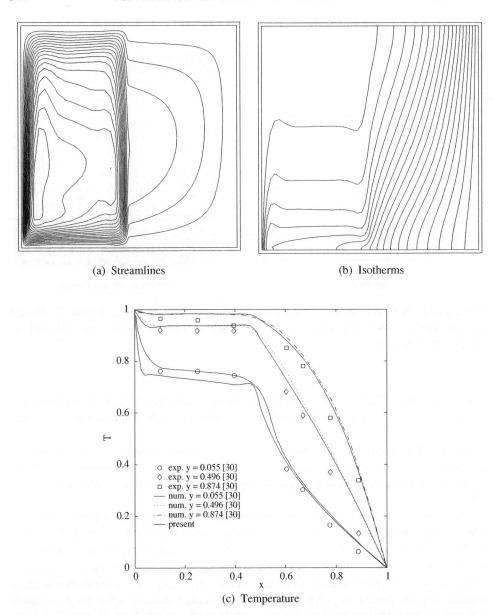

(a) Streamlines   (b) Isotherms

(c) Temperature

Figure 11.13 Natural convective heat transfer in a square cavity equally and vertically divided into saturated porous medium and free fluid. Streamlines, isotherms and temperature distribution.

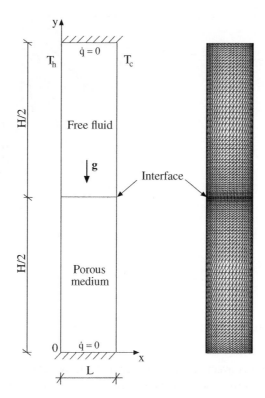

Figure 11.14 Natural convective heat transfer in a square cavity equally and horizontally divided into saturated porous medium and free fluid. Problem domain, boundary conditions and the finite element mesh used.

Example 11.5.2 *Natural convection – horizontally divided cavity*

The second type of porous-fluid interface problem considered is an enclosure divided horizontally as shown in Figure 11.14(a). The bottom half is filled with the saturated solid matrix. The mesh generated for this problem contains 2231 points and 4224 elements, is nonuniform, and is presented in Figure 11.14(b) (Massarotti et al. 2001). All the walls are assumed to obey no-slip conditions, and horizontal walls are insulated and vertical walls are placed at two different temperatures which trigger the buoyant flow.

In order to compare the results obtained with the present CBS procedure with some experiments, one case with a high Prandtl number was obtained run, and the results are compared in Figure 11.15 with those presented by Nishimura et al. (1986) for silicon oil as fluid and glass beads as solid matrix. The parameters used are: $Ra = 10^5, Da = 10^{-3}, Pr = 8000, \epsilon = 0.4$ and $k^* = 1.0$. The temperature is evaluated at five different horizontal sections, in the porous as well as in the fluid part of the cavity. The present results, in general, are in excellent agreement with the experimental data.

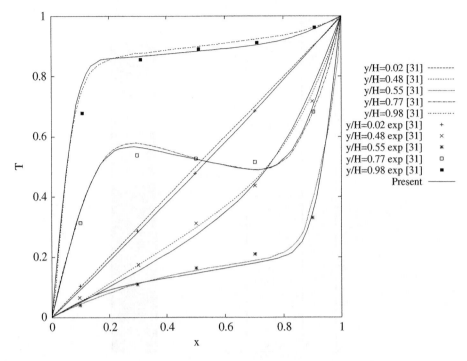

Figure 11.15 Natural convective heat transfer in a square cavity equally and horizontally divided into saturated porous medium and free fluid. Temperature distribution.

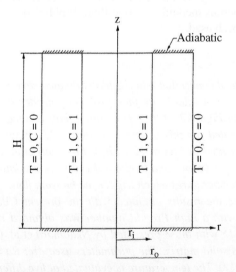

Figure 11.16 Double-diffusive natural convection in an axisymmetric, fluid saturated porous, square enclosure. Geometry and boundary conditions.

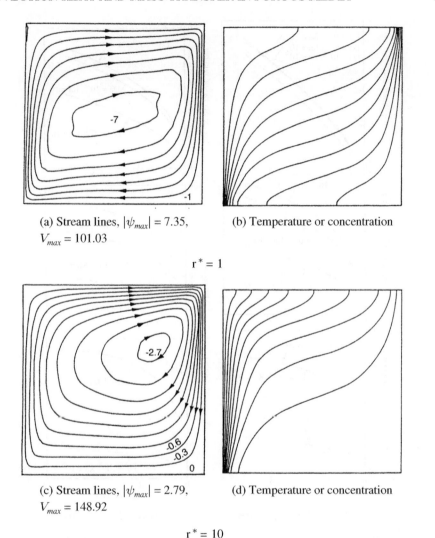

(a) Stream lines, $|\psi_{max}| = 7.35$,   (b) Temperature or concentration
$V_{max} = 101.03$

$r^* = 1$

(c) Stream lines, $|\psi_{max}| = 2.79$,   (d) Temperature or concentration
$V_{max} = 148.92$

$r^* = 10$

Figure 11.17 Double-diffusive natural convection in an axisymmetric, fluid saturated porous, square enclosure. Stream lines , isotherms and isoconcentration lines, $Pr = 0.71$, $Da = 10^{-6}$, $Ra = 10^8$, $\epsilon = 0.6$, $Le = 1$, $B = 1$.

11.6   Double-diffusive Convection

A basic understanding of double-diffusive natural convection in a fluid saturated porous media is important in many areas such as fibrous insulation, food processing and storage, contaminant transport in ground water, geophysical systems, electro chemistry, metallurgy etc. (Nithiarasu *et al.* 1996, 1997a,b).

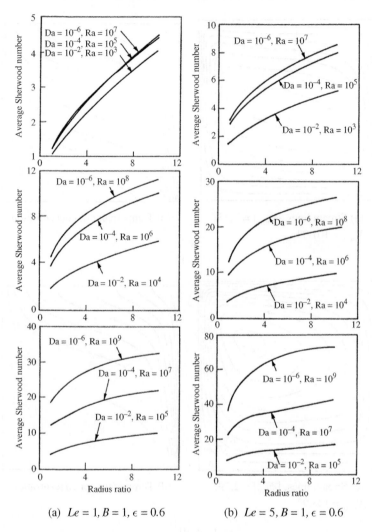

(a) $Le = 1, B = 1, \epsilon = 0.6$ (b) $Le = 5, B = 1, \epsilon = 0.6$

Figure 11.18   Double-diffusive natural convection in an axisymmetric, fluid saturated porous, square enclosure. Average Sherwood number distribution against the radious ratio r^*.

Example 11.6.1   *Double-diffusive convection – axisymmetric cavity*

A model problem of double-diffusive natural convection in an axisymmetric enclosed is discussed here. The geometry and boundary conditions for the problem are shown in Figure 11.16. A cavity filled with a saturated porous medium, whose inner vertical wall is maintained at a constant higher temperature and concentration than the outer wall, is considered. The horizontal walls are insulated. All the properties are assumed to be constant, except that of density. The generalized governing equations for double-diffusive natural convection inside the axisymmetric enclosure are used to obtain the solution here (Nithiarasu et al. 1997). The additional nondimensional parameter needed here is the radius ration $r^* = r_i/r_o$ (see Figure 11.10).

Table 11.3 Average Nusselt and Sherwood number distributions, $Le = 2, B = 1.0, r^* = 5, \epsilon = 0.6$

$Ra^* = RaDa$	Da	Nu	Sh
100	10^{-6}	8.29	13.76
500	10^{-6}	19.46	30.87
100	10^{-2}	3.85	5.39
500	10^{-2}	6.14	8.64

Figure 11.17 shows the flow, isothermal and iso-concentration patterns for different radius ratios at a Darcy number of 10×10^{-6}, Lewis number of 1 and Rayleigh number of 10^8. It is observed from the flow pattern that with increase in radius ratio, the eye of the vortex shifts towards the top-right corner of the cavity. Also, the isolines of temperature and concentration accumulate near the bottom-left corner. This can be attributed to flow acceleration towards the hot inner wall, due to reduction in the flow area. The crowded isotherms or iso-concentration lines lead to a thin thermal or solutal boundary layer near the inner wall, thus causing more heat and mass transfer. The packed stream lines near the top-right corner of the cavity also indicate thin velocity layers in that region.

The variations of inner wall Sherwood number with radius ratio at different Da, Ra and Le are shown in Figures 11.18(a) and (b). At low Darcy numbers the Sherwood number varies approximately as $r^{*1/2}$ with radius ratio. This feature can be attributed to the existence of thin thermal and solutal boundary layers adjacent to the inner wall, at the low Da (see Figure 11.1(a)). The boundary layer flow is predominant only when $Ra^* = RaDa$ is large and Da is very small. Therefore, the magnitude of the Sherwood number is high for the combination of high Ra^* and small Da values. Similar features are observed at different Lewis numbers, except that Sherwood number increases with Le due to the occurrence of a thinner mass transfer boundary layer. Table 11.3 shows the distribution of average Nusselt number (Nu) and Sherwood number (Sh).

11.7 Summary

In this chapter a brief summary of convection in porous media has been discussed. It is important to fully understand Chapter 7 before carrying out the porous medium flow calculations. Several details have deliberately not been included in this chapter in order to keep the discussion brief. It is important that readers who may be interested in carrying out further research on the topic read the books and papers listed in the bibliography to further enhance their knowledge.

References

Beckermann C, Ramadhyani S, and Viskanta R (1987) Natural convection flow and heat transfer between a fluid layer and a porous layer inside a rectangular enclosure. *Journal of Heat Transfer*, **109**, 363–370.

Berenati RF and Brosilow CB (1962) Void fraction distribution in packed beds. *AIChE Journal*, **8**, 359–361.

Brinkman HC (1947) A calculation of viscous force exerted by a flowing fluid on a dense swarm of particles. *Applied Science Research*, **1**, 27–34.

Darcy H (1856) Les Fontaines Publiques *de la ville de Dijon*, Dalmont, Paris.

David E, Lauriat G and Cheng P (1991) A numerical solution of variable porosity effects on natural convection in a packed sphere cavity. *ASME J Heat Transfer*, **113**, 391–399.

Ergun S (1952) Fluid flow through packed column. *Chemical Engineering Progress*, **48**, 89–94.

Forchheimer P (1901) Wasserbewegung durch bodem, *Z. Ver. Deutsch. Ing.*, **45**, 1782.

Hsu CT and Cheng P (1990) Thermal dispersion in a porous medium. *International Journal of Heat and Mass Transfer*, **33**, 1587–1597.

Inaba H and Seki N (1981) An experimental study of transient heat transfer characteristics in a porous layer enclosed between two opposing vertical surfaces with different temperatures. *International Journal of Heat and Mass Transfer*, **24**, 1854–1857.

Kaviany M (1991) *Principles of Heat Transfer in Porous Media*. Springer-Verlag, New York.

Lauriat G and Prasad V (1989) Non-Darcian effects on natural convection in a vertical porous enclosure. *International Journal of Heat Mass Transfer*, **32**, 2135–2148.

Lewis RW and Schrefler BA (1998) *The Finite Element Method in the Deformation and Consolidation of Porous Media*. John Wiley & Sons, Ltd, Chichester.

Lewis RW and Ferguson WJ (1990) The effect of temperature and total gas pressure on the moisture content in a capillary porous body. *International Journal for Numerical Methods in Engineering*, **29**, 357–369.

Lewis RW, Morgan K and Johnson KH (1984) A finite element study of 2-D multiphase flow with a particular reference to the five-spot problem. *Computer Methods in Applied Mechanics and Engineering*, **44**, 17–47.

Lewis RW, Morgan K, Pietlicki and Smith TJ (1983) The application of adaptive mesh methods to petroleum reservoir simulation. *Revue de Institut Francais de Petrol*, **36**, 751–761.

Lewis RW, Roberts PJ and Schrefler BA (1989) Finite element modeling of two phase heat and fluid flow in deforming porous media. *Transport in Porous Media*, **4**, 319–334.

Lewis RW and Sukirman Y (1993) Finite element modeling of three-phase flow in deforming saturated oil reservoirs. *International Journal for Numerical and Analytical Methods in Geomechanics*, **17**, 577–598.

Malan AG, Lewis RW and Nithiarasu P (2002) An improved unsteady, unstructured, artificial compressibility, finite volume scheme for viscous incompressible flows: Part I. Theory and implementation. *International Journal for Numerical Methods in Engineering*, **54**, 695–714.

Massarotti N (2001) The characteristic based split scheme for heat and mass transfer in porous media, PhD Thesis, Swansea University.

Massarotti, N., Nithiarasu, P. and Carotenuto, A (2003) Microscopic and macroscopic approach for natural convection in enclosures filled with fluid saturated porous medium. *International Journal of Numerical Methods for Heat and Fluid Flow*, **13**, 862–886.

Massarotti N, Nithiarasu P and Zienkiewicz OC (2001) Porous medium – fluid interface problems. The finite element analysis by using the CBS procedure, *International Journal of Numerical Methods for Heat Fluid Flow*, **11**, 473–490.

Murugesan K, Suresh HN, Seetharamu KN, *et al.* (2001) A theoretical model of brick drying as a conjugate problem *International Journal of Heat and Mass Transfer*, **44**, 4075–4086.

Nield DA and Bejan A (1992) *Convection in Porous Media*. Springer Verlag, New York.

Nishimura T, Takumi T, Shiraishi M, *et al.* (1986) Numerical analysis of natural convection in a rectangular enclosure horizontally divided into fluid and porous regions. *International Journal of Heat Mass Transfer*, **29**, 889–898.

Nithiarasu P (2001) A comparative study on the performance of two time stepping schemes for convection in a fluid saturated porous medium. *International Journal of Numerical Methods for Heat and Fluid Flow*, **11**, 308–328.

Nithiarasu P (2003) An efficient artificial compressibility (AC) scheme based on the characteristic based split (CBS) method for incompressible flows. *International Journal for Numerical Methods in Engineering*, **56**, 1815–1845.

Nithiarasu P, Seetharamu KN and Sundararajan T (1996) Double-diffusive natural convection in an enclosure filled with fluid saturated porous medium – a generalised non-Darcy approach. *Numerical Heat Transfer, Part A, Applications*, **30**, 413–426.

Nithiarasu P, Seetharamu KN and Sundararajan T (1997a) Non-Darcy double-diffusive convection in fluid saturated axisymmetric porous cavities. *Heat and Mass Transfer*, **32**, 427–434.

Nithiarasu P, Seetharamu KN and Sundararajan T (1997b) Double-diffusive natural convection in a fluid saturated porous cavity with a freely convecting wall. *International Communications in Heat and Mass Transfer*, **24**, 1121–1130.

Nithiarasu P, Seetharamu KN and Sundararajan T (1997c) Natural convective heat transfer in an enclosure filled with fluid saturated variable porosity medium. *International Journal of Heat and Mass Transfer*, **40**, 3955–3967.

Nithiarasu P, Seetharamu KN and Sundararajan T (2002) Finite element modeling of flow, heat and mass transfer in fluid saturated porous mdiea, *Archives of Computational Methods in Engineering, State of the Art Reviews*, **9**, 3–42.

Nithiarasu P and Ravindran K (1998) A new semi-implicit time stepping procedure for buoyancy driven flow in a fluid saturated porous media. *Computer Methods in Applied Mechanics and Enginnering*, **165**, 147–154.

Pao WKS, Lewis RW and Masters I (2001) A fully coupled hydro-thermo-poro-mechanical model for black oil reservoir simulation. *International Journal of Numerical and Analytical Methods for Geomechanics*, **25**, 1229–1256.

Prasad V, Kulacki FA and Keylhani M (1985) Natural convection in porous media. *Journal of Fluid Mechanics*, **150**, 80.

Sinha SK, Sundararajan T and Garg VK (1992) A variable property analysis of alloy solidification using the anisotropic porous medium approach. *International Journal of Heat Mass Transfer*, **35**, 2865–2877.

Tong TW and Subramanian E (1985) A boundary layer analysis for natural convection in vertical porous enclosures – use of Brinkmann – extended Darcy model. *International Journal of Heat and Mass Transfer*, **28**, 563–571.

Trevisan OV and Bejan A (1985) Natural convection with combined heat and mass transfer buoyancy effects in porous medium. *International Journal of Heat and Mass Transfer*, **28**, 1597–1611.

Vafai K and Tien CL (1981) Boundary and inertia effects on flow and heat transfer in porous media. *International Journal of Heat Mass Transfer*, **24**, 195–203.

Vafai K, Alkire RL and Tien CL (1984) An experimental investigation of heat transfer in variable porosity media. *ASME Journal of Heat Transfer*, **107**, 642–647.

Walker KL and Homsy GM (1978) Convection in a porous cavity. *Journal of Fluid Mechanics*, **87**, 449–474.

Whitaker S (1961) Diffusion and dispersion in porous media. *American Institute of Chemical Engineering Journal*, **13**, 420–427.

Zienkiewicz OC, Chan AHC, Pastor M, *et al.* (1999) *Computational Geomechanics with Special Reference to Earthquake Engineering*. John Wiley & Sons, Ltd, Chichester.

12

Solidification

12.1 Introduction

Solid–liquid phase change is one of the major topics of study in the heat and mass transfer literature. This is due to the fact that materials processing, metallurgy, purification of metals, growth of pure crystals from melts and solutions, solidification of casting and ingots, welding, electroslag melting, zone melting, thermal energy storage using phase change materials etc. involve melting and solidification. These phase change processes are accompanied by either absorption or release of thermal energy. A moving boundary exists, which separates the two thermo-physical states where the thermal energy is either absorbed or liberated. If we consider the solidification of a casting or ingot, the super heat in the melt and the latent heat liberated at the solid-liquid interface are transferred across the solidified metal, interface and the mold, encountering at each of these stages a certain thermal barrier. In addition, the metal shrinks as it solidifies and an air gap is formed between the metal and mold. Thus, additional thermal resistance is encountered. The heat transfer processes, that occur are complex. The cooling rates employed range from 10^{-5} to 10^{10} K/s and the corresponding solidification systems extend from depths of several meters to a few micrometers. These various cooling rates produce different micro structures and hence a variety of thermo-mechanical properties. During the solidification of binary and multi-component alloys, the physical phenomena becomes more complicated due to phase transformation taking place over a range of temperatures. During the solidification of an alloy, the concentrations vary locally from the original mixture, as material may have been preferentially incorporated, or rejected, at the solidification front. This process is called macro-segregation. The material between the solidus and the liquidus temperatures is partly solid and partly liquid and resembles a porous medium and is referred to as a "mushy zone." A complete understanding of the phase change phenomenon involves an analysis of the various processes that accompany it. The most important of these processes, from a macroscopic point of view, is the heat transfer process. This is complicated by the release, or absorption, of the latent heat of fusion at the solid–liquid interface. Several methods

Fundamentals of the Finite Element Method for Heat and Mass Transfer, Second Edition.
P. Nithiarasu, R. W. Lewis, and K. N. Seetharamu.
© 2016 John Wiley & Sons, Ltd. Published 2016 by John Wiley & Sons, Ltd.

have been used to take into account the liberation of latent heat. The following section gives a brief account of commonly employed methods, which deal with transient heat conduction during a phase change. In the latter sections, more complex cases with fluid flow are discussed. The topics covered in this chapter are meant to be an introduction to solidification. The readers are referred to the recent works listed at the end of this chapter to learn more about the state of the art in this area.

12.2 Solidification via Heat Conduction

12.2.1 The Governing Equations

The classical problem involves considering the conservation of energy in the domain, Ω, by dividing this into two distinct domains, Ω_l (liquid) and Ω_s (solid), where $\Omega_l + \Omega_s = \Omega$. The energy conservation equation is written for the one dimensional case, for simplicity, as

$$\left(\rho_l c_{pl}\right)\frac{\partial T}{\partial t} = k_l\frac{\partial^2 T}{\partial x^2} \quad \text{in} \quad \Omega_l, \tag{12.1}$$

where the subscript l denotes the liquid. Note that in the above equation, the convective motion is neglected. For details on convection, refer to latter sections of this chapter and Chapter 7. Similarly, the equation for the solid portion is written as

$$\left(\rho_s c_{ps}\right)\frac{\partial T}{\partial t} = k_s\frac{\partial^2 T}{\partial x^2} \quad \text{in} \quad \Omega_s, \tag{12.2}$$

where the subscript s represents the solid. The problem will be complete only if the initial and boundary conditions and the interface conditions are given. The interface conditions are

$$T_{sl} = T_f \tag{12.3}$$

and

$$-k_s\left(\frac{\partial T}{\partial x}\right)_s = \rho_s L\frac{ds}{dt} - k_l\left(\frac{\partial T}{\partial x}\right)_l \quad \text{on} \quad \Gamma_{sl}, \tag{12.4}$$

where sl represents the position of the interface, L is the latent heat, ds/dt the interface velocity and T_f is the phase change temperature. Equation (12.4) states that the heat transferred by conduction in the solidified portion is equal to the heat entering the interface by latent heat liberation at the interface and the heat coming from the liquid by conduction. The main complication in solving this classical problem lies in tracking the interface and applying the interface conditions.

12.2.2 Enthalpy Formulation

In the enthalpy method, one single equation is used to solve both the solid and liquid domains of the problem. A single energy conservation equation is written for the whole domain as

$$\frac{\partial H}{\partial t} = k\frac{\partial^2 T}{\partial x^2} \quad \text{in} \quad \Omega, \tag{12.5}$$

where H is the enthalpy function, or the total heat content, which is defined for an isothermal phase change as

$$H(T) = \int_{T_r}^{T} \rho c_s(T) dT \quad \text{if} \quad (T \le T_f)$$

$$H(T) = \int_{T_r}^{T_f} \rho c_s(T) dT + \rho L + \int_{T_f}^{T} \rho c_l(T) dT \quad \text{in} \quad (T \ge T_l) \tag{12.6}$$

and, for a phase change over an interval of temperature T_s to T_l, that is, the solidus and the liquidus temperatures respectively, we have the following

$$H(T) = \int_{T_r}^{T_s} \rho c_s(T) dT + \int_{T_s}^{T} \left[\rho \left(\frac{dL}{dT} \right) + \rho c_f(T) \right] dT \quad (T_s < T \le T_l)$$

$$H(T) = \int_{T_r}^{T_s} \rho c_s(T) dT + \rho L + \int_{T_s}^{T_l} \rho c_f(T) dT + \int_{T_l}^{T} \rho c_l(T) dT \quad (T \ge T_l), \tag{12.7}$$

where c_f is the specific heat in the freezing interval, L is the latent heat and T_r is a reference temperature, which is below T_s. One of the earliest and most commonly used methods for solving such problems has been the "effective heat capacity" method. This method is derived from writing

$$\frac{\partial H}{\partial t} = \frac{\partial H}{\partial T} \frac{\partial T}{\partial t} = k \frac{\partial^2 T}{\partial x^2} \quad \text{in} \quad \Omega, \tag{12.8}$$

We can rewrite the above equation as

$$c_{eff} \frac{\partial T}{\partial t} = k \frac{\partial^2 T}{\partial x^2}, \tag{12.9}$$

where $c_{eff} = \partial H / \partial t$ is the effective heat capacity. This can be evaluated directly from Equation (12.7) as

$$c_{eff} = \rho c_s \quad (T < T_s)$$

$$c_{eff} = \rho c_f + \frac{L}{T_l - T_s} \quad (T_s < T < T_l)$$

$$c_{eff} = \rho c_l \quad (T > T_l), \tag{12.10}$$

Figure 12.1 shows the effective heat capacity variation with respect to temperature. As seen, the effective heat capacity will become infinitely high if the liquidus and solidus temperatures are close to each other. In order to demonstrate the effective heat capacity method discussed above a one-dimensional phase change problem is considered in the following example.

Example 12.2.1   *A phase change problem with an initial temperature of 0.0°C as shown in Figure 12.2 is subjected to a cooling temperature of −45.0°C at the left face and the right-side face is subjected to a liquidus temperature of −0.15°C. The solidus temperature is −10.15°C. Determine the temperature distribution with respect to time if the latent heat of solidification is 70.26, $\rho c_p = 1.0$ and $k = 1.0$. Draw the temperature variation at a uint distance from the left side with respect to time.*

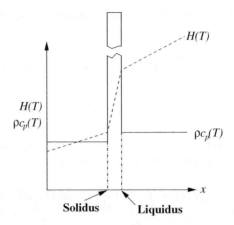

Figure 12.1   Variation of effective heat capacity and enthalpy across the solid–liquid interface.

The unstructured mesh used to solve this problem is shown in Figure 12.3(a). The temper-ature contours at a time of four units is shown in Figure 12.3(b) and the temperature variation at a point of unit length from the left face is shown in Figure 12.3c. These results show a close agreement with existing results (Lewis et al. 2004).

12.3   Convection During Solidification

For a rapid assessment of phase-change problems, the simplified, conduction procedure dis-cussed in the previous section may be used. However, for a more precise analysis of a solidifi-cation, or melting problem, including flow then the flow-driven heat transfer process inside a mold is essential. Both forced and natural convection are possible in a solidification problem but in general natural convection dominates the process. At the early stages of a solidification process, the molten metal is poured into a mold to fill the mold before solidification starts (Lewis *et al.* 1995, 1997; Ravindran and Lewis 1999; Usmani *et al.* 1992). At the mold filling stage, the heat transfer is mainly forced or mixed convection in nature. However, once the mold is filled and allowed to cool, the convection is dominated by buoyancy.

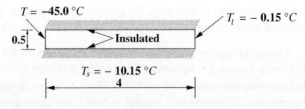

Figure 12.2   A one-dimensional solidification problem.

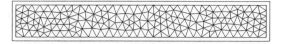

(a) Unstructured mesh, nodes: 202, elements: 328

(b)Temperature distribution at $t = 4$

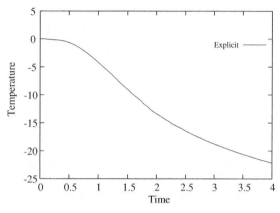

(c) Temperature distribution at point (1,0.25)
with respect to time

Figure 12.3   Solution for the phase change problem using effective heat capacity method.
(a) Unstructured mesh, nodes: 202, elements: 328; (b) Temperature distribution at $t = 4$; (c)
Temperature distribution at point (1,0.25) with respect to time.

Figure 12.4 shows a model problem of solidification in which the molten metal alloy is
allowed to solidify by cooling along the right vertical side. As seen, at any instant during the
solidification process, the domain contains a solid and liquid region and a mushy region. The
physical process governing the heat transfer in the solid region is normally heat conduction
and it is often assumed to be linear isotropic or anisotropic. In the liquid region, the buoyancy-
driven convection dominates the heat transfer process. The properties of liquid and solid
regions are normally different. The third and more complex region is the mushy region that
lies between the solid and liquid regions. A variety of structures in the form of dendrites within
the mushy region are possible (Davis 2001; Xu 2004) and discussing this in detail is not within
the scope of this chapter. However, it is important to realize that this region resembles a porous
medium with nearly zero porosity at the solid surface to a porosity of nearly unity close to
the liquid surface. Due to the directional and complex nature of the dendrite growth from the

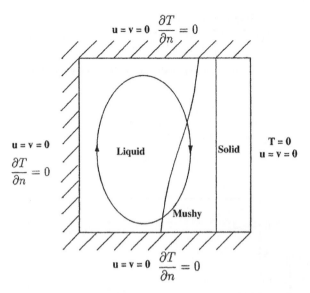

Figure 12.4   Solidification in a square cavity. Geometry and boundary conditions.

solid surface, the mushy region is often treated as a highly anisotropic porous region. In the
example shown, in addition to heat loss through the right vertical wall, the mass transfer in
the form of macro-segregation is also possible. In this chapter we limit the discussions to only
heat transfer.

If the solidifying material is pure, then the interface between the solid and liquid is normally
sharp without a mushy region (Example: Freezing of pure water) although a dendrite structure
is possible (Davis 2001; Xu 2004).

12.3.1   Governing Equations and Discretization

As mentioned in the previous section, convection in a solidification problem is extremely complex. Often precise modeling of mushy region can only be possible by including some experimental measurements on porosity and permeability of the mushy region (Zabaras and Samanta
2004). Thus, the governing equations can be written with varying complexity depending on
the modeling precision required. However, the basic equilibrium equations of incompressible
Navier-Stokes equations are still valid. Before introducing the flow equations, let us introduce
the solid drag terms introduced in the previous chapter to represent the mushy region. Since
the flow speed is really small in the mushy region, it is assumed that the linear drag term or
Darcy term is sufficient to represent the resistance here.

$$- \frac{\mu_f u_i}{\kappa},$$

(12.11)

where μ_f is the fluid viscosity, u_i are the velocity components and κ is the permeability. The
above drag term forms part of the momentum equation as explained in the previous chapter.

The permeability may be related to the porosity of the mushy region (liquid volume fraction), ϵ using

$$\kappa = \frac{\epsilon^3 d_p{}^2}{150(1 - \epsilon)^2}, \tag{12.12}$$

where d_p is the representative solid dendrite size in the mushy region. Since the above relationship was developed for packed beds, a more appropriate Kozeny-Carman relationship for the permeability is often preferred, that is,

$$\kappa(\epsilon) = \frac{\kappa_o \epsilon^3}{(1 - \epsilon)^2}, \tag{12.13}$$

where $\kappa_o = d^2/180$ with d being the secondary dendrite arm spacing. The permeability in the mushy region is directional and is often treated as a vector or tensor (Beckermann and Viskanta 1993; Sinha et al. 1992). The permeability distribution can also be expressed using experimental measurements (Incropera and Bennon 1987; Zabaras and Samanta 2004). A much simplified form of the drag term can be constructed using liquid volume fraction ϵ (Kim et al. 2002; Viswanath and Jaluria 1993) as

$$-C\frac{(1 - \epsilon)^2}{(\epsilon^3 + b)} u_i. \tag{12.14}$$

where C is a very large constant and b is a very small number to avoid division by zero. The simplest way of calculating the liquid volume fraction ϵ is

$$\epsilon = \frac{T - T_s}{T_l - T_s}. \tag{12.15}$$

The liquid volume fraction in the mushy region can also be calculated iteratively using more advanced models by computing the total enthalpy. Total enthalpy in the mushy region may be expressed as

$$\frac{H}{\rho} = \epsilon c_l(T - T_s) + (1 - \epsilon)c_s(T - T_s) + c_s T_s + \epsilon L. \tag{12.16}$$

The enthalpy H should be obtained by solving the energy equation reformulated in terms of enthalpy.

The equilibrium equations for solidification (or melting) may now be written as
Continuity:

$$\frac{\partial u_i}{\partial x_i} = 0. \tag{12.17}$$

Momentum:

$$\rho\left(\frac{\partial u_i}{\partial t} + u_j\frac{\partial u_i}{\partial x_j}\right) = -\frac{\partial p}{\partial x_i} + \mu\frac{\partial^2 u_i}{\partial x_j^2} - \rho_l g_i \beta(T - T_r) + S_i. \tag{12.18}$$

Temperature:

$$C_e \left[\frac{\partial T}{\partial t} + u_j \frac{\partial T}{\partial x_j} \right] = k_e \frac{\partial^2 T}{\partial x_j^2}. \tag{12.19}$$

where u_i are the velocity components, t is the time, x_i are the coordinate axes, p is the pressure, μ is the kinematic viscosity, ρ is the density, C_e is the effective heat capacity including the latent heat and k_e is the effective thermal conductivity.

The expression for C_e may be written as

$$C_e = (\rho c_p)_e + \frac{L}{T_l - T_s}, \tag{12.20}$$

where the subscript e represents effective. The source term S_i in Equation (12.18) may be represented either by Equation (12.11) or by Equation (12.14) depending on the details required. In order to estimate the liquid fraction in the mushy region, for alloy solidification, Equation (12.16) should be used. To close the enthalpy equation, the enthalpy should be calculated from a reformulated equation of energy in terms of the enthalpy, that is,

$$\frac{\partial H}{\partial t} + u_j \frac{\partial H}{\partial x_j} = k_e \frac{\partial^2 T}{\partial x_i^2}. \tag{12.21}$$

Along with the above equation and Equation (12.16), the liquid volume fraction ϵ may be estimated iteratively. The nondimensional form of Equations (12.16)–(12.19) may be obtained by using the following nondimensional scales.

$$x_i^* = \frac{x_i}{D}; u_i^* = \frac{u_i}{\alpha_l/D}; p^* = \frac{p}{\rho_l \alpha_l^2/D};$$

$$\rho^* = \frac{\rho}{\rho_l}; t^* = \frac{t}{D^2/\alpha_l}; T^* = \frac{T - T_r}{T_i - T_r}, \tag{12.22}$$

where D is a characteristic dimension, α is the thermal diffusivity, subscripts l, i and r represent liquid, initial and reference respectively. Substituting the nondimensional scales into Equations (12.16)–(12.19), we obtain

Continuity:

$$\frac{\partial u_i^*}{\partial x_i^*} = 0. \tag{12.23}$$

Momentum:

$$\rho^* \left(\frac{\partial u_i^*}{\partial t^*} + u_j^* \frac{\partial u_i^*}{\partial x_j^*} \right) = -\frac{\partial p^*}{\partial x_i^*} + \mu^* Pr \frac{\partial^2 u_i^*}{\partial x_j^{*2}} + \gamma_i RaPrT^* + S_i^*. \tag{12.24}$$

Temperature:

$$\sigma \left[\frac{\partial T^*}{\partial t^*} + u_j^* \frac{\partial T^*}{\partial x_j^*} \right] = k^* \frac{\partial^2 T^*}{\partial x_i^{*2}}. \tag{12.25}$$

where

$$\mu^* = \frac{\mu}{\mu_l}; v = \frac{\mu}{\rho}; Pr = \frac{v_l}{\alpha_l};$$

$$Ra = \frac{g_i\beta(T_i - T_r)D^3}{v_l\alpha_l}; \sigma = \frac{C_e}{(\rho c_p)_l}; k^* = \frac{k}{k_l}$$

(12.26)

and γ_i is the unit vector in the buoyancy direction. In Equation (12.24), the source term S_i takes different forms. If the mushy region is ignored, or not important, the source term is equal to zero. If Darcy's term is used, the nondimensional form of the source term becomes

$$S_i^* = -\mu^* \frac{Pr}{Da} u_i^*,$$

(12.27)

where Da is the Darcy number ($\frac{\kappa}{D^2}$). If Equation (12.14) is used for the source term, the nondimensional source becomes

$$S_i^* = -C^* \frac{(1 - \epsilon)^2}{(\epsilon^3 + b)} u_i^*.$$

(12.28)

C^* is a very large nondimensional constant. The finite element discretization of the continuity, momentum and energy equations follow the identical procedure discussed for natural convection in Chapter 7.

Example 12.3.2   *Solidification in a square cavity*

The model problem shown in Figure 12.4 is used here to demonstrate the finite element method. This is a simple problem and the finite element method discussed in Chapter 7 can directly be used to solve this problem. More complex problems with appropriate changes to the equations can also be solved using the algorithm discussed in Chapter 7. The objective of the problem considered here is to find the solidification pattern in a square cavity filled with a molten alloy with the simplest formulation possible with fluid flow. In this formulation, the source term S_i is assumed to be zero and the liquidus and solidus temperatures are predefined for an alloy solidification. In addition, the mushy region is represented only by introducing the latent heat in this region as discussed below. With these assumptions, the solidification is simplified to a very basic model with fluid flow.

In the problem shown in Figure 12.4, both the horizontal walls and left vertical wall are assumed to be insulated and the right vertical wall is assumed to be at a constant temperature that is less than that of the initial molten metal and solidification temperatures. The Rayleigh number is defined based on the temperature difference between the initial molten metal and the right vertical wall. To obtain a model solution, the following values for different parameters are used here. They are: $Pr = 0.02$; $\sigma = 5$ in the mushy region; $\sigma = 1$ in solid and fluid regions; $T_l^* = 0.5$; $T_s^* = 0.3$; $T_i^* = 1.0$; $T_c^* = 0$.

Figure 12.5 shows the flow and isotherm patterns with at different instances during the solidification process. The meshes used here are adaptively generated(Nithiarasu 2000; Nithiarasu and Zienkiewicz 2000; Nithiarasu 2002) (see Chapter 14). As seen, the natural convective flow is covering almost the whole part of the domain at the early stages of the solidification and the flow region progressively shrinks with respect to increase in time. It is also noticed that at the later stages of the solidification process, multi-cellular flow patterns

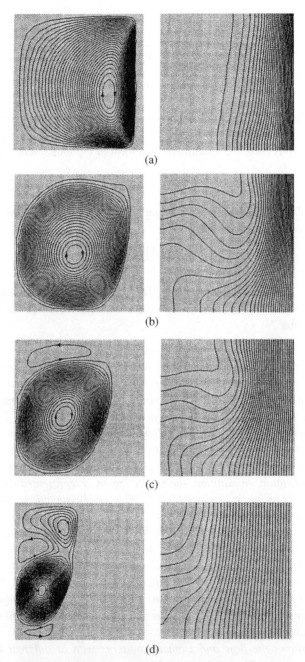

(a)

(b)

(c)

(d)

Figure 12.5 Solidification in a square cavity. Flow and isotherm patterns at different nondimensional times (a) $t = 0.0305$; (b) $t = 0.1945$; (c) $t = 0.4815$; (d) $t = 0.9275$.

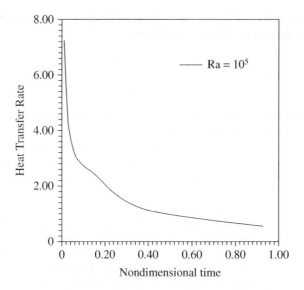

Figure 12.6   Solidification in a square cavity. Heat transfer rate variation with time.

are observed (Figure 12.5(d)) as the liquid region shrinks. The isotherms shown clearly shows the dominant conduction mode in the solid region and convective mode with highly nonlinear variation in the liquid part.

Figure 12.6 shows the rate of heat transfer through the cold wall (right wall) with respect to nondimensional time. The following relation is used to find the average rate of heat transfer

$$\frac{1}{D} \int_0^D \frac{\partial T}{\partial x} dy, \qquad (12.29)$$

where x is the direction normal to the wall, D is the height of the wall and y is the tangential direction along the wall. As expected, the heat transfer rate is high at the beginning and reduces exponentially with time to a more or less a fully conduction mode (heat transfer rate ≈ = 1). This is expected as the conduction mode becomes the predominant mode of heat transfer when the majority part of the domain is solidified.

12.4   Summary

This chapter provided a brief introduction to the finite element solution of solidification problems with only heat transfer aspects of the problem in mind. The amount of literature available on the numerical modeling of flow and heat transfer needs a comprehensive book to explain all aspects of solidification and melting. Some of the important topics not covered herein are dendrite growth, macro-segregation and multi-scale approach. These subjects need substantially more space to thoroughly explain them and this is not the objective of this chapter. We believe that the brief introduction given in this chapter allows a student, or researcher, to obtain a sufficient introductory knowledge to progress towards more comprehensive modeling of flow and heat transfer in solidification problems.

References

Beckermann C and Viskanta R (1993) Mathematical modeling of transport phenomena during alloy solidification. *Applied Mechanics Reviews*, **46**, 1–27.

Davis SH (2001) *Theory of Solidification*. Cambridge University Press, Cambridge, UK.

Incropera FP and Bennon WD (1987) A continuum model for momentum, heat and species transport in binary solid liquid phase change systemsI. Model formulation. *International Journal of Heat and Mass Transfer*, **30**, 2161–2170.

Kim S, Kim MC and Lee B (2002) Numerical analysis of convection-driven melting and solidification in a rectangular enclosure. *Journal of Inustrial Engineering Chemistry*, **8**, 185–190.

Lewis RW, Usmani AS and Cross JT (1995) Efficient mould filling simulation in castings by an explicit finite element method. *International Journal for Numerical Methods in Fluids*, **20**, 493–506.

Lewis RW, Navti SE and Taylor C (1997) A mixed Lagrangian-Eulerian approach to modelling fluid flow during mould filling. *International Journal for Numerical Methods in Fluids*, **25**, 931–952.

Lewis RW, Nithiarasu P and Seetharamu KN (2004) *Fundamentals of the Finite Element Method for Heat and Fluid Flow*. John Wiley & Sons, Inc., New York.

Nithiarasu P (2000) An adaptive finite element procedure for solidification problems. *Heat and Mass Transfer*, **36**, 223–229.

Nithiarasu P and Zienkiewicz (2000) Adaptive mesh generation for fluid mechanics problems. *International Journal for Numerical Methods in Engineering*, **47**, 629–662.

Nithiarasu P (2002) An adaptive remeshing technique for laminar natural convection problems. *Heat and Mass Transfer*, **38**, 243–250.

Ravindran K and Lewis RW (1999) Finite element modelling of solidification effects in mould filling. *Finite Elements in Analysis and Design*, **31**, 99–116.

Sinha SK, Sundararajan T and Garg VK (1992) A variable property analysis of alloy solidification using the anisotropic porous medium approach. *International Journal of Heat Mass Transfer*, **35**, 2865–2877.

Usmani AS, Cross JT and Lewis RW (1992) A finite element model for the simulations of mould filling in metal casting and associate heat transfer. *International Journal for Numerical Methods in Engineering*, **35**, 787–806.

Viswanath R and Jaluria Y (1993) A comparison of different solution methodologies for melting and solidification problems in enclosures. *Numerical Heat Transfer. Part B Fundamantals*, **24**, 77–105.

Xu JJ (2004) *Dynamical Theory of Dendritic Growth in Convective Flow*. Kluwer Academic Publishers, Norwell, MA.

Zabaras N and Samanta D (2004) A stabilized volume-averaging finite element method for flow in porous media and binary alloy solidification processes. *International Journal for Numerical Methods in Engineering*, **60**, 1103–1736.

13

Heat and Mass Transfer in Fuel Cells

13.1   Introduction

Fuel cells are electrochemical devices that directly convert chemical energy of a fuel into electrical energy. The advantages of such a nonconventional energy generation device include better efficiency and pollutant free operation. Figure 13.1 shows a schematic diagram of the operating principle of a typical fuel cell. As seen, a fuel cell consists of three major compartments. They are: fuel side on the left, oxidant on the right and electrolyte in the middle. The fuel side consists of an anode (negative electrode) and an appropriate passage for supplying a gaseous fuel. The oxidant side consists of a cathode (positive electrode) and an appropriate passage for the oxygen to flow. The middle part, the electrolyte layer, transports ionic charge to the cathode, and thereby completes the cell electric circuit as illustrated in Figure 13.1. It also provides a physical barrier between the fuel and oxidant to prevent direct mixing. The electrolyte is normally coated with a catalyst (platinum) layer, which triggers ionization of hydrogen.

In a typical fuel cell, a gaseous fuel (often hydrogen) is continuously supplied to the anode compartment and an oxidant (i.e., oxygen from air) is supplied to the cathode compartment. The hydrogen gas is ionized as soon as it is in contact with the catalyst, that is,

$$H_2 \rightarrow 2H^+ + 2e^-. \tag{13.1}$$

The positively charged ions from the above reaction (protons), permeate through the electrolyte towards the cathode side but the negatively charged electrons do not. As a result of proton migration, a potential difference between the two electrodes is generated. This electrical gradient drives the electrons through an external circuit between the two electrodes resulting in

Fundamentals of the Finite Element Method for Heat and Mass Transfer, Second Edition.
P. Nithiarasu, R. W. Lewis, and K. N. Seetharamu.
© 2016 John Wiley & Sons, Ltd. Published 2016 by John Wiley & Sons, Ltd.

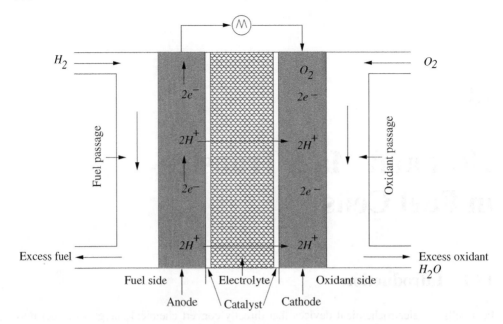

Figure 13.1   A schematic diagram of the typical proton exchange fuel cell principle.

power generation. The H^+ ions meet the electrons from the external circuit at the cathode. The electrons, the ions from the electrolyte and the oxidant induce the following electrochemical reaction at the cathode.

$$\frac{1}{2}O_2 + 2e^- + 2H^+ \rightarrow H_2O. \tag{13.2}$$

Summing Equations (13.1) and (13.2), the global electrochemical reaction of a hydrogen fuel cell is obtained as

$$H_2 + \frac{1}{2}O_2 \rightarrow H_2O. \tag{13.3}$$

In addition to producing water (H_2O), hydrogen fuel cells produce heat as another end product. In summary, the processes of a typical fuel cell may be listed as (a) continuous electric current is generated due to flow of electrons through the external circuit; (b) thermal energy is generated by the two electrochemical reactions (Equations (13.1) and (13.2)) that are highly exothermic; (c) water is produced at the cathode compartment as an end product. The performance of a fuel cell is measured in terms of power generated. The power generation can be obtained by multiplying the voltage generated and the current. A mathematical model of a fuel cell should be aimed at predicting the power of a fuel cell, if the material properties of various components of a fuel cell are provided. Before discussing the details of the mathematical model, the following section provides a brief summary on different type of fuel cells.

13.1.1   Fuel Cell Types

Fuel cells may be classified depending on the fuel-oxidant combination used, whether the fuel is processed outside (external reforming) or inside (internal reforming) the cell, the type of electrolyte employed, the operating temperature of the fuel cell, etc.

The most common classification of fuel cells is by the type of electrolyte used in the cells. Currently, six such types of fuel cells are available (Kirubakaran *et al.* 2009; Mauro 2009, 2011).

(i)   Proton exchange membrane fuel cell (PEMFC):

   (a) Direct formic acid fuel cell (DFAFC)

   (b) Direct Ethanol Fuel Cell (DEFC)

(ii) Alkaline fuel cell (AFC):

   (a) Proton ceramic fuel cell (PCFC)

   (b) Direct borohydride fuel cell (DBFC)

(iii) Phosphoric acid fuel cell (PAFC)

(iv) Molten carbonate fuel cell (MCFC)

(v)   Direct methanol fuel cell (DMFC)

(vi) Solid oxide fuel cell (SOFC).

The classifications under operating temperatures may be divided into low temperature fuel cells that operate in the temperature range of 50–250 °C and high temperature fuel cells that have a temperature range of 650–1000 °C. Examples of low temperature fuel cells include PEMFC, AFC and PAFC, while MCFC and SOFC are two good examples of high temperature fuel cells. A brief description of some fuel cell types is given below.

(i). The first type of fuel cell, PEMFC, uses a solid polymer electrolyte (Teflon-like membrane), which is an excellent conductor of protons and an insulator for electrons. The operating temperature of this fuel cell is as low as 100 °C. The first sub-classification of PEMFC is DFAFC, which uses formic acid ($HCOOH$) as the fuel. This fuel consists of small organic molecules and it is directly fed to the anode. The second subcategory of PEMFC, DEFC, uses nafion as a catalyst and ethanol as the fuel.

(ii). Alkaline Fuel Cells (AFC) were first employed in NASA's space missions. This fuel cell is also referred to as Bacon Fuel Cell, after its British inventor. It operates at low temperatures around 100 °C and is capable of working at an efficiency range of 60–70%. It uses an aqueous solution of potassium hydroxide (KOH) as the electrolyte, which transports negatively charged ions from the cathode to the anode and releases water as its byproduct. The water produced in the anode side migrates to the cathode to produce hydroxyl ions (OH^-). One of the advantages of this fuel cell is that it starts easily. Protonic ceramic fuel cell (PCFC) is a relatively new fuel cell, which is developed basically with the ceramic electrolyte material.

It can operate at temperatures as high as 750 °C and electrochemically oxidize gas molecules of the hydrocarbon supplied to the anode, without the aid of an additional reformer. Sodium borohydride ($NaBH_4$) is a potential fuel, if mixed with water to generate hydrogen. After releasing hydrogen, borohydride is oxidized at the cathode to produce $NaBO_2$ or borax. The second subclassification of AFC, DBFC fuel cell, operates at temperatures as low as 70 °C. The main advantages of this type of fuel cell are higher power density, no need for an expensive platinum catalyst and high open circuit cell voltage (about 1.64 V).

(iii). Phosphoric acid fuel cell (PAFC) operates at temperatures of about 175–200 °C. This operating temperature is almost twice the temperature at which PEM fuel cells operate. PAFC utilizes the liquid phosphoric acid as electrolyte. Unlike PEM and AFC, it is very tolerant to impurities in reformed hydrocarbon fuels.

(iv). Molten carbonate fuel cells (MCFC) operate at temperatures of about 600–700 °C. MCFC consists of two porous electrodes with good conductivity that are in contact with a molten carbonate electrolyte. Due to its internal reforming capability, it separates the hydrogen from unreformed fuel within the cell. The main advantages of MCFC are higher efficiency of 50–60%, no need of a metal catalyst and a separate reformer is also not required as its operating temperatures are quite high (Farooque and Maru 2001).

(v). DMFC technology is relatively new in comparison to the other type of fuel cells described. DMFC uses a polymer electrolyte and liquid methanol or alcohol as fuel instead of reformed hydrogen. At the anode, hydrogen is obtained by dissolving liquid methanol (CH_3OH) in water without the need for an external reformer. At the cathode, the recombination of the positive and negative ions takes place and in the presence of the oxidant, water is produced as a byproduct.

(vi). SOFCs belongs to the category of high temperature fuel cells. They use dense yttria stabilized zirconia, which is a solid ceramic material, as an electrolyte. In SOFCs, oxygen ions combine with hydrogen ions to generate water and heat. SOFCs produce electricity at a high operating temperature of about 1000 °C. A number of researchers have started working with Intermediate Temperature Solid Oxide Fuel Cells (ITSOFCs), which operate at a temperature between 550 and 800 °C. The ITSOFCs offer cost effective fabrication and they are potentially more reliable than very high temperature SOFCs. The main advantages of SOFCs are high efficiency of about 50–60% and no need for an external reformer to extract hydrogen from the fuel, thanks to its internal reforming capability. Waste heat can be recycled to make additional electricity by cogeneration operation (Farooque and Maru 2001). The slow start up, high cost and intolerance to sulphur content of the fuel cell are its main drawbacks.

In the following sections, the discussion of the mathematical models and numerical solution procedures are presented for Solid Oxide Fuel Cells (SOFCs). The extension of the procedures to other types of fuel cells is similar and the underlying numerical procedure can be employed for all fuel cell types.

13.2   Mathematical Model

A simple solid oxide fuel cell consists of an anode, cathode and electrolyte. The planar geometries of SOFCs are popular as they are easy to manufacture and they have the potential to generate higher power densities than tubular ones. Thus, the models presented here only represent the planar fuel cell structures. The electrochemical reactions of an SOFC depends on the fuel used. In general, the cathode and anode reactions may be written as:

$$O_2 + 2e^- \rightarrow O^= \tag{13.4}$$

and

$$H_2 + O^= \rightarrow H_2O + 2e^- \tag{13.5}$$

respectively. The electrons emitted at the anode travel via the external circuit to the cathode to produce $O^=$ ions. These ions are transported through the electrolyte towards the anode to produce electrons and water. Unlike the reaction explained in Section 13.1, the oxide ion travels across the electrolyte to complete the reaction and the water is produced as a product at the anodic side.

The ideal electromotive force associated with the electrochemical reactions (Equations 13.4 and 13.5) may be represented by the Nernst equation (Singhal and Kendall, 2003), that is,

$$E = E_o - \frac{RT}{2F} ln \left(\frac{p_{H_2O}}{p_{H_2} p_{O_2}^{0.5}} \right), \tag{13.6}$$

where E_o is the standard electrode potential (V), R is the universal gas constant (8.314 $Jmol^{-1}K^{-1}$), T is the temperature (K), F is the Faraday's constant (96 487 $Cmol^{-1}$) and p is the partial pressure (Pa). The standard electrode potential is given as $E_o = 1.2723 - 2.7645 \times 10^{-4}T_s$. This is the ideal voltage for hydrogen oxidation at ambient pressure as a function of temperature at the reaction sites.

Under normal operating conditions, the actual voltage of a fuel cell is lower than the ideal one given by Equation (13.6) due to losses. The voltage losses are often caused by the activation, concentration and ohmic over-potentials. Thus, the effective or corrected potential may be calculated as:

$$V = E - \eta_a - \eta_c - \eta_{ohm}, \tag{13.7}$$

where η_a is the activation over-potential, η_c the concentration over-potential and η_{ohm} the ohmic over-potential. The activation losses are the result of activation energy required by the chemical reactions, concentration losses are due to the concentration gradient across the electrodes and ohmic losses are due to the resistance to electron flow in the electrodes and to ion flow in the electrolyte. Equation (13.7) is the equation used in evaluating the performance of SOFCs. Thus the remaining part of the mathematical model should be developed to address the unknowns of Equations (13.6) and (13.7).

The activation over-potential η_a in Equation (13.7) is related to the cell current density via the Butler-Volmer equation (Singhal and Kendall 2003), that is,

$$i = i_o \left[exp \left(\alpha_a \frac{n_e F \eta_a}{RT} \right) - exp \left(-\alpha_c \frac{n_e F \eta_a}{RT} \right) \right], \tag{13.8}$$

Table 13.1   Available models for exchange current density, i_o, estimation

Reference	k_o		E_a	
	Anode	Cathode	Anode	Cathode
Campanari and Iora (2004)	$7 \times 10^9 \frac{p_{H_2}p_{H_2O}}{p_{ref}^2}$	$7 \times 10^9 \frac{p_{O_2}}{p_{ref}}$	110	120
Suwanwarangkul et al. (2006)	$\frac{RT}{3F} 6.2 \times 10^{11} \left(\frac{p_{H_2O}}{K_{eq \cdot H_2}p_{h2}} \right)^{0.266}$	$0.25 \times 10^{10} RTp_{O_2}^{0.5}$	120	130
Aguiar et al. (2004)	$\frac{RT}{2F} 2.35 \times 10^{11}$	$\frac{RT}{2F} 6.54 \times 10^{11}$	140	137
Suzuki et al. (2005)	$\frac{RT}{2F} 125.6 \times 10^{10} p_{O_2}^{0.15}$	$\frac{RT}{2F} 62.7 \times 10^6 p_{O_2}^{0.5}$	138	136

where i_o is the exchange current density at the electrode-electrolyte interface. The Arrhenius equation of the following type is often employed to compute i_o, that is,

$$i_o = k_o exp \left(-\frac{E_a}{RT} \right),$$ (13.9)

where k_o is a constant and E_a is the activation energy. Both these quantities are empirically determined. Some of the models used in the literature are given in Table 13.1.

The remaining undefined parameters in Equation (13.8) are n_e, the number of electrons transferred and α_a and α_c the transfer coefficients for anode and cathode respectively. Both the transfer coefficients are assumed to be 0.5. Equation (13.8) is used to calculate the activation losses. This is normally calculated iteratively by prescribing an average operating current density of the fuel cell.

The second loss, due to concentration gradient, is a result of varying concentration along an electrode as the fuel/oxidant are consumed when they travel along the reaction site. The difference in concentration between the flow channel and the reaction site is directly related to the concentration losses. These losses can be evaluated as the difference between the Nernst potential in the bulk flow of the channel and the reaction site, that is,

$$\eta_c = \eta_c^a + \eta_c^c = \frac{RT}{2F} ln \left(\frac{X_{H_2}^b X_{H_2O}^r}{X_{H_2}^r X_{H_2O}^b} \right) + \frac{RT}{4F} ln \left(\frac{X_{O_2}^b}{X_{O_2}^r} \right),$$ (13.10)

where X is the molar fraction. The third loss is a result of the resistance offered by the electrolyte to the ion movement and resistance of the circuit connecting the electrodes. This may be calculated as

$$\eta_{ohm} = IR_i,$$ (13.11)

where I is the current and R_i is the resistivity. The main contribution to resistivity is from the electrolyte which varies inversely with the operating temperature. The following relationship is commonly used in determining the temperature dependent resistivity (Mauro *et al.* 2010).

$$R_i = Aexp\left(\frac{B}{T}\right). \qquad (13.12)$$

The constants A and B in the above equation are determined via experiments.

Thus, to calculate the fuel cell voltage from Equation (13.7) we require the temperature and partial pressures of H_2, O_2 and H_2O (Equation (13.6)), prescribe a current density (Equation (13.8)) and determine the molar fractions of different species (Equation (13.10)). In order to calculate the pressure, temperature and concentrations (molar fractions), appropriate governing equations need to be solved throughout the domain. Such a thermofluid dynamic field in the anodic and cathodic compartments is described by using a generalized porous medium model. To make the presentation simple we may divide a fuel-cell domain into three compartments, that is, (i) the anodic compartment, that includes the fuel channel, the porous anode and the catalyst layer; (ii) the electrolyte; and (iii) the cathodic compartment, that includes the oxidant channel, the porous cathode and the catalyst layer. The first and last compartment of the fuel cell may be modeled using the generalized porous medium model and the second component, electrolyte, is modeled using a Laplace equation for the electric potential and energy equation with a source term generated by electric potential. The electrolyte part of the domain is often simplified as the thickness of this part is often orders of magnitude smaller than the electrode. In the following subsections, the equations governing the first and third part of the domain are reviewed before discussing the electrolytic comportment.

13.2.1   Anodic and Cathodic Compartments

The flow, heat and mass transfer in these compartments are governed by conservation of the mass, momentum and energy equations. They are:
Conservation of mass:

$$\frac{\partial u_i}{\partial x_i} = 0. \qquad (13.13)$$

Momentum conservation:

$$\frac{\rho}{\epsilon}\left(\frac{\partial u_i}{\partial t} + u_j\frac{\partial u_i}{\partial x_j}\right) = -\frac{\partial p}{\partial x_i} + \mu_e\frac{\partial}{\partial x_i}\left(\frac{\partial u_i}{\partial x_i}\right) - \frac{\mu u_i}{\kappa} - \frac{\rho F_o|\mathbf{u}|}{\sqrt{\kappa}}. \qquad (13.14)$$

Energy conservation:

$$\epsilon\left[(\rho c_p)_f + (1-\epsilon)(\rho c_p)_s\right]\frac{\partial T}{\partial t} + (\rho c_p)_f u_j\frac{\partial T}{\partial x_j} = k_e\frac{\partial}{\partial x_j}\left(\frac{\partial T}{\partial x_j}\right) + S_e. \qquad (13.15)$$

Species conservation:

$$\rho\epsilon\left(\frac{\partial c^k}{\partial t} + u_j\frac{\partial c^k}{\partial x_j}\right) = D^e_{km}\frac{\partial}{\partial x_j}\left(\frac{\partial c^k}{\partial x_j}\right) + S^m_k, \qquad (13.16)$$

where u_i are the velocity components, x_i coordinate axes, ρ density, ϵ porosity, t time, μ_e equivalent viscosity, μ fluid viscosity, κ permeability, F_o Forchheimer constant, c_p specific heat at constant pressure, k_e equivalent thermal conductivity, S_e energy source term, c^k is the concentration of species k, D^e_{km} equivalent mass transfer coefficient of species k and S^m_k is the mass transfer source term for species k.

Note that Equations 13.13–13.16 are valid both for the electrodes and for the channels carrying the fuel and oxidant. Inside the electrodes, the porosity assumes the value corresponding to the actual electrode type and microstructure (obtained through experiments), while in the fuel and oxidant channels the porosity approaches unity to recover incompressible Navier-Stokes equations. For more details on the generalized porous medium model, readers are referred to Chapter 11.

The equivalent mass transfer diffusion coefficient D^e_{km} is a crucial parameter and its calculation is important. In the absence of detailed experimental data, several models have been employed in the literature, which are functions of pressure, temperature and species concentration value. The most common method of theoretical estimation of gaseous diffusion is the one developed by Chapman and Cowling (2004). The effect of porous electrodes are taken into account through the Knudsen diffusion coefficient (Zho and Kee 2003) as

$$D_k = \frac{4}{3} r_p \sqrt{\frac{8RT}{\pi M_k}}, \tag{13.17}$$

where M_k is the molar weight of the kth species in $kg\ mol^{-1}$. The effective species diffusion coefficient may now be calculated as

$$\frac{1}{D^e_{km}} = \frac{\tau_g}{\epsilon} \left(\frac{1}{D_{km}} + \frac{1}{D_k} \right), \tag{13.18}$$

where τ_g is the tortuosity which is often assumed to be a constant and D_{km} is the multi-component diffusion coefficient $m^2\ s^{-1}$.

The source terms S_e and S^m_k in Equations 13.13–13.16 need defining. The heat generation term S_e is defined as

$$S_e = \frac{Ti}{2F} \left(s_{H_2O} - s_{H_2} - \frac{1}{2} s_{O_2} \right) + i(\eta_a + \eta_{ohm}). \tag{13.19}$$

In the above equation, the first term on the RHS represents the anodic and cathodic reversible heat generation per unit area and the second term represents the irreversible heat flux due to electrochemical reactions. The letter s above represents the entropies and the subscripts of s represent respective species (Mauro $et\ al.$ 2011).

The source term S^m_k in the species equation is the result of electrochemical reactions at the catalyst layer of the cell (see Figure 13.2). Thus the equation of the mass flux at the catalyst layer is given as

$$S^m_k = \frac{iM_k}{n_e F}, \tag{13.20}$$

where n_e is the number of transferred electrons.

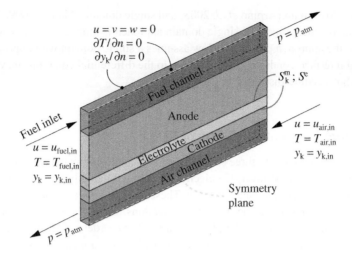

Figure 13.2 Heat and mass transfer in a planar SOFC. Schematic diagram and typical boundary conditions for a single domain approach. *Source*: Mauro (2009). Reproduced with permission of Mauro.

13.2.2 Electrolyte Compartment

Under normal and steady-state operating conditions, the oxide ions $O^=$ migrate from the cathode to the anode. As a result a static electric field is created which is governed by

$$\nabla.\mathbf{E} = 0. \tag{13.21}$$

The current flux $\mathbf{E}$ in the electrolyte may be expressed in terms of the static potential as $\mathbf{E} = -\lambda\nabla\phi = 0$, where λ is the electrical conductivity of the electrolyte and ϕ the electric potential. Thus, Equation 13.21 may be rewritten as

$$\lambda\nabla^2\phi = 0. \tag{13.22}$$

The above Laplace equation is the static potential equation. The generic energy equation for an electrolyte placed in between two electrodes may be written as

$$\rho c_p \left(\frac{\partial T}{\partial t} + \mathbf{u}.\nabla T \right) = \nabla.(k\nabla T) + \lambda|\nabla\phi.\nabla\phi|. \tag{13.23}$$

If the electrolyte is static, the velocity $\mathbf{u}$ in the above equation becomes zero, which results in the heat conduction equation with the heat generation term due to the static electric potential.

13.3 Numerical Solution Algorithms

There are a number of ways in which a finite element solution to the system of equations can be obtained. We provide a summary of modeling flow, heat and mass transfer in an anode supported SOFC below. The method explained below is derived from two distinctive approaches, multi domain (Arpino *et al.* 2008) and single domain (Mauro, 2009; 2011; Mauro *et al.* 2011) approaches. Since the single domain approach is more comprehensive and easier to implement, the approach provided below assumes a single domain with appropriate initial, boundary and interface conditions. The solution to the differential equations may be obtained using the CBS procedure discussed in Chapter 7.

13.3.1   Finite Element Modeling of SOFC

The device here is divided into three subdomains: (1) anodic compartment, including the fuel channel, anode and catalyst layer, (2) electrolyte, and (3) cathodic compartment, including oxidant channel, cathode and catalyst layer. These subdomains are combined into a single composite domain for solving the governing equations. Since the thickness of the catalyst layer, where chemical reactions take place, is at least two orders of magnitude smaller than the other parts of the domain, this is simply assumed to be an interface between the porous electrode and the electrolyte. This aspect allows us to model reactants and product consumption/production, as well as the thermal energy generation associated with the device operation, as Neumann boundary conditions are imposed at the interface between the porous electrode and the electrolyte. At the interfaces, continuity of temperature, species concentration and their fluxes should be maintained. The continuity in the concentration and temperature are automatically satisfied. The flux continuity may be approximately satisfied by refining the mesh at the interface. In addition, the following assumptions are also invoked:

- The flow is incompressible and laminar.

- The gases used are assumed to behave as ideas gases.

- The electrodes are assumed to be made of a homogeneous material with constant porosity.

- Transient effects are negligible. Thus, only a steady-state solution is sought.

Since flow within the fuel cell subdomains is independent of transport, the momentum and continuity equations may be solved independently of the temperature and species equations. However, the strong coupling between energy and species equations demands an iterative solution between them. The coupling between these two transport equations exists due to the temperature dependent species diffusion coefficients (Equations (13.17) and (13.18)) and source terms of the two equations (Equations (13.19) and (13.20)). Although only a steady-state solution is sought, the time-stepping procedure of the CBS scheme acts as an excellent iterative mechanism to balance the quantities. To obtain a satisfactory result, it is therefore important to converge one or more of the important quantities that impose a strong coupling between the two transport equations. The quantities of interest are the source terms, different over potentials (linked to energy source), current density (if not prescribed) and the diffusion coefficient. Bearing in mind all the assumptions and the coupling, the following solution sequence may be employed to obtain an approximate finite element solution to SOFC performance.

- Assume appropriate initial conditions for velocity, temperature, species concentration, average current density and diffusion coefficient.

- Solve continuity and momentum equations to steady state using the CBS method as described in Chapter 7.

- Using the initial conditions, compute the source terms of the species and energy equations and use them as the interface condition between anode and electrolyte and cathode and electrolyte respectively (see Figure 13.2).

- With the prescribed initial, boundary and interface conditions, obtain a steady-state solution to the energy and species equations.

- Solve the Butler-Volmer equation using the given average current density to compute activation losses.

- Compute the effective species diffusion coefficient.

- Check whether or not the diffusion coefficient value converges. If not, go to step 3 and continue until convergence.

As a final step of the calculation procedure, the over potentials, the actual voltage of the cell and other quantities required to estimate the performance of the cell are evaluated.

Restricting interest to planar solid oxide fuel cells, with particular reference to an anode supported configuration, it can be pointed out that the thickness of electrolyte and cathode is typically orders of magnitude smaller than that of the anode. As a consequence, the local current density distribution can be assumed to vary only in the flow direction (parallel to the catalyst layer) and not across the electrolyte. For the same reason, the temperature gradient across the electrolyte can be assumed to be negligible. As a consequence, the anodic and cathodic temperature distributions at the catalyst layer can be assumed to be identical, and the electrolyte compartment can be excluded from calculations. A typical example of such an approach is available in Figure 13.2, where domain definition and the boundary conditions employed are reported for the case of a planar anode-supported SOFC. In fact, in this example, the influence of the electrolyte is neglected as its thickness is several orders of magnitude smaller than the anode (Arpino *et al.*, 2008; Arpino and Massarotti, 2009; Mauro *et al.* 2011). The following example will provide further details on the problem solved using the proposed finite element approach.

Example 13.3.1 *Anode supported SOFC – planar configuration*

Figure 13.2 shows a schematic representation of an SOFC (not to scale) along with the boundary and interface conditions. At the inlet sections of the anodic/cathodic domain, a mixture of assigned species concentration values with given velocity and temperature is assumed to enter the domain. At the exit of the domain, constant pressure boundary conditions

Figure 13.3   A section of the three-dimensional finite element mesh. *Source*: Mauro (2009). Reproduced with permission of Mauro.

are assumed and natural boundary conditions are applied for species and concentration equations. No-slip velocity and adiabatic conditions are imposed on all other sides of the domain, which are also considered to be impermeable to the species in the mixture. The device is assumed to be supplied with hydrogen mixture and air. An example of computational grids used in the simulations is available in Figure 13.3. As seen, the hydrogen mixture channel and air channel are placed at the top and bottom of the domain respectively. All the interfaces and walls are refined to capture rapid changes in the variables. All the different parameters required to carryout the calculations are provided in Table 13.2.

To simulate a practical application of SOFC, heat and mass transfer in an anode supported case with the parameters given in Table 13.2 has been studied. The experimental data for this case are available in Yakabe et al. (2000). The fuel side in this case is supplied with a mixture of H_2, H_2O and Ar.

Figure 13.4 shows a comparison of over-potentials against experimental values Yakabe et al. (2000). The reference value η^A_{c0} is calculated at $H_2/(H_2 + H_2O + Ar) = 0.8$. Since the ratio H_2/H_2O is kept at 80/20, the reference value is calculated when no argon is added. The results clearly indicate that need for 3D simulations. While the 3D simulation gives results close to the experimental results, the 2D results are not very accurate (Mauro et al. 2011).

Table 13.2   SOFC parameters used in the simulation (Mauro *et al.* 2011)

Parameter	Symbol	Value
Operating temperature (K)	T	1023
Operating pressure (bar)	p	1.0
Fuel inlet velocity (m/s)	$u_{fuel,in}$	2.0
Air inlet velocity (m/s)	$u_{air,in}$	2.0
Current density (A/cm^2)	i	0.30, 0.70
Porosity	ε	0.46
Average pore size (?m)	r_p	2.6
Tortuosity	τ_g	4.5
Anode permeability (m^2)	μ_A	1.76×10^{-11}
Cathode permeability (m^2)	μ_C	1.76×10^{-11}
Anode thickness (mm)	t^A	2.0
Cathode thickness (mm)	t^C	0.050
Electrolyte thickness (mm)	t^E	0.050
Fuel channel height (mm)	h^F	1.0
Air channel height (mm)	h^A	1.0
Cell width (mm)	w	1.0

Source: Mauro (2009). Reproduced with permission of Mauro.

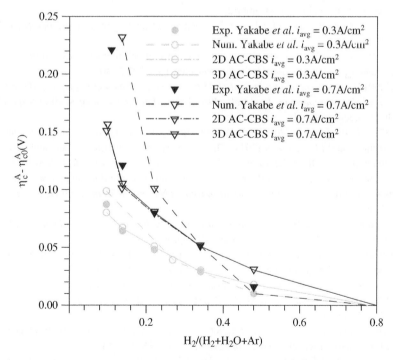

Figure 13.4   Concentration overpotentials for the H_2-H_2O-Ar fuel mixture, at $i_{avg} = 0.3 A/cm^2$ and $i_{avg} = 0.7 A/cm^2$: comparison with the experimental and numerical data available in Yakabe *et al.*, 2000. *Source*: Data from Yakabe *et al.* (2000).

13.4   Summary

This chapter is intended to provide a very brief overview on how to model heat and mass transfer in fuel cells using the finite element method. Although the discussion was focused around solid oxide fuel cells (SOFC), extending the procedure to other fuel cells may be carried out by appropriately changing the electrochemistry and boundary conditions. The core finite element procedure discussed in this chapter remains the same for different types of fuel cells.

References

Appleby AJ and Foulkes FR (1989) *Fuel Cell Handbook.* Van Nostrand Reinhold, New York.

Aguiar P, Adjiman CS and Brandon NP (2004) Anode-supported intermediate temperature direct internal reforming solid oxide fuel cell. I: model-based steady-state performance. *Journal of Power Sources*, **138**, 120–136.

Arpino F, Carotenuto A, Massarotti N and P Nithiarasu (2008) A robust model and numerical approach for solving solid oxide fuel cell (SOFC) problems. *International Journal of Numerical Methods for Heat and Fluid Flow*, **18**, 811–834.

Arpino F and Massarotti N (2009) Numerical simulation of mass and energy transport phenomena in solid oxide fuel cells. *Energy*, **34**, 2033–2041.

Campanari S and Iora P (2004) Definition and sensitivity analysis of a finite volume SOFC model for a tubular cell geometry. *Journal of Power Sources*, **132**, 113–126.

Chapman S and Cowling TG (1970) *The Mathematical Theory of Non-uniform Gases*, 3rd Edition. Cambridge Mathematical Library Cambridge.

Farooque M and Maru HC (2001) Fuel cells the clean and efficient power generators. *IEEE Proc*, **89**, 1819–1829.

Kirubakaran A Jain S and Nema RK (2009) A review on fuel cell technologies and power electronic interface. *Renewable and Sustainable Energy Reviews*, **13**, 2430–2440.

Mauro A (2009) *Development of a Fully Explicit Matrix Inversion Free Finite Element Algorithm for the Simulation of High Temperature Fuel Cells*, PhD Thesis, Universita Degli Studi di Napoli Federico II, Facolta di Ingegneria, Italy.

Mauro A (2011) *Finite Element Modeling of Solid Oxide Fuel Cells*. Giannini Editore, Naples.

Mauro A, Arpino F and Massarotti N (2011) Three-dimensional simulation of heat and mass transport phenomena in planar SOFCs. *International Journal of Hydrogen Energy*, **36**, 10288–10301.

Mauro A, Arpino F, Massarotti N, Nithiarasu P (2010) A novel single domain approach for numerical modeling solid oxide fuel cells. *International Journal of Numerical Methods for Heat and Fluid Flow*, 20(5), 587–612.

Singhal SC and Kendall K (2003) *High-temperature Solid Oxide Fuel Cells: Fundamentals, Design and Applications.* Elsevier Advanced Technology, Kidlington.

Suwanwarangkul R, Croiset E, Entchev E, *et al.* (2006) Experimental and modeling study of solid oxide fuel cell operating with syngas fuel. *Journal of Power Sources*, **161**, 308–322.

Suzuki M, Fukagata K, Shikazono, N and Kasagi N (2005) Numerical analysis of temperature and potential distributions in planar type SOFC. Paper presented at *6th KSME-JSME Thermal and Fluids Engineering Conference*, Jeju.

Yakabe H, Hishinuma M, Uratani M, *et al.* (2000) Evaluation and modeling of performance of anode-supported solid oxide fuel cell. *Journal of Power Sources*, **86**, 423–431.

Zho H and Kee RJ (2003) A general mathematical model for analysing the performance the performance of fuel-cell membrane-electrode assemblies. *Journal of Power Sources*, **141**, 79–95.

14

An Introduction to Mesh Generation and Adaptive Finite Element Methods

14.1 Introduction

It is important to realize that the finite element method is approximate and its accuracy depends on several aspects including domain and equation discretizations (refer to Chapter 1). In order to obtain an acceptable accuracy, the domain discretization, or mesh generation, should be carefully carried out. The mesh convergence studies, discussed in Chapter 7 are one way of improving confidence in the solution. Obtaining a mesh convergence is easy for geometrically and physically simple problems and it is also possible to design an accurate domain discretization *a priori* for such problems. For example, refining near a solid wall of a forced convection problem to capture high gradients of velocity and temperature is a way of increasing accuracy. Such intuition from the basic fluid dynamics and heat transfer knowledge may not always be easy or possible. One can refine the mesh everywhere to a degree to obtain a reasonably accurate solution, but this approach is not necessarily efficient and may not be cheap. Thus, alternatives such as an *adaptive mesh refinement* scheme is often considered. However, it is important to note here that the adaptive method can be more expensive than using a uniform fine discretization if not carried out carefully. Thus, practitioners should be aware of the available computing power and pros and cons of the adaptive strategies. The adaptive strategies generally contain two parts. The first part is linked to the estimation of solution error and the second part links the error indication to a mesh refinement strategy. Sometimes the error estimation is the only interest. In such cases, focussing on the design of a good error estimator is important. For geometrically and physically complex problems however, a

Fundamentals of the Finite Element Method for Heat and Mass Transfer, Second Edition.
P. Nithiarasu, R. W. Lewis, and K. N. Seetharamu.
© 2016 John Wiley & Sons, Ltd. Published 2016 by John Wiley & Sons, Ltd.

mesh refinement strategy may be of prime importance. Our main interest here is the second category of problems in which designing a suitable mesh for flow and heat transfer analysis is of primary concern. Before discussing adaptive meshing capabilities, a brief introduction to mesh generation is provided in the following sections.

14.2   Mesh Generation

In order to understand adaptive refinement, a fundamental understanding of mesh generation procedures may be necessary. In this section we provide a brief introduction to mesh generation (Frey and George 2008; Löhner 2001; Zienkiewicz *et al.* 2013a). The mesh generation, or domain discretization, may be classified under various categories such as topology, method of generation, element type, conformity, body alignment etc. The classification based on topology is shown in Figure 14.1. We classify the meshes under this category, in a broader sense, as structured and unstructured meshes. Within the structured and unstructured mesh classifications, the methods may further be subdivided into uniform and nonuniform categories as

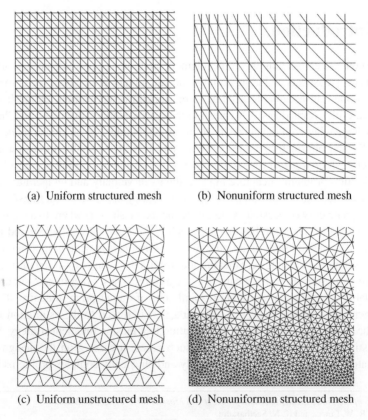

(a) Uniform structured mesh   (b) Nonuniform structured mesh

(c) Uniform unstructured mesh   (d) Nonuniformun structured mesh

Figure 14.1   Different type of meshes.

shown in Figure 14.1. Mesh generation methods may vary from fully manual, semi-automatic to fully automatic. The element types normally employed are triangles and quadrilaterals in two dimensions and tetrahedra and hexahedra (or bricks) in three dimensions. In addition to the shape of the elements, further classifications under the element type include the order of polynomials used within an element. Element types with interpolating polynomials of first order to very high orders are in use. The meshes may also be classified as conformal, or non-conformal, depending on how the elements are conformed within a domain. A mesh is said to be conformal if a perfect match between edges and nodes exist between two neighboring elements, that is, no elements overlap and no space exists between the elements. Otherwise, the mesh is a nonconformal one. Finally, a mesh may be classified as a body fitted mesh if it is aligning with the given geometry boundaries. Occasionally, meshes with no boundary alignment are used. In such cases, the solution algorithm need to make sure that the boundary of the domain is appropriately identified.

Our interest in this chapter is only to employ automatically generated unstructured meshes. Unstructured and automatic generation of meshes has been an active area of research since the early eighties. Due to their flexibility for boundary fitting and the speed at which an unstructured mesh can be generated, they are very popular for modeling complex problems. If the problem of interest is geometrically unchallenging, a researcher may spend a significant amount of time to generate a structured mesh. However, if the geometry involved is complex, then generating a structured mesh such as a multi-block mesh may take months. Thus, if the geometry is expected to change during an analysis and the geometry is complex in nature, then automatic and unstructured meshes are recommended, as long as appropriate precautions are taken to obtain a suitable mesh.

Most of the mesh generation methods require a discretization of the domain boundary first. In 2D cases, this means inserting boundary nodes which connect the boundary edges, that is, approximate the boundary by a polygon, or set of polygons if the domain is not simply connected. In 3D cases, the corresponding boundary discretization requires building a surface mesh, for example, a polyhedron with triangular faces on the surface. The two main automatic mesh generation methods, advancing front technique and Delaunay triangulation methods, start with such a boundary discretization. Both these methods are discussed in the following subsections.

14.2.1 Advancing Front Technique (AFT)

The AFT is one of the first developed automatic methods for mesh generation. In this method, a front is defined as a set of edges separating the existing mesh elements and a still nontriangulated part of the domain (see Figure 14.2). Initially, this front coincides with the boundary segments of the domain to be meshed or discretized. After defining the initial front, a new point is inserted aside a front-edge at an ideal location so that the front-edge and the new point form a triangle which is as close to being an equilateral triangle as possible. If the front-edge is essentially longer or shorter than the locally prescribed element size h then the ideal point location can be moved closer to or farther from the edge, respectively, along the normal passing through the edge mid-point (Löhner 2001; Peraire et. al. 1999). If other front nodes are closer to that ideal node, it can be (a) shifted towards the selected front-edge or (b) removed

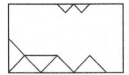 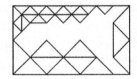 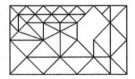

Figure 14.2   Stages in the progression of an advancing front algorithm.

and replaced by a new closer node to the front-edge. In all the cases, a new element is built and added to the set of elements as shown in Figure 14.2. Once a satisfactory new element is constructed, the front-edge used to form the element is excluded from the front and new edge/edges of the new element are added to the front. The selection of edges to build a new element may be carried out by using the shortest edge (Löhner 2001; Peraire *et. al.* 1999) or the oldest edge among the front edges. During triangulation, multiple fronts may be used to accelerate the triangulation. The algorithm stops when the domain is completely filled with triangles.

If the boundary of a 2D domain is smooth, the AFT produces a high quality mesh and a proper element sizing. Some poor elements can appear in the area where two parts of the front meet each other. These can be corrected by mesh cosmetic procedures (see below). The AFT algorithm is slow due to the fact that, for every new point, it is necessary to compute the distance between the point and all existing front points. In some versions, AFT is based on an analysis of the angle formed by two adjacent front edges (Schoberl 1997). An analogous technique to that of the 2D AFT may be applied for 3D mesh generation. Here, we have a front of triangular faces with the initial front coinciding with the boundary surface grid. An ideal position for a new point is along the normal passing through the centroid of a front-face (triangular surface) such that the distances from the new point to the face vertices are as close as possible to the prescribed local element size.

The AFT method requires multiple intersection and enclosure tests, especially when it comes to opposing fronts approaching each other, to ensure that the integrity of the final mesh is maintained. Although 2D AFT provides good quality meshes on simple geometries, a standard 3D version of the AFT, that is, generalization of standard 2D AFT, does not guarantee high quality elements, even near a smooth boundary with a high quality surface mesh (Sazonov *et al.* 2006). To obtain good quality elements close to the boundaries, a nonstandard method, or cosmetics, should be employed.

14.2.2   Delaunay Triangulation

A triangle in a triangulation is known as a Delaunay element if its circumcircle contains no nodes in its interior (Dellaunay 1934) as shown in Figure 14.3(a). A Delaunay triangulation is one in which every element is a Delaunay element. In two dimensions, a Delaunay triangulation is optimal in the sense that it maximizes the minimum angle found over the entire mesh (Weatherill 1992). Unfortunately, this property of a Delaunay mesh can not be extended to its 3D counterpart, that is, maximizing the minimum solid angle of the elements in the domain.

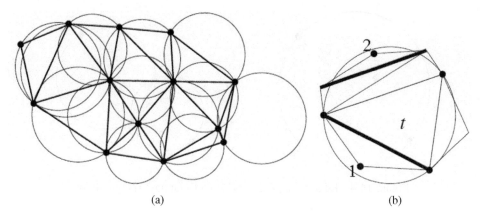

(a) (b)

Figure 14.3 Delaunay triangulation: (a) Delaunay triangulation of a given set of points; (b) constrained Delaunay element t: bold lines are the boundary edges, points 1,2 are within its circumcircle but not seen from element t.

It should be noted that the Delaunay method provides an algorithm to form topology of the mesh using existing nodes and does not supply any measure to position the nodes. Note also that a Delaunay triangulation is the triangulation of the domain that is a convex hull of the given set of points. Thus, if the domain boundary $\partial\Omega$ has concave parts, a Delaunay triangulation may generate "out of the domain" triangles which can be easily found and removed. If by chance four or more points lay on the same circumcircle, then we meet a degenerated case. A rectangular grid of points is an example of such a degenerated point set. In such a case, the connection is not unambiguous. However, an infinitesimally small displacement of points can remove this degeneration. All these definitions are valid for 2D and 3D Delaunay triangulation. In three dimensional domains, a circumsphere is used in place of a circumcircle for two-dimensional domain.

14.2.2.1 Delaunay Triangulation Algorithms

One of the simplest algorithms to make existing triangles Delaunay, or constrained Delaunay, is via edge swapping. Figure 14.4 gives such an example. In this case, a simple swapping of the diagonal makes the triangles Delaunay. This leads to a straightforward algorithm that states "construct any triangulation on a given number of points and then swap edges until no triangle is non-Delaunay". Unfortunately, this does not extend to three dimensions. The mesh thus obtained after the swapping algorithm is a Delaunay or constrained Delaunay mesh. The latter is a triangulation that contains as many Delaunay triangle as possible and the rest are constrained. The circumcircle of an element of a constrained Delaunay triangulation contains in its interior no nodes visible from the element as shown in Figure 14.3(b).

An alternative and widely used algorithm is based on the Bowyer-Watson point insertion algorithm (Bowyer 1981; Watson 1981). It works by adding points, one at a time, to a valid initial Delaunay triangulation. After every insertion, any triangles whose circumcircles contain

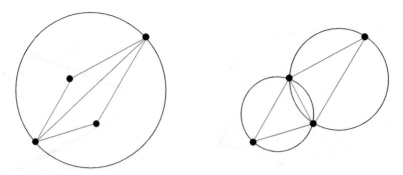

Figure 14.4   Edge swapping to obtain a Delaunay triangulation.

the new point are marked and their common edges are deleted to create a polygon. All the vertices of this polygon are then connected to the new point as shown in Figure 14.5. To make this procedure applicable to any domain, it starts with determining four auxiliary points forming a rectangle aligned along the horizontal and vertical directions that completely enclose the domain to be triangulated. This rectangular domain may be refereed to as the background region. An initial triangulation is obtained by connecting all four points of the rectangle. Now, the points are inserted one by one, first on the real boundaries of the domain and then the inner regions, re-triangulating the domain in accordance with the Bowyer-Watson algorithm. This gives a Delaunay triangulation with elements present both within and outside the real domain, that is, in-between the real domain and the background region. The outside elements should be detected and removed from the mesh along with the background region. Figure 14.6 shows a typical example of generating a mesh on a concave domain.

The Bowyer-Watson algorithm can be readily generalized to 3D domains. It starts with creating eight points constituting a hexahedron, that is, a right-angled parallelepiped aligned to the x, y and z directions, which encloses the domain to be meshed. The background parallelepiped region should be partitioned into 5 or 6 tetrahedra. This is an initial mesh into

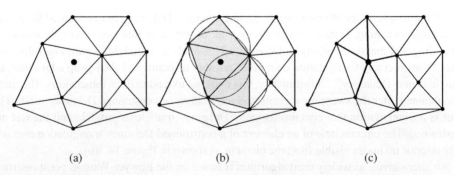

Figure 14.5   The Bowyer-Watson method of point insertion for generating Delaunay triangulation.

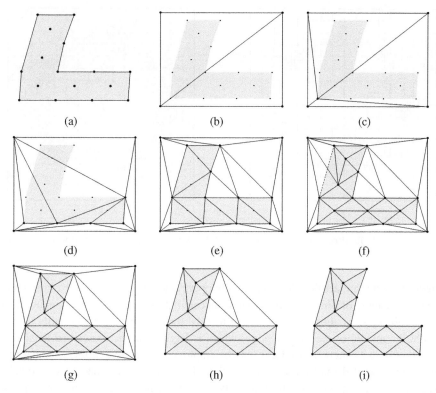

Figure 14.6   a: concave domain to be triangulated and all data points; b: surrounding box and initial mesh of 2 elements; c: one point is inserted; d: first 5 points are inserted; e: inserted all boundary points; f: all the points have been inserted (one boundary edge is not in the triangulation); g: boundary edge recovery by edge swapping; h: elements containing box vertices are removed (result: convex domain); i: final mesh—all out-of-domain elements are deleted.

which data points are inserted one-by-one, first on the boundaries of the real domain and then inside the problem domain.

14.2.2.2   Automatic Point Creation

In contrast to the AFT, Delaunay triangulation does not indicate how to place inner points to obtain a better element quality. However, it provides guidelines on how to optimally connect such points. Therefore, different approaches to optimally place new points have been developed to obtain good quality Delaunay triangulation.

One of the simplest approaches is to place a new point at the circumcenter of the element with the poorest quality, in terms of shape and size, among all the elements in the mesh. The process can be terminated when the size and quality have improved beyond a prescribed

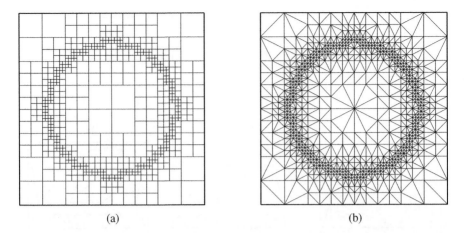

(a)   (b)

Figure 14.7   (a) Quadtree grid structure; (b) Delaunay triangulation generated on its nodes.

threshold. Despite the fact that this approach improves element quality, the resulting mesh is often not sufficiently of a high quality to carry out calculations. Thus, mesh cosmetics are used to improve the mesh quality further. Another method for targeting element quality is to insert a point at the middle of the longest edge, which is larger than the prescribed local element size. Again, mesh cosmetic procedures may be employed to improve further the quality.

To produce graded meshes, some researchers have resorted to quadtrees. A quadtree is a recursive data structure used to efficiently manipulate multiscale geometric objects in a plane. Quadtrees recursively partition a region into axis-aligned squares. A top-level square called the root encloses the entire input planar straight line graph. Each quadtree square can be divided into four child squares, which can be divided in turn, as illustrated in Figure 14.7(a). Octrees are the generalization of quadtrees to three dimensions; each cube in an octree can be subdivided into eight cubes. The points generated by quadtrees and octrees may be used in a Delaunay triangulation as shown in Figure 14.7(b).

14.2.2.3   Boundary Recovery

Delaunay triangulation guarantees a consistent mesh within the interior. However, it does not guarantee to preserve the original boundary edges (2D) or faces (3D) in the final mesh. Hence, it is necessary to check the mesh and, if necessary, modify the elements to ensure that the elements conform to the boundary. In some cases for a planar triangulation, it is possible to recover the boundary edge by simple edge swapping (see Figure 14.8).

For more general cases, edge recovery through node creation can be applied. In this approach, an additional node/nodes should be added at points of intersection of a boundary and triangulation edges (Weatherill 1990; Borouchaki and George 1997). If short boundary edges and low quality boundary triangles are generated as shown in Figure 14.9(a-b), the new node is moved to one of the end points of the recovered boundary edge and the corresponding short edge is contracted as shown in Figure 14.9(c-d). This results in the same configuration as

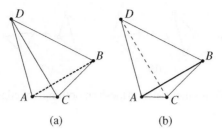

(a)   (b)

Figure 14.8   Boundary recovery by swapping edge: (a) boundary edge *AB* is not in the triangulation, (b) swapping edge *CD* → *AB* we recover the boundary edge but the triangulation becomes not Delaunay (only constrained Delaunay).

obtained by an edge swap as shown in Figure 14.9(d) and Figure 14.8(b). Note that inserting nodes and edge contraction is a more robust procedure than edge swapping.

In three dimensions, the existence of a constrained Delaunay triangulation that matches with a prescribed surface triangulation is not guaranteed. There are many examples of surface configurations, which cannot be triangulated without inserting an additional point. It is also not clear whether a given triangulation can be transformed to another one using the same points by a finite number of 3D flips. Thus, a robust procedure that works for the overwhelming majority of configurations should be used (Weatherill and Hassan 1994). This procedure resembles that described for 2D and comprises inserting additional points and edge/face contraction.

14.2.3   Mesh Cosmetics

14.2.3.1   Mesh Quality

Mesh quality can have a considerable impact on the computational analysis in terms of the accuracy of the solution and the CPU time needed. A high quality mesh is one in which all edge lengths are close to that of a prescribed value and the element shapes reflect the requirement. This implies that the mesh does not contain elements which can cause large discretization errors and at the same time the mesh should not be finer than the necessary requirement. The

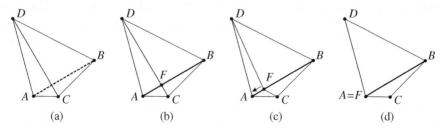

(a)   (b)   (c)   (d)

Figure 14.9   Boundary recovery by inserting additional node *F* in the intersection of *AB* and *CD* (b) and forming boundary edges *AF* and *FB*. Contraction of edge *AF* (c) and obtaining configuration shown in Figure 14.8(b).

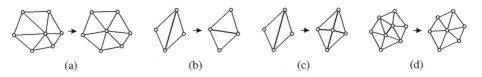

(a)	(b)	(c)	(d)

Figure 14.10   Mesh cosmetics methods: (a) mesh smoothing, (b) edge swapping (2D), (c) edge splitting, (d) edge contraction.

global mesh quality is defined through the quality of individual elements. Thus, histograms of element qualities are often generated to compare the qualities of meshes.

There are several ways of computing the quality of individual elements and how to quantify the overall quality of a mesh. For a triangular element, a ratio of geometrical characteristics is taken which achieves a maximum for a perfect triangle. There are many ways to define this ratio. It may be taken as the ratio of inscribed and circumscribed circles or the ratio of the inscribed circle and perimeter. For three-dimensional problems more quality measures are often used.

14.2.3.2   Mesh Smoothing

The smoothing of a mesh is achieved by displacing nodes in order to improve the element quality without changing the connection between the nodes, that is, without changing the mesh topology. One of the most popular methods is Laplace smoothing, in which every inner node p is displaced to the centroid of its contiguous nodes $p_i : i \in P_p$ as shown in Figure 14.10(a). Its smoother version reads as

$$\mathbf{p}^{\text{new}} = (1 - \alpha)\mathbf{p} + \alpha \frac{1}{d_p} \sum_{i \in P_p} \mathbf{p}_i \qquad (14.1)$$

where $\mathbf{p}$ represents the coordinates of node p; α is tuning parameter ($\alpha = 1$ for the standard Laplace smoothing); d_p is the nodal index, that is, the number of contiguous nodes to node p. The procedure can be applied to all inner nodes simultaneously or to some selected nodes. The procedure can be repeated several times and in many cases five iterations are sufficient to increase the quality of most of the elements of a planar mesh. If the initial mesh is highly irregular and non-smooth, Laplace smoothing may result in entangled elements. This method may also be applied to 3D meshes, but performance may not be as good as for 2D meshes. Thus, element quality measures should be used to accept or reject Laplace smoothing results.

14.2.3.3   Edge Swapping

Swapping an edge, shared by two triangles, is often used to improve mesh quality. The swapping criterion is based on an analysis of the angles of the triangles before and after swapping. If the angles are greater after smoothing than before smoothing, the common edge between two triangles is swapped to improve quality as shown in Figure 14.10(b). A better approach to the angle based approach is a topology based technique. In this approach, a beneficial effect on the mesh quality over the angle based approach can be obtained if used in conjunction with a smoothing procedure (Frey and Field 1991; Sazonov and Nithiarasu 2012).

In the topology based edge swapping method, in order to decide whether or not an edge needs swapping, we should compute the change of the mesh relaxation index U caused by swapping. Let 1 and 2 be the end points of an edge before swapping and d_1 and d_2 are then respective nodal indices. Let 3 and 4 are end points of the edge after swapping and d_3 and d_4 are respective nodal indices of them. Now, the change in the mesh relaxation index ΔU, is evaluated as

$$\Delta U = [(m_1-1)^2 + (m_2-1)^2 + (m_3+1)^2 + (m_4+1)^2] - \left[m_1^2 + m_2^2 + m_3^2 + m_4^2\right] \quad (14.2)$$

where $m_i = d_i - d_{opt}$ is the deviation from the ideal nodal index (Frey and Field 1991). An edge should be swapped if $\Delta U < 0$. The optimal nodal index $d_{opt} = 6$ for a 2D triangulation, but it is smaller for boundary nodes (it depends on the angle formed by two boundary edges sharing the node). This procedure can be applied to all edges in the mesh and repeated until swapping cannot cause any further changes of ΔU (Sazonov and Nithiarasu 2012).

The relaxation method may be started based on some edge properties (Borouchaki and George 1997). Topology based swapping may be started with an edge that induces a maximum change in ΔU after swapping. Note that in a high quality mesh almost all inner nodes will have a nodal index of 6 with a small percentage of nodes having indices 5 or 7. Nodes with $d > 7$ or $d < 5$ can be corrected by swapping.

In 3D, swapping is used to improve the quality of the elements. In contrast to 2D mesh, where there is a single swap transformation: $2 \rightarrow 2$ triangle, in 3D there are several such transformations: $2 \rightarrow 3, 3 \rightarrow 2, 4 \rightarrow 4$ and so on. The optimal nodal index in a 3D situation is 14. This is the nodal index of a mesh of high quality with tetrahedra filing the space.

14.2.3.4   Edge Splitting/Contraction

To provide a proper element size edge splitting/contraction procedures can be applied both in 2D and 3D triangulations. A short edge can be contracted, together with some elements sharing it, to its midpoint. If one of the edge points is a boundary node then it should be contracted to a boundary node. The edges that are longer than the prescribed element size can be split by introducing a midpoint. This results in additional elements (Figures 14.10(c)-(d)).

After these transforms the quality of elements attached to the new node may not be sufficient. Therefore, a local cosmetic procedure may be required. The mesh cosmetics may first start with a swap of edges of all elements associated with the new node or contracted point before a Laplace smoothing is applied.

14.2.3.5   Centroidal Voronoi Tesselation

Another cosmetic approach is based on the concept of Centroidal Voronoi tessellation (CVT). If a mesh is Delaunay, then a polygon formed by connecting the circumcenters of elements sharing node p is a boundary of a Voronoi cell of node p. These Voronoi cells are elements of the Voronoi tessellation or diagram, which is the dual to the Delaunay triangulation, that is, element circumcenters of Delaunay triangulation are vertices of the Voronoi diagram and

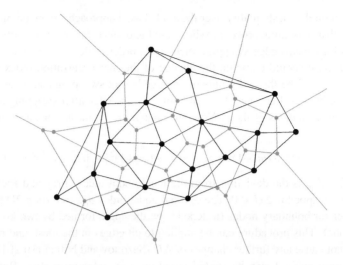

Figure 14.11   Dual Voronoi diagram (gray) to a Delaunay triangulation.

generators of the Voronoi diagrams are nodes of the Delaunay triangulation, as shown in Figure 14.11.

The Voronoi tessellation is called a Centroidal Voronoi if its every generator point coincides with the centroid of the corresponding Voronoi cell. One of the methods to generate CVT is to apply Lloyd's iterations in which generator points are moved to barycenters of the Voronoi cells as shown in Figure 14.12. If a Voronoi cell surrounding an inner point intersects a boundary of the domain Ω then the barycenter of the polygon forming the intersection of the domain and Voronoi cell should be used. The CVT method can convert an absolutely random distribution of initial points into a high-quality mesh with a very good point distribution. The method essentially improves mesh topology thus diminishing the mesh relaxation index. The Laplace smoothing applied after Lloyd's iteration may slightly improve the mesh quality in 2D. The method is easily generalized for three dimensions.

14.3   Boundary Grid Generation

For a computational domain Ω, the boundary $\partial\Omega$ should first be accurately defined. The boundary of simple geometries may be described analytically or piecewise analytically using cubic spline, Bezier curves/patches, or nonuniform rational basis splines (NURBS), etc.

14.3.1   Boundary Grid for a Planar Domain

If the object boundary contains corners, the nodes should be placed on them first. If the boundary contains high curvature parts, then it is useful to place nodes at local places where the curvature is greatest. Between these initial nodes, new nodes can be placed one by one in

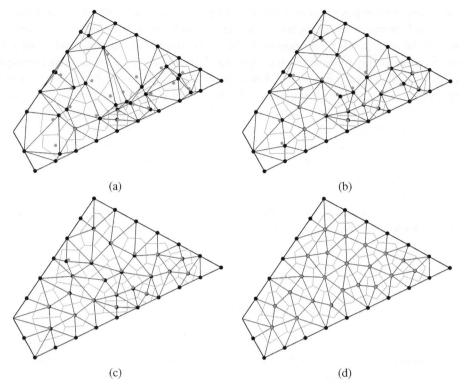

(a) (b)

(c) (d)

Figure 14.12   CVT: Lloyd's iterations. Grey polygons indicate Voronoi cells for inner nodes, grey circles – their centroids. (a): innitial mesh (inner nodes are placed randomly), (b): 1st iteration, (c): 3rd iteration, (d): 20th iteration.

accordance with the locally prescribed element size. The prescribed local size, if not uniform, can be described analytically or piecewise analytically using the so-called background mesh. It may also be computed automatically based on boundary features such as local boundary curvature or/and the local domain width or/and corner angles etc. After points are inserted they should be connected one by one to form boundary edges to describe the geometry as a closed polygon or several polygons if the domain is not simply connected.

14.3.2   NURBS Patches

The surfaces of turbine blades, car bodies and boat hulls, and other industrial objects designed by CAD, are split by curvilinear quadrilateral patches matching each other at edges and vertices. The surface of every patch is described analytically as a tensor product of Bezier curves or nonuniform rational basis splines (NURBS). The latter ones allow an exact description of cylindrical, conical and spherical surfaces. Due to high-order approximations, the number of patches can be minimized for an accurate description of complicated objects.

If an object boundary is set by such NURBS patches and the locally prescribed element size is small compared to the patch size, then every patch can be triangulated independently (but with the same patch border nodes). If the patch does not deviate too much from a 2D plane, it can be simply projected onto the plane and triangulated by a 2D meshing technique. For cases where the patch is not very close to a 2D plane, it can be projected using a metric such a way that the distance between two closely located points in the plane approximately equals the distance between the corresponding points on the surface. In this case, the local prescribed element size should be adjusted to account for the local metric used.

The 2D mesh cosmetic methods can be applied directly to a 3D surface mesh if its surface is well defined. The Laplace smoothing, or edge splitting/contraction, should be carried out outside the surface and then projected back to the surface by the shortest path.

If a surface can be described analytically without dividing into patches, for example, a sphere or ellipsoid, then it cannot be mapped on to a planar domain. To mesh such an analytical surface, it may be (a) divided into patches, then meshed, or (b) AFT adopted for surface meshing may be used for the whole surface.

14.4   Adaptive Refinement Methods

Any adaptive procedure needs a powerful refinement strategy to properly refine the mesh. Two major classifications of adaptive refinement are: (1) that of p-refinement, in which the order of interpolation is varied and the other is (2) h-refinement, in which the element sizes are varied. Within h-refinement, three distinctive methods are possible. They are: (a) r-refinement, in which the structure and connectivity of the mesh remain unaltered and so does the number of nodes; (b) mesh enrichment (subdivision), where individual elements are subdivided without altering their original position; and (c) complete regeneration by adaptive remeshing.

The order of the interpolation polynomial used to form the finite elements is used in a p-adaptivity process to control the error measured by a suitable indicator. This process known as p-refinement is often coupled with a h-refinement procedure. This method essentially increases the order of the interpolation polynomial in order to reduce the error in the solution.

In the r-refinement method, the spring analogy is often used to redistribute the nodes in the existing mesh. Mesh movement is accomplished by advancing the solution towards steady state and, at certain stages, replacing the mesh sides by springs of a certain stiffness. This stiffness is usually based on the local error in the solution. The nodes are moved until the spring system is in equilibrium. Since new nodes are not added, the accuracy of the solution derived by this method is limited by the initial number of nodes and elements.

The basic principle of mesh enrichment is that as the solution proceeds towards a steady state, then the nodes are added locally based on an error indicator. The portion to be refined is identified locally and the nodes are added in this region. The process of enrichment is continued until the solution reaches the required accuracy. Even though this method can lead to accurate solutions, the process can be tedious for complex problems. Further difficulties arise in the coupling of refined and original elements.

The above problems are avoided in adaptive remeshing, in which the whole domain is remeshed in a manner based on the error indicator computed from the previous solution. In this process, one can use directionally stretched elements wherever the one-dimensional feature dominates and de-refinement can be introduced when the error is small. Obviously, such a procedure will be more economical in the number of nodes used at the expense of additional remeshing. Among the procedures discussed above, mesh enrichment and adaptive remeshing are suitable for most heat transfer problems. When these procedures are coupled with a powerful automatic mesh generator, they can yield excellent results.

14.5   Simple Error Estimation and Mesh Refinement

The error in a solution is the difference between the exact and approximate solutions, which can be written for temperature as

$$e_T = T - T^h,$$ (14.3)

where T is the exact value of the temperature and the superscript h represents a finite element solution. In a similar fashion, the error in heat flux may be written as

$$\mathbf{e}_q = \mathbf{q} - \mathbf{q}^h.$$ (14.4)

As in all problems requiring numerical analysis, since we do not know the exact solution, we need an alternative approximation to represent the analytical solution. In the problems considered here the trial functions (linear with C^0 continuity) result in a discontinuous approximation of the fluxes. The acceptable continuous solution can be obtained by an averaging or projection process. In an averaging process, the element heat fluxes are averaged over a node connected to a patch of elements. The projection process, on the other hand, assumes that the flux is interpolated using the same function as the temperature. Now, the error indication in the heat flux may be obtained as (Zienkiewicz and Zhu 1987):

$$\mathbf{e}_q \approx \mathbf{q}^* - \mathbf{q}^h,$$ (14.5)

where the superscript $*$ is the heat flux obtained via projection or averaging. Generally, we write the errors in terms of some error norms. The energy and L_2 norms are frequently used as measures in the literature. Here, we use the L_2 norm of the flux error measure for heat conduction problems

$$\|\mathbf{e}_q\| = \left(\int_\Omega \mathbf{e}_q^T \mathbf{e}_q d\Omega \right)^{\frac{1}{2}}.$$ (14.6)

The above error can be calculated by summing up all the elemental errors as

$$\|\mathbf{e}_q\|^2 = \Sigma_i^m \|\mathbf{e}_q\|_i^2,$$ (14.7)

where m is the number of elements. It is convenient to express the error as a percentage, that is,

$$\eta = \frac{\|e_q\|}{\|q^h\|} \times 100\%. \tag{14.8}$$

This percentage of error may be used in the refinement strategy. Using the error indicator, the following form of refinement strategy may be used to refine and de-refine a mesh.

$$h_{new} = h_{old} \text{ (aimed error)}/\text{(element error indicator)}$$

where h_{new} is the new element size and h_{old} is the original element size. The above formula gives a mesh with equal error over every element. The element error indicator may be given by

$$E = \frac{\|e_q\|}{\|q^h\|} m^{1/2} \tag{14.9}$$

with m being the number of elements. Thus, $h_{new} = h_{old}\zeta/E$ with ζ being the target error.

If the new element size as predicted by the above procedure is bigger than the old element size, then the new element size may be restricted to a prescribed maximum size. To avoid very small elements, a minimum allowable size may also be given as input. An alternative refinement procedure may also be written using the error calculated as

$$h_{new}E_e = C \tag{14.10}$$

with $E_e = \sqrt{\|e_q\|^2}$. In the above equation, C is an equilibration constant. By changing C and other allowable quantities a suitable adapted mesh may be generated.

14.5.1 Heat Conduction

In this section two simple examples of two-dimensional heat conduction are given. In both examples an advancing front technique is employed to adaptively refine the mesh (Lewis *et al.* 1991; Nithiarasu and Zienkiewicz 2000; Peraire *et al.* 1987).

Example 14.5.1 *A square plate size 100 cm, as shown in Figure 5.4, is subjected to isothermal boundary conditions of 100°C on all sides except the top side which is subjected to 500°C. If the thermal conductivity of the material is constant and equal to 10 W/m°C, implement the adaptive strategy explained in the previous section and refine/de-refine the mesh for an error of 0.05. Control the element size to improve accuracy if required.*

To start the process of adaptive refinement, an initial mesh and initial solution should be obtained first as shown in Figures 14.13(a) and 14.13(d) respectively. As seen, the initial mesh is a uniform unstructured mesh and gives a central temperature value of 199.7°C against

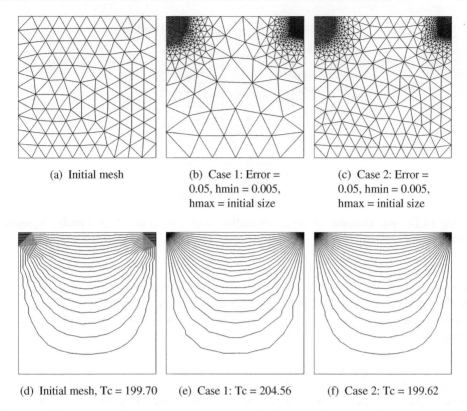

(a) Initial mesh

(b) Case 1: Error =
0.05, hmin = 0.005,
hmax = initial size

(c) Case 2: Error =
0.05, hmin = 0.005,
hmax = initial size

(d) Initial mesh, Tc = 199.70   (e) Case 1: Tc = 204.56   (f) Case 2: Tc = 199.62

Figure 14.13   Heat conduction in a square plate. Initial and adapted meshes and respective solutions.

the analytical solution value of 200°C (see Chapter 5). With the maximum and minimum element sizes constrained at twice the initial mesh size and 0.005 respectively, the adaptive meshes produced are shown in Figure 14.13(b) and (c). As seen the mesh is refined to a very high degree at both top corners of the plate. This is due to the fact that heat flux at these locations is very high due to the sudden temperature change at the boundaries from 500 to 100°C. It is also noticed that the element size increases at locations away from these two corners due to only moderate values of temperature gradient. This may have an adverse effect on the solution away from the two corners as shown in Figure 14.13(e). As seen, the temperature at the middle of the domain is increased to 204.56°C when the maximum allowed element size is twice as large as the original size. This indicates a dramatic decrease in accuracy at the mid point of the domain. Although the adaptive refinement has increased the accuracy of the calculations at the top corners of the domain, it decreased the accuracy at the center of the domain due to the constraints placed on the maximum element size. If this constraint is changed to the initial size of the elements, the error at the center is reduced and the temperate values are reduced to 199.62, close to the analytical solution as

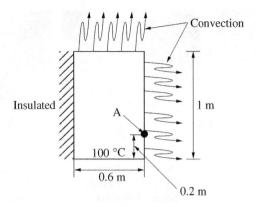

Figure 14.14   An example of heat conduction with convective heat transfer boundary conditions.

shown in Figures 14.13(c) and (f). A further reduction in maximum element size will make the solution even closer to the analytical solution at all points of the domain. This example clearly demonstrates that some practice and experience is required to obtain an accurate solution.

Example 14.5.2   *Use adaptive refinement method to reduce the error step by step to 0.01 for the problem shown in Figure 14.14. The left vertical side of the wall is insulated with a zero heat flux along this side. The bottom horizontal side is subjected to a temperature of 100 °C and the remaining sides are assumed to convect heat to the ambient maintained at a temperature of 0 °C. The convective heat transfer coefficient is 750 W/m °C and the thermal conductivity of the material is 52 W/m °C. The analytical solution to this problem is given by Carslaw (1959) and computed value for the point A in Figure 14.14 is 18.2535 °C (Huang and Usmani 1994).*

Figure 14.15 shows the meshes and solutions for the step by step reduction in error. As seen the initial mesh solution is uniform and reasonably accurate to produce temperature value of 18.32 °C against the analytical solution value of 18.2535 °C. The point A in Figure 14.14 is chosen deliberately away from the corner. Since a very high value of flux is anticipated at the bottom right corner due to the sudden change in temperature at the corner point, the adaptive refinement is expected to be intensive in the vicinity of this corner to reduce the overall error of the problem. As a result, element sizes away from this corner is expected to increase in size. The idea here is to analyze the impact of such de-refinement. Figures 14.15 shows the refinements and solutions with progressive decrease in error. Note that the constraints on the element sizes are not changed here but the error is continuously decreased. As seen, for the first and second refinements, the solution error against the analytical solution is increased. This is due to the fact that the element sizes are increased away from the bottom right corner of the plate. However, at the third refinement with 1% error the solution at point A reached a value very close to the analytical solution.

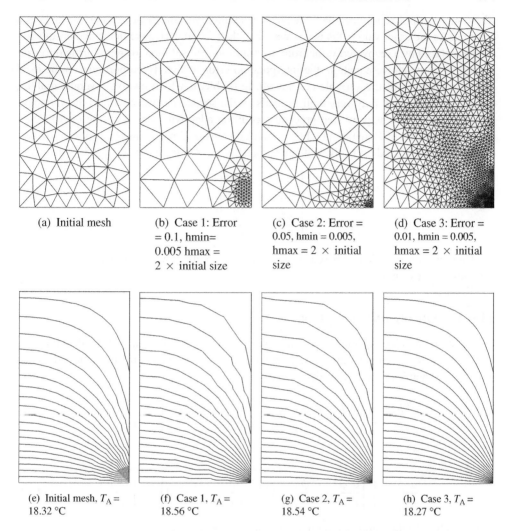

(a) Initial mesh

(b) Case 1: Error = 0.1, hmin= 0.005 hmax = 2 × initial size

(c) Case 2: Error = 0.05, hmin = 0.005, hmax = 2 × initial size

(d) Case 3: Error = 0.01, hmin = 0.005, hmax = 2 × initial size

(e) Initial mesh, T_A = 18.32 °C

(f) Case 1, T_A = 18.56 °C

(g) Case 2, T_A = 18.54 °C

(h) Case 3, T_A = 18.27 °C

Figure 14.15   Heat conduction in a rectangular plate. Initial and adapted meshes.

14.6   Interpolation Error Based Refinement

Another mesh refinement method that is more suitable for fluid flow problems is based upon the interpolation error. The objective here is to reduce and *equally distribute* the interpolation error throughout the domain. For one-dimensional problems, with linear finite element approximations, the local interpolation error for a scalar variable ϕ may be expressed in terms of the second derivative as (Davies *et al.* 2007, 2008; Nithiarasu 2000, 2002; Nithiarasu and Zienkiewicz 2000; Peraire *et al.* 1987; Zienkiewicz *et al.* 2013b,a)

$$e = \phi - \phi^h = ch^2 \frac{d^2\phi}{dx^2}, \tag{14.11}$$

where superscript h represents the finite element solution, h is the local element size and c is a constant. As the exact value of ϕ is not available, the second derivative in the above equation can be replaced by a recovered value from the finite element solution to approximate the error. As shown in the above equation, the interpolation error is directly proportional to the local second derivatives of the variable considered. In problems which involve several variables, the interpolation error may very well depends on the second derivatives of all the variables. Before developing a step by step procedure of mesh adaptivity, based on the interpolation error, an extension of Equation (14.11), to multi dimensional problems is considered. In two dimensions, the second derivatives of any variable is a 2×2 tensor (in 3D 3×3) is given as

$$
\begin{bmatrix}
\dfrac{\partial^2 \phi}{\partial x^2} & \dfrac{\partial^2 \phi}{\partial x \partial y} \\[2ex]
\dfrac{\partial^2 \phi}{\partial y \partial x} & \dfrac{\partial^2 \phi}{\partial y^2}
\end{bmatrix}.
\tag{14.12}
$$

Thus, the error indicator based mesh refinement boils down to computing the second derivatives of a variable.

14.6.1 Anisotropic Adaptive Procedure

Once the nodal values of the second derivative tensor (see Appendix F for calculating the nodal values of second derivatives), Equation (14.12), are calculated, the local maximum and minimum principle second derivatives can be calculated as

$$
\frac{\partial^2 \phi}{\partial X_1^2} = \frac{1}{2}\left[\frac{\partial^2 \phi}{\partial x^2} + \frac{\partial^2 \phi}{\partial y^2}\right] + \sqrt{\left[\frac{1}{2}\left(\frac{\partial^2 \phi}{\partial x^2} - \frac{\partial^2 \phi}{\partial y^2}\right)\right]^2 + \left[\frac{\partial^2 \phi}{\partial x \partial y}\right]^2}
\tag{14.13}
$$

and

$$
\frac{\partial^2 \phi}{\partial X_2^2} = \frac{1}{2}\left[\frac{\partial^2 \phi}{\partial x^2} + \frac{\partial^2 \phi}{\partial y^2}\right] - \sqrt{\left[\frac{1}{2}\left(\frac{\partial^2 \phi}{\partial x^2} - \frac{\partial^2 \phi}{\partial y^2}\right)\right]^2 + \left[\frac{\partial^2 \phi}{\partial x \partial y}\right]^2}
\tag{14.14}
$$

respectively. The direction of minimum principle value is

$$
\tan 2\gamma = 2\left(\frac{\partial^2 \phi}{\partial x \partial y}\right)\left(\frac{\partial^2 \phi}{\partial x^2} - \frac{\partial^2 \phi}{\partial y^2}\right)^{-1},
\tag{14.15}
$$

where γ is the angle between the x direction and the minimum principle direction. From the above relations, for the equal distribution of interpolation error, the following condition needs to be satisfied:

$$
h_{min}^2 \left|\frac{\partial^2 \phi}{\partial X_1^2}\right| = h_{max}^2 \left|\frac{\partial^2 \phi}{\partial X_2^2}\right| = C.
\tag{14.16}
$$

Thus, the maximum and minimum element sizes h_{max} and h_{min} can be calculated locally. The two element sizes in the above equation indicate that the elements regenerated will not be equilateral triangles in places where the minimum and maximum principal second derivatives are not equal. The mesh thus obtained is *anisotropic*. Although such an anisotropic mesh has limited use in an incompressible flow context, it does reduce the total number of elements when used. If an isotropic, unstructured mesh is of interest, only h_{min} needs to be calculated. In both the isotropic and anisotropic cases, the value of the constant C needs to be established.

14.6.2 Choice of Variables and Adaptivity

In incompressible heat convection problems, strong discontinuities are rarely seen. Also, the direction of rapid variation of the variables, temperature and velocity, is not always the same, that is the thermal and momentum boundary layers do not necessarily coexist along the same wall. In places where momentum boundary layers exist, the thermal boundary layer may be absent and vice-versa. It is also important to note that high flux and hot spots can occur at any arbitrary locations when complicated geometries are considered. It is thus obvious to consider the variation of all influential variables together to generate an adapted mesh for heat convection problems. Such a mesh will represent the physics of the problem and deliver an accurate solution. The following procedure may be used for natural (and forced) convection problems.

Step 1 Solve the natural convection heat transfer problem on a coarse, initial mesh for a fixed number of time iterations (the solver used in the present study is based on the CBS algorithm and is transient, see Chapter 7).

Step 2 Calculate the local second derivatives of variables at all nodes of the initial mesh and normalize them using the corresponding maximum values calculated in the whole problem domain. The normalized second derivatives of the variables ensure the use of common "equilibration" constant C for all variables. For example, consider two variables, temperature and magnitude of total velocity, the maximum and minimum values of principal second derivatives are

$$\frac{\partial^2 |V|}{\partial X_1^2}, \frac{\partial^2 T}{\partial X_1^2} \tag{14.17}$$

and

$$\frac{\partial^2 |V|}{\partial X_2^2}, \frac{\partial^2 T}{\partial X_2^2} \tag{14.18}$$

respectively. Now the rule of "equilibration" of error can be applied to the first variable as

$$h_{min}^2 \frac{\partial^2 |V|}{\partial^2 X_1^2}^* = h_{max}^2 \frac{\partial^2 |V|}{\partial X_2^2}^* = C, \tag{14.19}$$

where the superscript $*$ indicates normalized second derivatives. The above rule is applied also to the temperature and local maximum and minimum element sizes at nodes are calculated for both variables and then stored. The local stretching ratios (ratio between local maximum and minimum sizes) at nodes are also calculated from each variable and stored. Here the calculation of C values need to be explained further. The procedure proposed relates C with

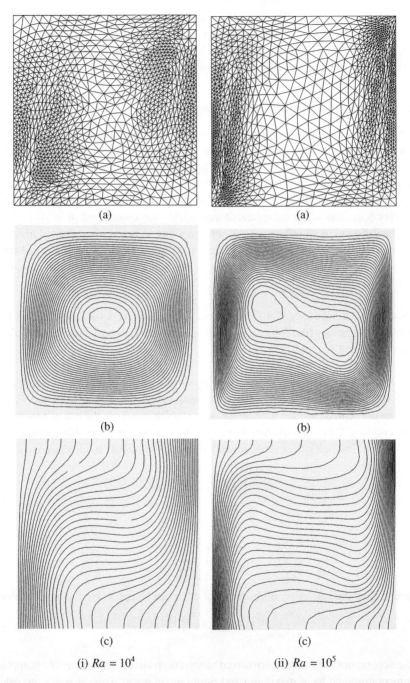

(a)   (a)

(b)   (b)

(c)   (c)

(i) $Ra = 10^4$ (ii) $Ra = 10^5$

Figure 14.16   Adaptive finite element meshes for natural convection in a differentially heated square cavity.

the major governing nondimensional number of the problem. The procedure can be extended to any number of nondimensional numbers but in the present study, only the Rayleigh number (Ra) is considered. The following relation is successfully used to determine C:

$$C = \frac{1}{Ra^m}, \tag{14.20}$$

where m is an index which is determined by solving natural convection in a square cavity for which a benchmark solution is available. In the laminar range m values are found to vary between 0.3 and 0.5.

Step 3 The minimum element sizes and the stretching ratios calculated from nodal temperatures and velocities are compared locally at nodes and the minimum values among the variables found are fixed as the new local element sizes and stretching ratios.

Step 4 The stretching direction of the strongest second derivatives among the second derivatives of velocity and temperature (before normalizing) is taken as the local stretching direction of the new mesh.

Step 5 From the new local element sizes, stretching ratios and directions a new mesh is generated. The variables are interpolated from the previous mesh and time stepping is continued.

Steps 2 to 5 are repeated until steady state. If no significant change in the total number of nodes is observed between the two meshes, then mesh regeneration is stopped and time stepping continued until steady state. For unstructured mesh generation, the advancing front type of procedure based upon background mesh principle may be used.

Example 14.6.1 *Natural convection in a cavity*

Figure 14.16 shows the adaptive finite element meshes generated for natural convection in a differentially heated square cavity. These meshes are generated using the interpolation based error indicator. Both the temperature and velocity values are used in the calculation of the error indicator.

14.7 Summary

This chapter provided a brief overview on finite element mesh generation and how to adapt the meshes for specific heat transfer problems. We believe that the adaptive mesh procedure has a role to play in heat transfer calculations. However, the ample availability of computing power significantly reduces the need for adaptive meshing nowadays. The fast growth in processor speed and parallel computing have reduced the need for adaptive meshing for heat transfer problems. However, the ability of adaptive meshing in reducing computing costs for a prescribed accuracy can be demonstrated in certain problems(Davies *et al.* 2008). Thus, adaptive meshing may still play an important role in certain heat transfer problems as the complexity of problems increase.

References

Borouchaki H and George PL (1997) Aspects of 2-D Delaunay mesh generation. *International Journal for Numerical Methods in Engineering*, **40**, 1957–1975.

Bowyer A (1981) Computing Dirichlet tessellations. *The Computer Journal*, **24**(2), 162–166.

Carslaw HS and Jaeger JC (1959) *Conduction of Heat in Solids*. Clarendon Press, Oxford.

Davies R, Davies H, Hassan O, *et al.* (2007) Investigations into the applicability of adaptive finite element methods to two-dimensional infinite Prandtl number thermal and thermochemical convection. *Geochemistry Geophysics Geosystems*, **8**, 1–25.

Davies R, Davies H, Hassan O, *et al.* (2008) Applicability of adaptive mesh generation for mid-ocean ridge and subduction zone problems. *International Journal of Numerical Methods for Heat and Fluid Flow*, **18**, 1015–1035.

Du Q and Wang D (2004) Boundary recovery for three-dimensional conforming Delaunay triangulation. *Computer Methods in Applied Mechanics and Engineering*, **193**, 2547–2563.

Delaunay B (1934) Sur la sphere vide, *Izvestia Akademii Nauk SSSR, Otdelenie Matematicheskikh i Estestvennykh Nauk*, **7**, 793–800.

Frey WH and Field DA (1991) Mesh relaxation: A new technique for improving triangulation. *International Journal for Numerical Methods in Engineering*, **31**, 1121–1133.

Frey P and George PL (2008) *Mesh Generation*. John Wiley & Sons Ltd, Chichester.

Huang H-C and Usmani A (1994) *Finite Element Analysis for Heat Transfer, Theory and Software*. Springer-Verlag, London.

Vasanth Kumar KS, Ramesh Babu AV, Seetharamu KN, *et al.* (1997) A generalized Delaunay triangulation algorithm with adaptive grid size control. *Communications in Numerical Methods in Engineering*, **13**, 941–948.

Lewis RW, Huang H-C, Usmani AS, and Cross JT (1991) Finite element analysis of heat transfer and flow problems using adaptive remitting including application to solidification problems. *International Journal of Numerical Methods in Engineering*, **32**, 767–781.

Löhner R (2001) *Applied CFD Techniques*. John Wiley & Sons, Inc., New York.

Nithiarasu P (2000) An adaptive finite element procedure for solidification problems. *Heat and Mass Transfer*, **36**, 223–229.

Nithiarasu P (2002) An adaptive remeshing scheme for laminar natural convection problems. *Heat and Mass Transfer*, **38**, 243–250.

Nithiarasu P and Zienkiewicz OC (2000) Adaptive mesh generation for fluid mechanics problems. *International Journal for Numerical Methods in Engineering*, **47**, 629–662.

Peraire J, Vahdati M, Morgan K and Zienkiewicz OC (1987) Adaptive remeshing for compressible flow computations. *Journal of Computational Physics*, **72**, 449–466.

Peraire J, Peiro J and Morgan K (1999) Advancing front grid generation. In JF Thompson, BK Soni and NP Weatherill (eds), *Handbook of Grid Generation, chapter 17*. CRC Press, London.

Sazonov I and Nithiarasu P (2012) Semi-automatic surface and volume mesh generation for subject-specific biomedical geometries. *International Journal for Numerical Methods in Biomedical Engineering*, **28**, 133–157.

Sazonov I, Hassan O, Morgan K and Weatherill WP (2006) Smooth Delaunay-Voronoi dual meshes for co-volume integration schemes. In PP Pébay (ed.), *Proceedings of 15 International Meshing Roundtable*, 529–541, Springer, New York.

Schoberl J (1997) NETGEN An advancing front 2D/3D-mesh generator based on abstract rules, *Computing and Visualization in Science*, **1**, 41.

Watson DF (1981) Computing the n-dimensional tessellation with application to Voronoi polytopes, *The Computer Journal*, **24**(2):167–172.

Weatherill NP (1990) The integrity of geometrical boundaries in the two-dimensional Delaunay triangulation. *Communications in Applied Numerical Methods*, **6**, 101–109.

Weatherill NP (1992) Delaunay triangulation in computational fluid dynamics. *Computers and Mathematics with Applications*, **24**, 129–150.

Weatherill NP and Hassan O (1994) Efficient 3-dimensional Delaunay triangulation with automatic point creation and imposed boundary constraints. *International Journal for Numerical Methods in Engineering*, **37**, 2005–2039.

Zienkiewicz OC and Zhu JZ (1987) A simple error estimator and adaptive procedure for practical engineering analysis. *International Journal for Numerical Methods in Engineering*, **24**, 337–357.

Zienkiewicz OC, Taylor RL and Nithiarasu P (2013) *The Finite Element Method, Vol. 3, Fluid Dynamics*. Elsevier, Amsterdam.

Zienkiewicz OC, Taylor RL and Zhu JZ (2013) *The Finite Element Method, Vol. 1, The Basis*. Elsevier, Amsterdam.

15

Implementation of Computer Code

15.1 Introduction

In this chapter a brief introduction to the implementation of computer code is given. It is assumed that readers are familiar with Fortran programming (Smith and Griffiths 1998; Wille 1995). The whole chapter is based upon the CBS scheme and the time-stepping algorithm discussed in the previous chapters. The discussion is limited to the essential aspects of CBSflow code. However, discussion on pre- and post-processing technique is common to many other schemes. Although CBSflow is a heat convection code, heat conduction may also be solved if the velocity calculations are suppressed. The codes may be downloaded from the website, www.zetacomp.com.

The following discussion will be limited to linear triangular elements, which has been discussed in detail in Chapters 3 and 7. The basic source codes for simple mesh generation and analysis are freely available for the readers to carry out two-dimensional studies.[1]

In general all the numerical programs contain three parts, that is, preprocessing, the main processing unit and postprocessing. The preprocessing part includes mesh generation, data structure and most of the element related data, which are constant for an element. The main processing unit is responsible for the computational effort and often most of the computing (CPU) time during a calculation. Efficient programming can reduce the CPU time, which is especially important in three dimensions. The details of an efficient data structure are not discussed here

---

[1] All the source codes available from the authors are copyrighted to the authors who developed the code. None of the material available within the code should be reproduced/copied in any form for commercial purposes without the written permission of the authors of the source codes. Readers are expected to acknowledge by citing the book in their publications if the full/part of the code is used for producing results

---

Fundamentals of the Finite Element Method for Heat and Mass Transfer, Second Edition.
P. Nithiarasu, R. W. Lewis, and K. N. Seetharamu.
© 2016 John Wiley & Sons, Ltd. Published 2016 by John Wiley & Sons, Ltd.

but readers may obtain information on such issues in various other relevant items of literature (Löhner 2001). In this chapter the basic implementation procedure are given so that the readers can understand the basics of the computer implementation of the finite element method.

The final part of a finite element code is the postprocessing unit. This unit can either be a coupled postprocessor, which directly gives the solution in graphical form or may be linked to an external postprocessor via an interface. The latter option is chosen in this text and the readers can then prepare their own interface and link to a postprocessing unit. Often it is necessary to extract data along a line within a domain. In such a situation one can either use other available software or employ an interpolation routine to compute the data along an arbitrary line or at a point.

The CBSflow code has been used for various applications in the past. The overall procedure of time-stepping CBSflow code for thermal problems can be summarized as

```
call preprocessing  ! preprocessing
do itime = 1,ntime  ! time loop
   call timestep     ! time step calculation
   call step1        ! intermediate momentum
   call step2        ! pressure calculation
   call step3        ! momentum(velocity) correction
   call step4        ! temperature calculation
   call check        ! check for steady state
enddo !
call postprocessing !postprocessing (output)
```

More details are given in the following sections.

15.2   Preprocessing

As mentioned previously the preprocessing operation normally takes place before the main solution unit. Often, the mesh generation solution is kept separately from the rest of the routines in order to simplify the data preparation. Such an approach is followed here and the mesh generation algorithm is kept separate from the rest of the program.

15.2.1   Mesh Generation

As mentioned in previous chapter, there are two main types of meshes, viz., structured and unstructured meshes. Structured meshes are generally simple in form and follow a certain pattern, which may either be uniform or nonuniform. Alternatively, unstructured meshes follow no particular pattern and are generated by dividing a domain into an arbitrary number of triangles or other finite element shapes. Since unstructured meshes follow no fixed pattern, the control of the solution accuracy in those sections of the domain which are dominated by high gradients is difficult. Structured meshes on the other hand results in more accurate solutions. However, the generation of a structured mesh for a complex geometry, especially in three dimensions, is both time-consuming and difficult. Therefore unstructured meshes, which are generated by a suitable unstructured mesh generator will be used in this text.

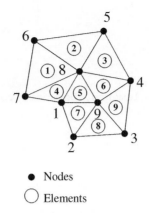

● Nodes

○ Elements

Figure 15.1   A typical unstructured mesh.

There are several methods available of generating unstructured meshes. Two of the most prominent methods are the "advancing front" (Löhner 2001; Peraire and Morgan 1997; Peraire *et al.* 1987) and "Delaunaly triangulation" (Kumar *et al.* 1997; Lewis *et al.* 1995; Thompson *et al.* 1999; Weatherill *et al.* 1994) techniques. Most of the unstructured meshes used in this book are generated by either one of these methods. Controlling the quality of elements, e.g, aspect ratio, is much easier in the Delaunay approach than in the advancing front method (chapter 14).

It is common practice to store finite element data in terms of the nodal coordinates and element connectivity. In addition to these, some convenient form of boundary condition specification is also necessary. It is therefore important that a mesh generator enables the coordinates of discrete points, the nodal connectivity of the finite elements and some form of boundary node/side information. A typical mesh is shown in Figure 15.1 and the typical input from a mesh generator is given by:

```
no of nodes, no elements and no of boundary sides
9   9   7
Element number and connectivity
1   7   8   6
2   6   8   5
3   8   4   5
4   1   8   7
5   1   9   8
6   9   4   8
7   2   9   1
8   2   3   9
9   9   3   4
Node number and xy-coordinates
1   1.1   1.2
2   1.6   0.0
3   3.3   0.1
4   3.4   1.9
```

```
5   2.1   3.3
6   0.4   3.0
7   0.0   1.0
8   1.8   2.1
9   2.3   1.1
Boundary side nodes and elements, boundary condition code
1   2   7   1
2   3   8   1
3   4   9   1
4   5   3   1
5   6   2   2
6   7   1   2
7   1   4   2
```

In the above mesh data, the total number of linear triangular elements is nine, the number of nodes is also nine and the number of boundary sides is seven. The element connectivity of all the elements is numbered in the anticlockwise direction. The node numbering follows no particular pattern. For simply connected domains the outer boundary sides are numbered in the anticlockwise direction and in a multiply connected domain the inner boundary is numbered in the clockwise direction.

The above-mentioned data structure of the element connectivity and the boundary side numbering are essential to make sure that the areas of the triangular elements are positive and that the appropriate boundary normals are determined from the boundary side data.

Note that the boundary condition code, that is, last column in the boundary side data, is used to represent an appropriate boundary condition on a side. For example, 1 in the above data can be used to represent an inlet condition and 2 may be used to represent a solid wall condition (no-slip). The third column in the boundary side data is the element to which the corresponding side belongs. This information is useful in evaluating the boundary integral terms and helpful in applying Neumann boundary conditions. The above data is normally prepared by a mesh generator, and once available from the mesh generator, these data may be read into the main analysis code by the following arrays

```
intma(i,j) - Connectivity array. i = 1,2,3
 and j = 1,2...number of elements
coord(i,j) - Coordinates array. i = 1,2
 and j = 1,2 ... number of nodes.
isido(i,j) - Boundary side array. i=1,2,3,4
 and j = 1,2, ..number of boundary sides.
```

15.2.2   Linear Triangular Element Data

As mentioned before, only linear triangular elements will be considered in this chapter. The essential data, including the mesh data and any other relevant data, are read from various input files at the preprocessing stage. Once all the external data are available, the remaining preprocessing procedure is carried out by the program. Some of the important preprocessing aspects of the finite element program are given in the following subsections.

Figure 15.2 A triangular element.

15.2.3 Element Area Calculation

The areas of the triangular elements are necessary for any finite element calculation and these areas are constant if the mesh is unchanged throughout the analysis. With reference to Figure 15.2, the area of an element may be determined from the following expressions:

$$A = \int dx_1 dx_2 = \frac{1}{2} \begin{vmatrix} 1 & x_{1i} & x_{2i} \\ 1 & x_{1j} & x_{2j} \\ 1 & x_{1k} & x_{2k} \end{vmatrix}. \tag{15.1}$$

Note that i, j and k are the nodes and the subscripts 1 and 2 indicate the coordinate directions. A sample routine, which calculates the area of the elements and the derivatives of the shape functions is given below

```
c-------------------------------------------------------------------
        subroutine getgeo(mxpoi,mxele,npoin,nelem,coord,intma,geome)
c-------------------------------------------------------------------
c       Derivatives of shape functions and 2A are calculated and

c       stored in the array geome(7,mxele). First six entries  are

c       derivatives of the shape functions and the last one

c       (seventh) is two times the area of an element

        implicit  none

        integer   mxpoi, mxele, npoin, nelem, ielem, inode, in

        integer   intma(3,mxele)

        real*8    x21,x31,y21,y31,rj,rj1,xix,xiy,etx,ety
        real*8    rnxi,rnet

        real*8    geome(7,mxele),   coord(2,mxpoi)
```

```
real*8    x(3),y(3),pnxi(3),pnet(3) !local arrays

data pnxi/-1.0d00, 1.0d00, 0.0d00/
data pnet/-1.0d00, 0.0d00, 1.0d00/

do ielem      = 1,nelem !loop over number of elements
  do inode    = 1,3
    in        = intma(inode,ielem)
    x(inode)  = coord(1,in)
    y(inode)  = coord(2,in)
  enddo !inode
  x21         = x(2)-x(1)
  x31         = x(3)-x(1)
  y21         = y(2)-y(1)
  y31         = y(3)-y(1)
  rj          = x21*y31-x31*y21
  rj1         = 1.0d+00/rj
  xix         =  y31*rj1
  xiy         = -x31*rj1
  etx         = -y21*rj1
  ety         =  x21*rj1
  do in                 = 1,2
    rnxi                = pnxi(in)
    rnet                = pnet(in)
    geome(in,ielem)     = xix*rnxi + etx*rnet
    geome(in+3,ielem)   = xiy*rnxi + ety*rnet
  enddo !in
  geome(3,ielem)        = -( geome(1,ielem) + geome(2,ielem) )
  geome(6,ielem)        = -( geome(4,ielem) + geome(5,ielem) )
  geome(7,ielem)        = rj ! two times area

enddo !ielem
end
```
!--

As stated previously, if the mesh is unchanged during the analysis then the above calculation is carried out once only and all the values are stored in the arrays for use in the main unit of the program.

15.2.4   Shape Functions and Their Derivatives

For linear elements an explicit calculation of the shape functions is not necessary as these may integrated directly. However, it is necessary to calculate the derivatives of the shape functions, which are constant for a linear element. Therefore, these derivatives can be evaluated at the preprocessing stage and stored in an appropriate array. For a linear triangular element we require six derivatives of the shape functions, that is,

$$\frac{\partial N_i}{\partial x_1}; \frac{\partial N_j}{\partial x_1}; \frac{\partial N_k}{\partial x_1}; \frac{\partial N_i}{\partial x_2}; \frac{\partial N_j}{\partial x_2} \quad \text{and} \quad \frac{\partial N_k}{\partial x_2}. \tag{15.2}$$

These derivatives are calculated and stored in the first six entries of an array

```
geome(7,mxele)
```

as mentioned in the previous subsection. Further details on the shape function derivatives are given in Chapter 3. Once the derivatives of the shape functions are stored, a calculation of the derivatives of any function/variable is straightforward. For example, the x_1 and x_2 derivatives of a nodal variable,

```
unkno(2,ip)
```

within the elements are calculated as

```
do ie         = 1,nelem !loop over elements
   dpdx(ie)   = 0.0d00   !x_1 derivative
   dpdy(ie)   = 0.0d00   !x_2 derivative
   do i       = 1,3
      ip       = intma(i,ie)
      dpdx(ie) = dpdx(ie) + geome(i,ie)*unkno(2,ip)
      dpdy(ie) = dpdy(ie) + geome(i+3,ie)*unkno(2,ip)
   enddo !i
enddo !ie
```

These derivatives will be constant over an element for linear triangular elements.

15.2.5   Boundary Normal Calculation

The unit boundary outward normal, **n** is shown in Figure 15.3. The components n_1 and n_2 are calculated and stored in an array at the preprocessing stage if the mesh is unchanged during the calculation. In addition to the normal components the boundary side lengths are also computed

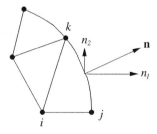

Figure 15.3   Outward normal from a boundary side.

and stored in the same array. The sample routine, which calculates the normal components
and side lengths, is given below.

```
c-----------------------------------------------------------------
      subroutine getnor(mxpoi,mxbou,npoin,nboun,coord,isido,rsido)
c-----------------------------------------------------------------
c     Boundary normal calculation

      implicit   none

      integer    mxpoi, mxbou, npoin, nboun,ib,ipoi0,ipoi1
      integer    isido(4,mxbou)

      real*8     dx,dy,rl
      real*8     rsido(3,mxbou),   coord(2,mxpoi)

      call rfillm(rsido,3,nboun,0.0d00) !fill with zeros
      do ib          = 1, nboun !loop over boundary sides
        ipoi0        = isido(1,ib) !first node of a side
        ipoi1        = isido(2,ib) !second node of a side
        dx           = coord(1,ipoi1) - coord(1,ipoi0)
        dy           = coord(2,ipoi1) - coord(2,ipoi0)
        rl           = dsqrt(dx*dx+dy*dy) ! length of a side
        rsido(1,ib) =  dy/rl   ! cos(theta)
        rsido(2,ib) = -dx/rl   ! sin(theta)
        rsido(3,ib) =  rl      ! side length
      enddo !ib
      end
c-----------------------------------------------------------------
```

   Readers are reminded that the above routine will be applicable only if the outer boundary
sides are numbered in an anticlockwise fashion for simply connected domains. For multiply
connected domains the inner boundary sides should be numbered in a clockwise direction in
order to ensure that the normals point outwards in the analysis domain as shown in Figure 15.4.
   In the routine considered above, the term

```
rsido(3,mxbou)
```

is the array used to store the normal components and the side lengths. The first two entries are
the x_1 and x_2 components of the normals and the third entry is the side length.

15.2.6   Mass Matrix and Mass Lumping

The calculation of the mass matrices is requires at many stages during the solution of a heat
transfer problem. For example all the transient terms, if solved in an explicit mode, lead to
mass matrices after spatial and temporal discretizations . These mass matrices can be "lumped"

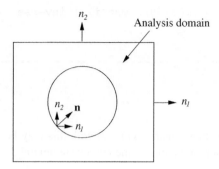

Figure 15.4   Multiply connected domain. Outward normal.

using a standard row summing approach if the steady-state solution is only of interest. In such situations, the mass matrix is lumped, inverted and stored in an array during the pre-processing stage if the mesh is unchanged during the calculation. For details of mass matrices and the lumping procedure, refer to Chapter 7. The following Fortran routine gives the details of how the inverse of the mass matrix is calculated and then stored into an array.

```
!--------------------------------------------------------------------
      subroutine getmat(mxpoi,mxele,npoin,nelem,intma,geome,dmmat)
!--------------------------------------------------------------------
c     This routine calculates inverse lumped mass matrix

c     and stores in an array dmmat(mxpoi)

      implicit    none

      integer     mxpoi, mxele, npoin, nelem,ielem,inode,i,in
      integer     intma(3,mxele)

      real*8      rj,rj6
      real*8      geome(7,mxele), dmmat(mxpoi)

      call rfillv(dmmat, npoin, 0.0d00) !fill with zeros

      do ielem = 1, nelem
        rj              = geome(7,ielem) ! 2A
        rj6             = rj/6.0d+00      ! A/3
        do inode = 1, 3
          in            = intma(inode,ielem)
          dmmat(in) = dmmat(in) + rj6 ! assembly
        enddo !inode
      enddo !ielem
      do i = 1, npoin
```

```
      dmmat(i)     = 1.0d+00/dmmat(i)  ! inverse
    enddo !i
    end
c--------------------------------------------------------------------
```

Note that

```
dmmat(mxpoi)
```

is the lumped and inverted mass matrix. Once stored, this may be used during the solution update of an explicit solution procedure in the main program unit.

15.2.7   Implicit Pressure or Heat Conduction Matrix

Often the pressure calculation in fluid dynamics or pure heat conduction calculations are carried out using implicit procedures. For instance the pressure Poisson equation of an incompressible flow calculation may have the following form:

$$\frac{\partial^2 p}{\partial x_1^2} + \frac{\partial^2 p}{\partial x_2^2} = \frac{1}{\Delta t}\left(\frac{\partial u_1^*}{\partial x_1} + \frac{\partial u_2^*}{\partial x_2}\right). \tag{15.3}$$

If a standard Galerkin weighting procedure and linear triangular elements are used then this will lead to the following discrete form of the LHS of the above equation (integration by parts) for a triangular element:

$$\frac{1}{4A}\begin{bmatrix} b_i^2 & b_i b_j & b_i b_k \\ b_j b_i & b_j^2 & b_j b_k \\ b_k b_i & b_k b_j & b_k^2 \end{bmatrix}\begin{Bmatrix} p_i \\ p_j \\ p_k \end{Bmatrix} + \frac{1}{4A}\begin{bmatrix} c_i^2 & c_i c_j & c_i c_k \\ c_j c_i & c_j^2 & c_j c_k \\ c_k c_i & c_k c_j & c_k^2 \end{bmatrix}\begin{Bmatrix} p_i \\ p_j \\ p_k \end{Bmatrix}, \tag{15.4}$$

where i, j and k are the three nodes of a triangle. The terms $\frac{b_i}{2A}, \frac{b_j}{2A}$ and $\frac{b_k}{2A}$ are the x_1 derivatives of the shape functions and $\frac{c_i}{2A}, \frac{c_j}{2A}$ and $\frac{c_k}{2A}$ are the x_2 derivatives of the shape functions (see Equation (15.2) and Chapters 3 and 7). The above equation needs to be assembled in order to obtain a global LHS matrix. As mentioned previously, the derivatives of the shape functions are constants and do not change if the mesh is fixed during the calculation. It is therefore convenient to calculate the matrices of the above equation at the pre-processing stage so that they may be used whenever necessary in the main unit of the code. A sample calculation of the pressure matrix for a banded (direct) matrix solver is given below.

```
c--------------------------------------------------------------------
      subroutine pstiff(mxpoi,mxele,mbw,npoin,nelem,nbw,intma,
    &                  geome,theta,gsm)
c--------------------------------------------------------------------
c *** calculates global LHS matrix for pressure

      implicit   none

      integer    mxpoi,mxele,mbw,npoin,nelem,nbw,i
```

```
      integer     ie,ip1,ip2,ip3,j,ielem,i3,j3,ii,i1,jj,i2,j1,j2

      integer     intma(3,mxele)

      real*8      area,thett

      real*8      geome(7,mxele), theta(2), gsm(mbw,mxpoi)
      real*8      s(3,3) !local

       do i = 1, npoin
         do j = 1, nbw
           gsm(j,i) = 0.0d00 !initialise
         enddo !j
       enddo !j

       do ielem = 1, nelem
         area     = geome(7,ielem)*0.5d00 ! area of an element
         thett    = theta(1)*theta(2) ! theta parameters (see
                                      ! Chapter 7 for details)
         do i    = 1, 3
           i3    = i + 3
           do j      = 1, 3
             j3      = j + 3
c      Element by element calculation of the shape function

c      derivatives and summation
             s(i,j) = thett*area*(geome(i,ielem)*geome(j,ielem)
     &                            + geome(i3,ielem)*geome(j3,ielem))
           enddo !j
         enddo !i
         do ii = 1, 3
           i1   = intma(ii,ielem)
           do jj   = ii, 3
             i2     = intma(jj,ielem)
             if(i2.lt.i1) then !banded arrangement
               j1 = i2
               j2 = i1
               j2 = j2 - j1 +1
               gsm(j2,j1) = gsm(j2,j1) + s(jj,ii)!assembly
             else
               i2 = i2 - i1 + 1 !banded arrangement
               gsm(i2,i1) = gsm(i2,i1) + s(jj,ii)!assembly
             endif
           enddo !jj
```

```
      enddo !ii
    enddo !ielem
  end
```
C--

In this case, the term

`gsm(mbw,mxpoi)`

is the global LHS matrix, which is unchanged during the calculation if the mesh is unaltered.

15.3   Main Unit

The following important list of parameters and quantities are normally available from the preprocessing unit.

```
intma(3,mxele) - connectivity; coord(2,mxpoi) - nodal coordinates;
isido(4,mxbou) - boundary side information; geome(7,mxele) -
derivatives of shape functions and element area; rsido(3,mxbou) -
boundary side normals and its length; dmmat(mxpoi) - lumped and
inversed mass matrix; gsm(mbw,mxpoi) - LHS matrix (only for
implicit solution); nelem - number of elements; npoin - number of
nodes, nboun - number of boundary sides
```

In addition to the above, several other quantities and parameters need to be either read from an input file or developed within a the preprocessing unit. Readers are asked to consult the source codes and manuals to understand these additional auxiliary parameters, which are available to download.

The discussion on the main unit of the program is provided here by assuming that a time-stepping approach is adopted for the solution of heat transfer problems and that the above listed parameters are available from the preprocessing unit.

15.3.1   Time-step calculation

As stated previously if a steady-state solution is obtained, via a time-stepping approach, an appropriate stable time step should be employed in the calculations. The time-step magnitude for a convection heat transfer problem may be stated as

$$\Delta t = min \left(\frac{h}{|u|}, \frac{h^2}{2v}, \frac{h^2}{2\alpha} \right), \tag{15.5}$$

where h is the element size, u is the velocity, v is the kinematic viscosity of the fluid and α is the thermal diffusivity. For Prandtl numbers of unity, the time-step values due to the kinematic viscosity and thermal diffusivity are equal. If the Prandtl number is greater than unity, then the time step calculated using the thermal diffusivity is greater than that of the one due to the kinematic viscosity. Assuming that the magnitude of the thermal time step, that is,

$h^2/2\alpha$ is greater than that of the viscous time step then the following routine may be utilized to calculate the value.

```
c-------------------------------------------------------------
      subroutine alotim( mxpoi, mxele, npoin, nelem, intma, geome,
     &              unkno, number, dtfix, ilots, csafm, ani  , deltp,
     &              delte )
c-------------------------------------------------------------
c     calculates the critical time step for all the elements

c     and nodes. iopt = -1 - fixed user specified global time step

c     (dtfix). iopt = 0 - global time step calculated as minimum

c     from all nodal values. iopt = 1 - local time step nodally

c     varies

      implicit   none

      integer    mxpoi,mxele,npoin,nelem,ilots,ip,ie,ip1,ip2,ip3

      integer    intma(3,mxele), number(mxpoi)

      real*8     u1,u2,u3,v1,v2,v3,vn1,vn2,vn3,veln,anx,any
      real*8     alen1,alen2,alen3,alen,dm,dtfix,csafm
      real*8     ani,aloti1,aloti2,tiny

      real*8     geome(7,mxele), unkno(4,mxpoi), deltp(mxpoi)
      real*8     delte(mxele)

c     global user specified fixed time step

      if(ilots.le.-1) then
         call rfillv(deltp, npoin, dtfix) !fill with fixed value
         call rfillv(delte, nelem, dtfix) !fill with fixed value
         return
      endif

      tiny  = 0.1d-05
      do ip = 1, npoin
         deltp(ip) = 1.0d06 !nodal value initialise
      enddo !ip
      do ie   = 1, nelem !loop over elements
```

```
        ip1   = intma(1,ie) !node1
        ip2   = intma(2,ie) !node1
        ip3   = intma(3,ie) !node3
        u1    = unkno(2,ip1) !u1 node1
        u2    = unkno(2,ip2) !u1 node2
        u3    = unkno(2,ip3) !u1 node3
        v1    = unkno(3,ip1) !u2 node1
        v2    = unkno(3,ip2) !u2 node2
        v3    = unkno(3,ip3) !u2 node3
        vn1   = dsqrt(u1**2 + v1**2) ! |V| node1
        vn2   = dsqrt(u2**2 + v2**2) ! |V| node2
        vn3   = dsqrt(u3**2 + v3**2) ! |V| node3
        veln  =  max(vn1,vn2,vn3)     ! Maximum |V|
        anx   = geome(1,ie)
        any   = geome(4,ie)
        alen1 = 1.0d+00/dsqrt(anx**2 + any**2) !element size (h1)
        anx   = geome(2,ie)
        any   = geome(5,ie)
        alen2 = 1.0d+00/dsqrt(anx**2 + any**2) !element size (h2)
        anx   = geome(3,ie)
        any   = geome(6,ie)
        alen3 = 1.0d+00/dsqrt(anx**2 + any**2) !element size (h3)
        alen  = min(alen1,alen2,alen3) !mimimum h

c    local time step

        aloti1     = alen/(veln+tiny) ! convection limit
        aloti2     = 0.5*alen**2/ani  ! viscous limit
        deltp(ip1) = min(deltp(ip1), aloti1,aloti2) !nodes
        deltp(ip2) = min(deltp(ip2), aloti1,aloti2) !nodes
        deltp(ip3) = min(deltp(ip3), aloti1,aloti2) !nodes
        delte(ie)  = min(deltp(ip3), aloti1,aloti2) !elements
      enddo !ie
      do ip = 1,npoin
        deltp(ip) = csafm*deltp(ip) !multiply by safety factor
      enddo !ip
      do ie = 1,nelem
        delte(ie) = csafm*delte(ie) !multiply by safety factor
      enddo !ie

c    global minimum time step

      if(ilots.eq.0)then
       dm = 5.0d03
```

```
      do ip = 1,npoin
        dm = min(deltp(ip),dm)
      enddo !ip
      do ip = 1, npoin
        deltp(ip) = dm
      enddo !ip
      do ie = 1, nelem
        delte(ie) = dm
      enddo!ie
      endif
      end
c- - - - - - - - - - - - - - - - - - - - - - - - - - - - - - - - - - - - - - - - - - - - - - - - - - - - -
```

The element size at a node is calculated in the routine using the sizes represented by Figure 15.5 as

$$h_i = min(h_1, h_2, h_3, h_4, h_5). \tag{15.6}$$

Again the above element size will be unchanged if the mesh is unaltered during a calculation. It is therefore possible to calculate and store the element sizes into an array at the preprocessing stage. A more accurate representation of an element size is possible by determining the element size in the streamline direction. However, such a calculation will lead to a variation in the element size at each time step, if a time-stepping scheme is employed, or it will vary at each iteration if a steady-state equation system with an iterative procedure is employed.

15.3.2   Element Loop and Assembly

A loop over the number of elements is the commonly employed form of LHS matrix/RHS vector construction in finite element codes. The assembly process is normally associated with the element loop. An example of such a loop when assembling the full viscous terms of the momentum equations is

```
      do ia = 1, nelem !loop over number of elements
        do lok      = 1, 3!loop over three nodes of an element
          in        = intma(lok,ia) !nodes of an element
```

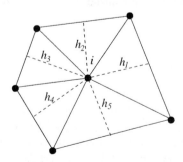

Figure 15.5   Element size calculation.

```
        lok1         = lok + 3
        velo1        = unkno(2,in) ! velocity component1
        velo2        = unkno(3,in) ! velocity component2
        sigxx(ia) = sigxx(ia) + ( ani )*
   &                  ( geome(lok,ia)*2.0*velo1 ) !stress 11
        sigyy(ia) = sigyy(ia) + ( ani )*
   &                  ( geome(lok1,ia)*2.0*velo2 )!stress 22
        sigxy(ia) = sigxy(ia) + ( ani )*
   &                  ( geome(lok,ia)*velo2
   &                  + geome(lok1,ia)*velo1 )    !stress 12
      enddo !lok
      do lok         = 1, 3
        lok1         = lok + 3
        rh1p(1,lok) = -geome(7,ia)*( sigxx(ia)*geome(lok,ia)
   &                  + sigxy(ia)*geome(lok1,ia) )*0.5d00
        rh1p(2,lok) = -geome(7,ia)*( sigxy(ia)*geome(lok,ia)
   &                  + sigyy(ia)*geome(lok1,ia) )*0.5d00
      enddo !lok
      do lok            = 1, 3
        in              = intma(lok,ia)
        do ja           = 1, 2
          ja1           = ja + 1
          rhs0(ja1,in) = rhs0(ja1,in) + rh1p(ja,lok) !assembly
        enddo !ja
      enddo !lok
    enddo !ia
```

The stress components, τ_{11}, τ_{22} and τ_{12} are determined element by element and assembled into the RHS vector

```
rhs0(4,mxpoi)
```

Both the stress arrays

```
sigxx(mxele); sigyy(mxele); sigxy(mxele)
```

and the RHS vector array have to be initialized to a value of zero at every time step of the calculation.

15.3.3   Updating Solution

Two types of solution updating are possible when a time-stepping procedure is employed. In the first type, a solution is updated after solving a simultaneous system of equations. In the second type, the solution is updated by multiplying a lumped and inverted mass matrix. In the latter procedure, the lumped mass matrix is a diagonal matrix and requires no simultaneous equation solution, as shown in the following portion of the code for the momentum equations.

```
c       add advection and diffusion RHS and multiply

c       by inversed mass

        do ip         = 1, npoin ! nodal loop
           dt         = dmmat(ip)
           rhs2(2,ip) = ( rhs2(2,ip) + rhs0(2,ip) )*dt
           rhs2(3,ip) = ( rhs2(3,ip) + rhs0(3,ip) )*dt
        enddo !ip


c       update the solution.

        do ip         = 1, npoin
           unkno(2,ip) = unkno(2,ip) + deltp(ip)*rhs2(2,ip) !update u_1
           unkno(3,ip) = unkno(3,ip) + deltp(ip)*rhs2(3,ip) !update u_2
        enddo !ip
```

Note that the time step is multiplied only at the end. The solution in the above part of the routine is updated as follows:

$$u_1^{n+1} = u_1^n + \Delta t * RHS * dmmat. \tag{15.7}$$

The matrix solution procedures for updating the analysis is carried out by either a direct or iterative solver. Direct solvers, such as the Gausian elimination technique, are employed when the simultaneous system is small and structured. However, for unstructured meshes and large systems, it is difficult to employ such direct solvers. It is therefore necessary to employ the iterative solvers, for example, conjugate gradient solver, in such situations. A typical LHS matrix is discussed in Section 15.2.7 for a banded direct solver. A RHS vector needs to be constructed before the solver can be used to obtain a solution. The RHS vector is constructed at each time step and is subjected to boundary conditions during the simultaneous solution procedure (see Chapter 3). The complete details of the solvers used are available along, with the source codes, from www.zetacomp.com.

15.3.4   Boundary Conditions

The boundary conditions are imposed after each time step by allotting an appropriate boundary condition code to a side (see mesh data). For instance the no velocity flux condition, or normal velocity zero condition, is imposed using the following routine during an explicit calculation. Note that the boundary code for such a condition is assumed to be 4.

```
c-----------------------------------------------------------------
        subroutine corsym( mxpoi, mxbou, npoin, nboun, unkno,
      &                    isido, rsido )
c-----------------------------------------------------------------
c *** Applies the zero velocity flux boundary conditions

        implicit   none
```

```
integer      mxpoi,mxbou,npoin,nboun,is,in,ip

integer      isido(4,mxbou)

real*8       anx,any,us

real*8       unkno(4,mxpoi), rsido(3,mxbou)

do is = 1, nboun
  if(isido(4,is).eq.4) then
     anx = rsido(1,is) !boundary normal
     any = rsido(2,is) !boundary normal
     do in = 1, 2
        ip = isido(in,is)
        us = -unkno(2,ip)*any + unkno(3,ip)*anx
        unkno(2,ip) = - us*any
        unkno(3,ip) =   us*anx
     enddo !in
  endif
enddo !is
end
```
c---

Note that

```
unkno(4,mxpoi)
```

is the unknown array. The first entry is the temperature, the second is the velocity component u_1, the third is the velocity component u_2 and fourth is the pressure. As seen in the above routine, the "no mass flux" condition is applied only to the velocity components.

15.3.5   Monitoring Steady State

The steady-state may be monitored via a fixed prescribed tolerance of the difference in a variable between two consecutive time steps. For example

$$max(\phi_i^{n+1} - \phi_i^n) \le 10^{-10}, \tag{15.8}$$

where ϕ is any variable such as velocity components, temperature etc. and the subscript i varies from 1 to the total number of nodes. Other ways of monitoring whether the steady state has been reached are discussed in Chapter 7. The following portion of the code explains how such a steady-state check is carried out between two consecutive time steps. In addition to screening the maximum difference, the following section of code stores the node at which such a maximum occurs.

```
do ip        = 1, npoin
   adel1      = unkno(1,ip) - unkn1(1,ip) !temperature
   adel2      = unkno(2,ip) - unkn1(2,ip) !u_1
```

```
      adel3          = unkno(3,ip) - unkn1(3,ip)  !u_2
      adel4          = pres1(ip) - pres(ip)         !pressure
      cder           = dabs(adel1)
      if(cder.gt.ha(1)) then
         icount(1) = ip                  !node
         ha(1)       = cder              !maximum value
      endif
      cder           = dabs(adel2)
      if(cder.gt.ha(2)) then
         icount(2) = ip                  !node
         ha(2)       = cder              !maximum value
      endif
      cder           = dabs(adel3)
      if(cder.gt.ha(3)) then
         icount(3) = ip                  !node
         ha(3)       = cder              !maximum value
      endif
      cder           = dabs(adel4)
      if(cder.gt.ha(4)) then
         icount(4) = ip                  !node
         ha(4)       = cder              !maximum value
      endif
   enddo !ip
   print*, (ha(ia),ia = 1,4)        !printing on screen
                                     max value
   print*, (icount(ia),ia = 1,4)  !printing on screen
                                     the node
```

Note that the array

```
unkn1(4,mxpoi)
```

stores the variables at the previous time step n. The array

```
unkno(4,mxpoi)
```

stores the variable values at the current time step of $n + 1$. The maximum difference between these two time levels forms the criterion for the steady-state condition.

15.4 Postprocessing

The postprocessing unit is mainly employed after a solution to a problem has been achieved. An interface to another graphical package may be linked to the main program unit so that the output from the main unit can be directly loaded into a postprocessor to visualize the data. For beginners it is important to asses the accuracy of the calculations by investigating the qualitative distribution of any quantity. The choice of graphical package is left to the user. The source code available on the web includes some standard packages.

15.4.1 Interpolation of data

It is often necessary to plot the quantities along a straight line within a domain or at an arbitrary point within a domain. If the nodes are not placed along the line of interest, or no node coincides with the point of interest, the variable required has to be interpolated using the shape functions. Such an interpolation routine may be used either as part of the main program unit or may be employed externally.

Once the data is obtained via interpolation, the plots may be generated using any standard package. Plots of interest can be of a spatial variation and or a temporal variation of the fluid flow and heat transfer variables.

15.5 Summary

In this chapter we have provided the readers with a brief introduction to the computer implementation of the finite element method for heat and fluid flow applications. Several advanced issues, such as the edge based data structure, parallel implementation and multi-grid acceleration procedure have not been discussed in this chapter. However, some appropriate references are provided for those who would like to read about such advanced topics. Further details on the programming and how to use the source codes are available at www.zetacomp.com.

References

Kumar KVS, Babu AVR, Seetharamu KN, *et al.* (1997) A generalised Delaunay triangulation algorithm with adaptive grid size control. *Communications in Numerical Methods in Engineering*, **13**, 941–948.

Lewis RW, Zhang Y and Usmani AS (1995) Aspects of adaptive mesh generation based on domain decomposition and delaunay triangulation. *Finite Elements in Analysis and Design*, **20**, 47–70.

Löhner R (2001) *Applied CFD Techniques*. John Wiley & Sons, Inc., New York.

Löhner R and Baum JD (1992) Adaptive h-refinement on 3D unstructured grid for transient problems. *Internationl Journal for Numerical Methods in Fluids*, **14**, 1407–1419.

Nithiarasu P (2000) Computer implementation of the CBS algorithm. In OC Zienkiewicz and RL Taylor, *The Finite Element Method, Vol 3, Fluid Dynamics, 5th Edition*. Butterworth Heinemann, Oxford.

Peraire J and Morgan K (1997) Unstructured mesh generation including directional refinement for aerodynamic flow simulation. *Finite Elements in Analysis and Design*, **25**, 343–356.

Peraire J, Vahdati M, Morgan K and Zienkiewicz OC (1987) Adaptive remeshing for compressible flow computations. *Journal of Computational Physics*,**72**, 449–466.

Smith I and Griffiths DV (1998) *Programming the Finite Element Method*, 3rd Edition, John Wiley & Sons, Ltd, Chichester.

Thompson JF, Soni BK and Weatherill NP (1999) *Handbook of Grid Generation*. CRC Press, London.

Weatherill NP, Eiseman PR, Hause J and Thompson JF (1994) *Numerical Grid Generation in Computational Fluid Dynamics and Related Fields*. Pinridge Press, Swansea.

Wille DR (1995) *Advanced Scientific Fortran*. John Wiley & Sons, Ltd, Chichester.

Appendix A

Gaussian Elimination

A system of simultaneous equations may be given as

$$\begin{aligned} a_{11}\phi_1 + a_{12}\phi_2 + a_{13}\phi_3 &= f_1 \\ a_{21}\phi_1 + a_{22}\phi_2 + a_{23}\phi_3 &= f_2 \\ a_{31}\phi_1 + a_{32}\phi_2 + a_{33}\phi_3 &= f_3 \end{aligned} \qquad (A.1)$$

In matrix form, this system of equations may be written as

$$\begin{bmatrix} a_{11} & a_{12} & a_{13} \\ a_{21} & a_{22} & a_{23} \\ a_{31} & a_{32} & a_{33} \end{bmatrix} \begin{Bmatrix} \phi_1 \\ \phi_2 \\ \phi_3 \end{Bmatrix} = \begin{Bmatrix} f_1 \\ f_2 \\ f_3 \end{Bmatrix}. \qquad (A.2)$$

The set of simultaneous equations given in Equation (A.1) or (A.2) may be solved using Gaussian elimination. The following step-by-step procedure may be followed for a 3×3 matrix. A similar procedure may be applied to larger matrices.

I. Eliminate ϕ_1 from second and third equations. (a) Multiply the first equation by a_{21}/a_{11} and subtract from the second equation, that is,

$$\begin{bmatrix} a_{11} & a_{12} & a_{13} \\ 0 & \left(a_{22} - a_{12}\frac{a_{21}}{a_{11}}\right) & \left(a_{23} - a_{13}\frac{a_{21}}{a_{11}}\right) \\ a_{31} & a_{32} & a_{33} \end{bmatrix} \begin{Bmatrix} \phi_1 \\ \phi_2 \\ \phi_3 \end{Bmatrix} = \begin{Bmatrix} f_1 \\ f_2 - \left(f_1\frac{a_{21}}{a_{11}}\right) \\ f_3 \end{Bmatrix} \qquad (A.3)$$

and (b) now multiple first equation by a_{31}/a_{11} and subtract from the third equation, that is,

$$\begin{bmatrix} a_{11} & a_{12} & a_{13} \\ 0 & \left(a_{22} - a_{12}\frac{a_{21}}{a_{11}}\right) & \left(a_{23} - a_{13}\frac{a_{21}}{a_{11}}\right) \\ 0 & \left(a_{32} - a_{12}\frac{a_{31}}{a_{11}}\right) & \left(a_{33} - a_{13}\frac{a_{31}}{a_{11}}\right) \end{bmatrix} \begin{Bmatrix} \phi_1 \\ \phi_2 \\ \phi_3 \end{Bmatrix} = \begin{Bmatrix} f_1 \\ f_2 - \left(f_1\frac{a_{21}}{a_{11}}\right) \\ f_3 - \left(f_1\frac{a_{31}}{a_{11}}\right) \end{Bmatrix}. \qquad (A.4)$$

Fundamentals of the Finite Element Method for Heat and Mass Transfer, Second Edition.
P. Nithiarasu, R. W. Lewis, and K. N. Seetharamu.
© 2016 John Wiley & Sons, Ltd. Published 2016 by John Wiley & Sons, Ltd.

II. Eliminate ϕ_2 from third equation, that is, multiply second equation by $\dfrac{\left(a_{32}-a_{12}\frac{a_{31}}{a_{11}}\right)}{\left(a_{22}-a_{12}\frac{a_{21}}{a_{11}}\right)}$ and

subtract from the third equation, that is,

$$
\begin{bmatrix}
a_{11} & a_{12} & a_{13} \\
0 & \left(a_{22}-a_{12}\frac{a_{21}}{a_{11}}\right) & \left(a_{23}-a_{13}\frac{a_{21}}{a_{11}}\right) \\
0 & 0 & \left(a_{33}-a_{13}\frac{a_{31}}{a_{11}}\right) - \left[\left(a_{23}-a_{13}\frac{a_{21}}{a_{11}}\right)\frac{\left(a_{32}-a_{12}\frac{a_{31}}{a_{11}}\right)}{\left(a_{22}-a_{12}\frac{a_{21}}{a_{11}}\right)}\right]
\end{bmatrix}
\begin{Bmatrix}
\phi_1 \\
\phi_2 \\
\phi_3
\end{Bmatrix}
$$

$$
= \left\{
\begin{matrix}
f_1 \\
f_2-\left(f_1\frac{a_{21}}{a_{11}}\right) \\
f_3-\left(f_1\frac{a_{31}}{a_{11}}\right)-\left(f_2-f_1\frac{a_{21}}{a_{11}}\right)\left[\frac{a_{32}-a_{12}\frac{a_{31}}{a_{11}}}{a_{22}-a_{12}\frac{a_{21}}{a_{11}}}\right]
\end{matrix}
\right\}. \quad \text{(A.5)}
$$

III. Use back substitution to find the values of ϕ_3, ϕ_2 and ϕ_1, i.e, ϕ_3 may be calculated as

$$
\phi_3 = \frac{f_3-\left(f_1\frac{a_{31}}{a_{11}}\right)-\left(f_2-f_1\frac{a_{21}}{a_{11}}\right)\left[\frac{a_{32}-a_{12}\frac{a_{31}}{a_{11}}}{a_{22}-a_{12}\frac{a_{21}}{a_{11}}}\right]}{\left(a_{33}-a_{13}\frac{a_{31}}{a_{11}}\right)-\left[\left(a_{23}-a_{13}\frac{a_{21}}{a_{11}}\right)\frac{\left(a_{32}-a_{12}\frac{a_{31}}{a_{11}}\right)}{\left(a_{22}-a_{12}\frac{a_{21}}{a_{11}}\right)}\right]}. \quad \text{(A.6)}
$$

The value of ϕ_3 can now be substituted into the second equation of Equation (A.5) to compute ϕ_2. After computing ϕ_3 and ϕ_2 respectively from the third and second equations, ϕ_1 can be calculated from the first equation of Equation (A.5).

The above description of Gaussian elimination is useful for programming the method.

Reference

Hamming RW (1987) *Numerical Methods for Scientists and Engineer.* Dover Books on Mathematics, New York.

Appendix B

Green's Lemma

Green's lemma states that, for differentiable functions α_1 and α_2, we can write (for a two-dimensional problem)

$$\int_\Omega \alpha_1 \frac{\partial \alpha_2}{\partial x_1} d\Omega = -\int_\Omega \frac{\partial \alpha_1}{\partial x_1} \alpha_2 d\Omega + \int_\Gamma \alpha_1 \alpha_2 n_1 d\Gamma \tag{B.1}$$

Similarly

$$\int_\Omega \alpha_1 \frac{\partial \alpha_2}{\partial x_2} d\Omega = -\int_\Omega \frac{\partial \alpha_1}{\partial x_2} \alpha_2 d\Omega + \int_\Gamma \alpha_1 \alpha_2 n_2 d\Gamma, \tag{B.2}$$

where n_1 and n_2 are the components of the outward normals on the enclosed curve Γ (see Figure B.1) and Ω is the two-dimensional domain. Let us consider the integration of a second-order term weighted by the shape function. The following form is common in finite element formulations:

$$\int_\Omega N_k \frac{\partial^2 T}{\partial x_1^2} d\Omega. \tag{B.3}$$

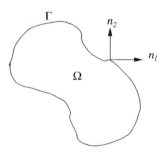

Figure B.1 Domain, boundary and outward normals.

Fundamentals of the Finite Element Method for Heat and Mass Transfer, Second Edition.
P. Nithiarasu, R. W. Lewis, and K. N. Seetharamu.
© 2016 John Wiley & Sons, Ltd. Published 2016 by John Wiley & Sons, Ltd.

Applying Green's lemma, the above equation becomes

$$-\int_{\Omega}\frac{\partial N_k}{\partial x_1}\frac{\partial T}{\partial x_1}d\Omega+\int_{\Gamma}N_k\frac{\partial T}{\partial x_1}n_1 d\Gamma. \tag{B.4}$$

In a similar fashion the x_2 direction can also be simplified using Green's lemma.

Appendix C

Integration Formulae

C.1 Linear Triangles

Let i, j and k be the nodes of a triangular element. Integrating over the triangular area gives

$$A = \int dx_1 dx_2 = \frac{1}{2} \begin{vmatrix} 1 & x_{1i} & x_{2i} \\ 1 & x_{1j} & x_{2j} \\ 1 & x_{1k} & x_{2k} \end{vmatrix}, \tag{C.1}$$

where A is the area of the triangle. For a linear triangular element, the integration of the shape functions can be written as

$$\int_\Omega N_i^a N_j^b N_k^c d\Omega = \frac{a! b! c! 2A}{(a + b + c + 2)!}. \tag{C.2}$$

On the boundaries

$$\int_\Gamma N_i^a N_j^b d\Gamma = \frac{a! b! l}{(a + b + 1)!}. \tag{C.3}$$

Note that $i - j$ is assumed to be the boundary side. The above equation is identical to the integration formula of a one-dimensional linear element. In the above equation l is the length of a boundary side.

C.2 Linear Tetrahedron

Let i, j, k and m be the nodes of a linear tetrahedron element. Integrating over the volume gives

$$V = \int dx_1 dx_2 dx_3 = \frac{1}{6} \begin{vmatrix} 1 & x_{1i} & x_{2i} & x_{3i} \\ 1 & x_{1j} & x_{2j} & x_{3j} \\ 1 & x_{1k} & x_{2k} & x_{3k} \\ 1 & x_{1m} & x_{2m} & x_{3m} \end{vmatrix}, \tag{C.4}$$

Fundamentals of the Finite Element Method for Heat and Mass Transfer, Second Edition.
P. Nithiarasu, R. W. Lewis, and K. N. Seetharamu.
© 2016 John Wiley & Sons, Ltd. Published 2016 by John Wiley & Sons, Ltd.

where V is the volume of a tetrahedron. For linear shape functions, the integration formula can be written as

$$\int_\Omega N_i^a N_j^b N_k^c N_m^d d\Omega = \frac{a!b!c!d!6V}{(a+b+c+3)!}. \tag{C.5}$$

On the boundaries

$$\int_\Gamma N_i^a N_j^b N_k^c d\Gamma = \frac{a!b!c!2A}{(a+b+c+2)!}. \tag{C.6}$$

Note that the above formula is identical to the integration formula of triangular elements within the domain. In the above equation A is the area of a triangular face.

Appendix D

Finite Element Assembly Procedure

Consider the two-dimensional linear triangular elements shown in Figure D.1. Let us assume the following elemental LHS matrix for the variable ϕ:

For the element 1

$$\mathbf{K_1} = \begin{bmatrix} a_{11} & a_{12} & a_{13} \\ a_{21} & a_{22} & a_{23} \\ a_{31} & a_{32} & a_{33} \end{bmatrix} \qquad (D.1)$$

and for the element 2

$$\mathbf{K_2} = \begin{bmatrix} b_{22} & b_{23} & b_{24} \\ b_{32} & b_{33} & b_{34} \\ b_{42} & b_{43} & b_{44} \end{bmatrix}, \qquad (D.2)$$

where the subscripts indicate the node numbers.

The elemental RHS vectors are:

For the element 1

$$\mathbf{f_1} = \begin{Bmatrix} c_1 \\ c_2 \\ c_3 \end{Bmatrix} \qquad (D.3)$$

and for the element 2

$$\mathbf{f_2} = \begin{Bmatrix} d_2 \\ d_3 \\ d_4 \end{Bmatrix}. \qquad (D.4)$$

Fundamentals of the Finite Element Method for Heat and Mass Transfer, Second Edition.
P. Nithiarasu, R. W. Lewis, and K. N. Seetharamu.
© 2016 John Wiley & Sons, Ltd. Published 2016 by John Wiley & Sons, Ltd.

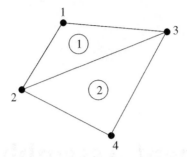

Figure D.1   A domain with two linear triangular elements.

Assembling the above elemental contributions gives the global equations as

$$[\mathbf{K}]\{\phi\} = \{\mathbf{f}\}, \tag{D.5}$$

where $[\mathbf{K}]$ and $\{\mathbf{f}\}$ are the global LHS matrix and RHS vector respectively and the unknown vector is $\{\phi\}$ is the unknown vector given for the system shown in Figure D.1 as follows:

$$\{\phi\} = \begin{Bmatrix} \phi_1 \\ \phi_2 \\ \phi_3 \\ \phi_4 \end{Bmatrix}. \tag{D.6}$$

The global LHS matrix is assembled as follows. The entries with the same subscripts in Equations (D.1) and (D.2) are added together to form an assembled global LHS matrix, that is,

$$[\mathbf{K}] = \begin{bmatrix} a_{11} & a_{12} & a_{13} & 0 \\ a_{21} & a_{22} + b_{22} & a_{23} + b_{23} & b_{24} \\ a_{31} & a_{32} + b_{32} & a_{33} + b_{33} & b_{34} \\ 0 & b_{42} & b_{43} & b_{44} \end{bmatrix}. \tag{D.7}$$

In a similar fashion, the RHS vector is assembled as

$$\{\mathbf{f}\} = \begin{Bmatrix} c_1 \\ c_2 + d_2 \\ c_3 + d_3 \\ d_4 \end{Bmatrix}. \tag{D.8}$$

The global system of equations is written as follows:

$$\begin{bmatrix} a_{11} & a_{12} & a_{13} & 0 \\ a_{21} & a_{22} + b_{22} & a_{23} + b_{23} & b_{24} \\ a_{31} & a_{32} + b_{32} & a_{33} + b_{33} & b_{34} \\ 0 & b_{42} & b_{43} & b_{44} \end{bmatrix} \begin{Bmatrix} \phi_1 \\ \phi_2 \\ \phi_3 \\ \phi_4 \end{Bmatrix} = \begin{Bmatrix} c_1 \\ c_2 + d_2 \\ c_3 + d_3 \\ d_4 \end{Bmatrix}. \tag{D.9}$$

As seen, there are four simultaneous equations, each of them associated with a node. The first equation, which is associated with node 1, is

$$a_{11}\phi_1 + a_{12}\phi_2 + a_{13}\phi_3 = c_1. \tag{D.10}$$

In the above equation, the contributions are from node 1 and the nodes connected to node 1. As seen, node 1 receives contributions from 2 and 3. Similarly, the second nodal equation receives contributions from all other nodes, which is obvious from Equation (D.9).

Appendix E

Simplified Form of the Navier–Stokes Equations

To derive the Navier–Stokes equations in their nonconservative form, we start with the conservative form. Conservation of mass:

$$\frac{\partial \rho}{\partial t} + \frac{\partial(\rho u_i)}{\partial x_i} = \frac{\partial \rho}{\partial t} + \rho \frac{\partial u_i}{\partial x_i} + u_i \frac{\partial \rho}{\partial x_i} = 0. \tag{E.1}$$

Conservation of momentum:

$$\frac{\partial(\rho u_i)}{\partial t} + \frac{\partial(u_j \rho u_i)}{\partial x_j} - \frac{\partial \tau_{ij}}{\partial x_j} + \frac{\partial p}{\partial x_i} = 0. \tag{E.2}$$

Conservation of energy:

$$\frac{\partial(\rho E)}{\partial t} + \frac{\partial(u_j \rho E)}{\partial x_j} - \frac{\partial}{\partial x_i}\left(k\frac{\partial T}{\partial x_i}\right) + \frac{\partial(u_j p)}{\partial x_j} - \frac{\partial(\tau_{ij} u_j)}{\partial x_j} = 0. \tag{E.3}$$

Rewriting the momentum equation with terms differentiated as

$$\rho \frac{\partial u_i}{\partial t} + u_i\left(\frac{\partial \rho}{\partial t} + \rho\frac{\partial u_j}{\partial x_j} + u_j\frac{\partial \rho}{\partial x_j}\right) + \rho u_j \frac{\partial u_i}{\partial x_j} - \frac{\partial \tau_{ij}}{\partial x_j} + \frac{\partial p}{\partial x_i} = 0 \tag{E.4}$$

and substituting the equation of mass conservation (Equation (E.1)) into the above equation gives the reduced momentum equation

$$\frac{\partial u_i}{\partial t} + u_j\frac{\partial u_i}{\partial x_j} - \frac{1}{\rho}\frac{\partial \tau_{ij}}{\partial x_j} + \frac{1}{\rho}\frac{\partial p}{\partial x_i} = 0. \tag{E.5}$$

Fundamentals of the Finite Element Method for Heat and Mass Transfer, Second Edition.
P. Nithiarasu, R. W. Lewis, and K. N. Seetharamu.
© 2016 John Wiley & Sons, Ltd. Published 2016 by John Wiley & Sons, Ltd.

The above momentum equation can be further simplified if the fluid is incompressible. For an incompressible fluid, the conservation of mass equation becomes

$$\frac{\partial u_i}{\partial x_i} = 0. \tag{E.6}$$

The deviatoric stresses in Equation (E.5) are written as

$$\tau_{ij} = \mu \left(\frac{\partial u_i}{\partial x_j} + \frac{\partial u_j}{\partial x_i} - \frac{2}{3} \frac{\partial u_k}{\partial x_k} \delta_{ij} \right). \tag{E.7}$$

Note that the last term in the above equation is zero from the continuity equation for incompressible flows. The devitoric stresses become

$$\tau_{ij} = \mu \left(\frac{\partial u_i}{\partial x_j} + \frac{\partial u_j}{\partial x_i} \right). \tag{E.8}$$

Substituting the above equation into Equation (E.5) we have (assuming μ is a constant)

$$\frac{\partial u_i}{\partial t} + u_j \frac{\partial u_i}{\partial x_j} - \frac{\mu}{\rho} \frac{\partial}{\partial x_j} \left(\frac{\partial u_i}{\partial x_j} + \frac{\partial u_j}{\partial x_i} \right) + \frac{1}{\rho} \frac{\partial p}{\partial x_i} = 0. \tag{E.9}$$

If we substitute $i = 1$ and $j = 1, 2$ we get the x_1 component of momentum equation as (in two dimensions)

$$\frac{\partial u_1}{\partial t} + u_1 \frac{\partial u_1}{\partial x_1} + u_2 \frac{\partial u_1}{\partial x_2} = -\frac{1}{\rho} \frac{\partial p}{\partial x_1} + 2v \frac{\partial^2 u_1}{\partial x_1^2} + v \frac{\partial^2 u_1}{\partial x_2^2} + v \frac{\partial}{\partial x_2} \left(\frac{\partial u_2}{\partial x_1} \right). \tag{E.10}$$

Rewriting the above equation as

$$\frac{\partial u_1}{\partial t} + u_1 \frac{\partial u_1}{\partial x_1} + u_2 \frac{\partial u_1}{\partial x_2} = -\frac{1}{\rho} \frac{\partial p}{\partial x_1} + v \frac{\partial^2 u_1}{\partial x_1^2} + v \frac{\partial^2 u_1}{\partial x_2^2} + v \frac{\partial}{\partial x_1} \left(\frac{\partial u_1}{\partial x_1} + \frac{\partial u_2}{\partial x_2} \right). \tag{E.11}$$

Applying the conservation of mass, we get

$$\frac{\partial u_1}{\partial t} + u_1 \frac{\partial u_1}{\partial x_1} + u_2 \frac{\partial u_1}{\partial x_2} = -\frac{1}{\rho} \frac{\partial p}{\partial x_1} + v \frac{\partial^2 u_1}{\partial x_1^2} + v \frac{\partial^2 u_1}{\partial x_2^2}. \tag{E.12}$$

In a similar fashion other components of momentum and energy equations can be simplified.

Appendix F

Calculating Nodal Values of Second Derivatives

In this method we assume that the second derivative is interpolated in exactly the same way as the main function and write the approximation as

$$\left(\frac{\partial^2 \phi}{\partial x_i \partial x_j}\right)^h = \mathbf{N} \overline{\left(\frac{\partial^2 \phi}{\partial x_i \partial x_j}\right)}^*. \tag{F.1}$$

This approximation is made to be a weighted residual approximation to the actual distribution of curvatures, that is,

$$\int_\Omega \mathbf{N}^{\mathrm{T}} \left[\mathbf{N} \overline{\left(\frac{\partial^2 \phi}{\partial x_i \partial x_j}\right)}^* - \frac{\partial^2 \phi^h}{\partial x_i \partial x_j} \right] d\Omega = 0 \tag{F.2}$$

and integrating by parts to give

$$\overline{\left(\frac{\partial^2 \phi}{\partial x_i \partial x_j}\right)}^* = \mathbf{M}^{-1} \left(\int_\Omega \mathbf{N}^{\mathrm{T}} \frac{\partial^2 \phi^h}{\partial x_i \partial x_j} \right) d\Omega = -\mathbf{M}^{-1} \left(\int_\Omega \frac{\partial \mathbf{N}^{\mathrm{T}}}{\partial x_i} \frac{\partial \mathbf{N}}{\partial x_j} d\Omega \right) \tilde{\phi}. \tag{F.3}$$

where $\mathbf{M}$ is the mass matrix given by

$$\mathbf{M} = \int_\Omega \mathbf{N}^{\mathrm{T}} \mathbf{N} \, d\Omega, \tag{F.4}$$

which of course can be "lumped."

Fundamentals of the Finite Element Method for Heat and Mass Transfer, Second Edition.
P. Nithiarasu, R. W. Lewis, and K. N. Seetharamu.
© 2016 John Wiley & Sons, Ltd. Published 2016 by John Wiley & Sons, Ltd.

Index

κ-ε model, 259
  eddy viscosity, 260

Rayleigh number, 188, 328, 401

Activation loss, 369
Activation over-potential, 369
Adaptive meshing, 379, 392
  anisotropic, 398
  choice of variables, 399
  de-refinement, 393
  error estimation, 393
  error estimator, 379
  error indicator, 392
  flux averaging, 393
  flux projection, 393
  h-refinement, 392
  heat conduction, 393, 394, 396
  interpolation error, 397
  mesh refinement, 379, 380
  p-refinement, 392
  r-refinement, 392
  regeneration, 392
  remeshing, 393
  stretching direction, 401
  stretching ratios, 401
Advancing front technique(AFT), 381, 392, 394, 407
  front, 381
  front-edge, 381
  front-face, 382

  poor elements, 382
  quality, 382
Alkaline fuel cell (AFC), 367
Analytical method, 158, 166
Analytical solution, 15, 106, 115, 123, 137, 166, 308, 309
Anode, 365, 374
Anodic compartment, 371
Area coordinate, 53
Arrhenius equation, 370
Artificial compressibility method, 213
  artificial compressibility parameter, 214
  dual time-stepping, 214
  real time solution, 214
Aspect ratio, 407
Assembly, 88, 122
Axisymmetric, 148
  element radius, 151
  gradient matrix, 150
  load vector, 151
  property matrix, 151
  stiffness matrix, 150
Axisymmetric problems, 243, 340
  conservation of energy, 244
  conservation of mass, 243
  conservation of momentum, 244

Backward Euler, 169
Backward facing step, 225, 235
  forced convection, 235
  turbulent flow, 266
  turbulent velocity profiles, 267

Fundamentals of the Finite Element Method for Heat and Mass Transfer, Second Edition.
P. Nithiarasu, R. W. Lewis, and K. N. Seetharamu.
© 2016 John Wiley & Sons, Ltd. Published 2016 by John Wiley & Sons, Ltd.

Benchmark solution, 401
Bezier curves, 390
Bilinear, 68
Bilinear element, 58, 142
Biot number, 159
Body force, 305
Boundary and initial conditions, 211
Boundary condition
    heat flux, 160
Boundary conditions, 13, 98, 106, 134, 160,
        407, 421
    convective, 134, 146, 151, 160, 163
    heat flux, 134, 146, 151, 163
    convective, 109
    heat flux, 109
Boundary mesh generation, 390
    planar domain, 390
Boundary normal, 411
Boundary side, 408
Boussinesq assumption, 255
Bowyer-Watson algorithm, 383
Brinkman, 324
Buoyancy, 305, 306
Buoyancy-driven convection, 186
Buoyancy-driven flow
    Nusselt number, 274
    turbulent, 272
Buoyancy ratio, 316
Butler-Volmer equation, 369

Capacitance matrix, 161, 162, 164, 165
Catalyst, 365, 371, 374
Cathode, 365, 366, 369, 374
Cathodic compartment, 371
CBS algorithm, 399, 405
CBS procedure, 374
CBSFlow, 406
Central difference, 169
Centroidal Voronoi tessellation (CVT),
        389
    Lloyd's iteration, 390
Channel flow, 219
    forced convection, 232
    turbulent velocity profile, 263
    velocity profile, 219

Characteristic based split (CBS) method,
        202
    boundary conditions, 211
    first step, 204
    fourth step, 206
    intermediate velocity, 210
    pressure, 210
    second step, 205
    spatial discretization, 207
    temperature calculation, 210
    third step, 206
    time-step calculation, 211
    velocity correction, 210
Characteristic dimension, 176, 360
Characteristic Galerkin, 189, 330
    convection-diffusion, 190
    simple, 190
    smoothing operator, 192
Circuit resistance, 370
Circular cylinder, 228, 269
    flow past, 228
    three-dimensional, 230
Circumcenter, 385
Colburn j-factor, 286
Collocation method, 83
Compact heat exchanger, 281
Composite slab, 20
Composite wall, 107, 108
Computational fluid dynamics (CFD), 175
Computer implementation, 405, 406
    boundary condition, 421
    element size, 419
    mesh generation, 406
Concave domain, 384
Concentration, 307
Concentration loss, 369, 370
Concentration over-potential, 369
Conduction, 2, 6, 10, 26, 89, 105
    analytical, 158
    axisymmetric, 148
    multi-dimensional, 131
    three-dimensional, 169
    transient, 157
    two-dimensional, 131
    unsteady, 157

solidification, 354
steady-state, 105
Conduction-convection systems, 123
Conjuate gradient, 421
Conjugate heat transfer
  identification of solid and fluid nodes,
    295
Conservation of energy, 182, 371
Conservation of mass, 359, 371
  derivation of, 177
  mass flux, 177
Conservation of momentum, 179, 371
  components, 182
  indicial form, 182
Conservation of species, 371
Continuity equation, 179, 203, 324, 327,
    328, 371
  derivation of, 177
  solidification, 359
Control volume, 302
Convection, 2, 6, 26, 28, 175
  axisymmetric, 243
  buoyancy-driven, 238, 356
  forced, 231
  mixed, 239
  natural, 238, 356
  porous media, 321
  solidification, 356, 361
Convection heat transfer, 321
Convection matrix, 194, 208
  one-dimensional, 194
Convection-diffusion
  characteristic Galerkin, 189, 190
  explicit, 192
  Lax-Wendroff, 189
  Multi-dimensions, 197
  steady-state, 201
  time-step restriction, 202
  two-dimensional, 201
Convection-diffusion equation, 188
Coordinate transformation, 60
Corrugated passages, 289
CPU, 405
Crank-Nicolson, 169
Cubic spline, 390

Current, 371
Current density, 370
Cylindrical coordinates, 12, 121
  heat source, 120
  hollow cylinder, 118
  one-dimensional, 118, 120
  solid cylinder, 120

Darcy, 361
  solidification, 358
Darcy number, 328
Darcy's law, 322
De-refinement, 393
Delaunay triangulation, 381, 382, 407
  background region, 384
  boundary recovery, 386
  Bowyer-Watson algorithm, 383
  circumcircle, 383
  circumsphere, 383
  concave domain, 384
  constrained, 383, 387
  convex hull, 383
  edge contraction, 386
  edge recovery, 386
  edge swapping, 383
  initial triangulation, 383
  minimum angle, 382
  out of the domain triangles, 383
  point creation, 385
  quadtree, 386
Detached eddy simulation, 278
Deviotoric stress, 180, 255
Diffusion coefficient
  mass, 372
Diffusion matrix, 195, 209
  one-dimensional, 195
Direct borohydride fuel cell (DBFC), 367,
    368
Direct Ethanol Fuel Cell (DEFC), 367
Direct formic acid fuel cell (DFAFC),
    367
Direct methanol fuel cell (DMFC), 367,
    368
Direct numerical simulation, 278
Dirichlet conditions, 304

Discrete systems, 19
  flow network, 23
  heat sink, 26
  matrix form, 22, 25
  slab, 20
  steady-state, 20
  steps, 20
  transient problem, 28
Divergence free velocity field, 179
Domain discretization, 379
Double diffusive, 306
Double-diffusive convection, 312, 347, 348
  axisymmetric cavity, 348
  square cavity, 347
Drag force, 215

Edge contraction, 389
Edge splitting, 389, 392
Edge swapping, 388
  relaxation index, 389
Effective heat capacity, 360
Effective thermal conductivity, 360
Effectiveness-NTU method, 285
Electric potential, 371, 373
Electrochemical device, 365
Electrochemical reaction, 366, 372
Electrodes, 365
Electrokinetic, 373
Electrolyte, 365, 367, 369, 374
Electron, 365, 366, 370, 372
Element characteristics, 22, 24, 28, 76, 87
Element connectivity, 407
Element loop, 419
Element size
  local maximum, 399
  local minimum, 399
  maximum, 398
  minimum, 398
Element size calculation, 419
Elemental matrix, 162
Emissivity, 4
Energy
  thermal, 366
Energy balance, 6, 8, 9, 11, 28, 89, 137,
      141, 145

Energy conservation, 354
  energy balance, 184
  indicial form, 184
Energy equation, 204, 307, 326–328, 371,
      373
  solidification, 359
Enthalpy, 354, 360
  total, 359
Enthalpy formulation, 354
Equilibrium, 311
Ergun, 322
Error (Gauss) function, 309
Error equilibration constant, 394
Error estimation, 393
Error estimator, 379
Error indicator, 392–394, 401
Error norms, 393
Exact solution, 106
Exercise, 16, 31, 100
Exothermic, 366
Experimental data
  fuel cell, 376
Explicit scheme, 308
Extended surface, 123
External circuit, 366
External reforming, 367

Faraday's constant, 369
Fick's law, 13, 301
Fin, 123
Finite difference method, 39, 163
Finite element assembly, 419
Finite element method
  assembly, 42
  discretization, 40
  domain, 40
  element, 40
  formulation, 42, 76
  history, 39
  identification of solid and fluid nodes,
      295
  linear element, 193
  node, 40
  numerical model, 40
  porous media, 329

post processing, 42
shape function, 42
simultaneous equations, 42
spatial discretization, 331
weighted residual form, 40
Finite element solution, 189
convection-diffusion, 189
fuel cell, 373
Navier-Stokes equations, 202
Finite volume method, 39
Fins, 26, 77, 96, 97
efficiency, 99
First law of thermodynamics, 5
Flow past sphere
forced convection, 233
Flow resistance, 23
Fluid dynamics, 175
Fluid flow
backward facing step, 225
circular cylinder, 228
isothermal, 218, 263, 266
laminar, 175, 218
lid-driven cavity, 221
nonisothermal, 265, 272
nonisothermal flow, 231
rectangular channel, 219
transient, 228
Flux boundary, 312
Force vector, 109, 120
Forced convection, 185, 305, 311, 334,
    379
backward facing step, 235
channel, 232
Nusselt number, 235
sphere, 233
Forced mass convection, 311
Forchheimer, 322, 372
Forcing vector, 109, 115, 120, 122, 125,
    195
element, 134
Formulation, 76
FORTRAN, 405
Forward difference, 166
Forward Euler, 169
Fourier's law, 3, 301

Fuel cell, 365
Alkaline fuel cell (AFC), 367
anode, 365
catalyst, 365
cathode, 365
continuity, 371
current, 366
Direct borohydride fuel cell (DBFC), 367
Direct Ethanol Fuel Cell (DEFC), 367
Direct formic acid fuel cell (DFAFC), 367
Direct methanol fuel cell (DMFC), 367
electrolyte, 365
energy, 371
experimental, 376
fuel, 365
heat transfer, 365
hydrogen, 366
ions, 365
loss, 369
mass transfer, 365
material properties, 366
mathematical model, 366, 368
Molten carbonate fuel cell (MCFC), 367
momentum, 371
operating temperature, 367
Phosphoric acid fuel cell (PAFC), 367
power, 366
Proton ceramic fuel cell (PCFC), 367
Proton exchange membrane fuel cell
    (PEMFC), 367
reaction, 365
Solid oxide fuel cell (SOFC), 367
species, 371
types, 366
voltage, 371
water, 366
Fuel cell types, 367

Galerkin method, 84, 86, 93, 150, 160, 168,
    198, 330
Gauss error function, 309
Gaussian elimination, 421
Generalized porous medium flow approach,
    324
Generalized porous medium model, 371

Gradient, 46
Gradient matrix, 133, 162
Grashof number, 188, 328
Green's lemma, 192

h-refinement, 392
Heat balance, 137, 141, 145
Heat conduction, 393, 394, 396, 414
Heat convection, 175, 401
　adaptive meshing, 399
　buoyancy-driven convection, 177, 186
　forced convection, 177, 185
　mixed convection, 177, 188
　natural convection, 177, 186
　types, 176
Heat dissipated, 98, 99
Heat dissipation, 123, 126
Heat exchangers, 281
　challanges, 297
　Colburn j-factor, 286
　compact heat exchanger, 281
　computational approach, 286
　conjugate heat transfer, 292
　corrugated passage, 289
　effectiveness-NTU, 283
　element characteristics, 288
　finite element mesh, 295
　finite element solution, 289
　identification of solid and fluid nodes,
　　295
　LMTD, 283
　Nusselt number, 292
　overall heat transfer coefficient, 281,
　　287
　passages, 289
　shell and tube heat exchanger, 287
　Stanton number, 286
　system approach, 286
　temperature difference, 282
　tubular heat exchanger, 281
Heat flow
　radial, 118
Heat flux, 5, 52, 125, 393
　element, 140
Heat generation, 8, 109, 112, 113, 121, 151,
　372

Heat sink, 26, 123
Heat source, 112, 113, 117, 121, 125, 134,
　146
Heat transfer
　conduction, 2, 10, 105, 354
　convection, 2, 175, 231, 356
　forced convection, 231, 334
　fuel cell, 365
　importance, 1
　laws, 3
　mixed convection, 239
　modes, 2
　natural convection, 238, 337, 401
　radiation, 3
　solidification, 354, 361
Heat transfer coefficient, 115, 125, 158
Heat transfer devices, 1
Hexahedron, 75, 384
Hollow cylinder
　conduction, 118
Horizontally divided enclosure, 345
Hydrogen, 366
Hydrogen fuel, 366
Hydrogen fuel cell, 366

Incompressible flow, 177, 182, 305, 325,
　372, 374
Infinitesimal control volume, 177, 324
Initial conditions, 6, 13, 160
Integration by parts, 160
Interface, 354
Interface velocity, 354
Intermediate velocity, 210
Internal energy, 158
Internal reforming, 367
Interpolation error, 397, 398
Interpolation function, 94
Interpolation of data, 424
Isoconcentration lines, 308
Isoparametric element, 63
　eight-node, 67
　one-dimensional, 64
　triangle, 69
　two-dimensional, 66
Isotherm, 52
Isotropic, 308

Jacobian, 70
Jacobian matrix, 65, 66

Kinematic viscosity, 329, 416
Knudsen diffusion coefficient, 372
Kolmogorov scale, 254
Kozeny–Carman relationship, 359
Kroneker delta, 181

Lagrange interpolation, 49
Laminar
    backward facing step, 225
    channel flow, 219
    circular cylinder, 228
    lid-driven cavity, 221
    nonisothermal flow, 231
Lamp, 7
Laplace equation, 308, 373
Laplace smoothing, 388, 390, 392
Large eddy simulation (LES), 275
    continuity, 275
    filtration, 275
    Kolmogorov constant, 277
    momentum, 275
    Smagorinsky model, 277
    standard subgrid scale, 277
    subgrid scale(SGS), 275
Latent heat, 354
Laws of heat transfer, 3
    first law of thermodynamics, 5
    Fourier's law, 3
    Newtons's law of cooling, 4
    Stefan-Boltzmann law, 4
Lax-Wendroff, 189
Least-squares method, 85
Lewis number, 329
Lid-driven cavity, 221
    velocity distribution, 222
Linear element, 45, 50, 86, 96, 119, 161,
        193
    one-dimensional, 121, 193
Linear temperature, 150
Linear variation, 133
Liquid volume fraction, 359, 360
Liquidus, 355
Lloyd's iteration, 390

LMTD method, 283
Load vector, 93, 120, 125, 133, 161, 162,
        165
    element, 134, 146
Local coordinate, 55, 108
Lumped heat capacity, 157

Main processing, 405, 416
Mass balance, 324
Mass diffusion, 303, 307
Mass diffusivity, 309
Mass flux, 13, 24, 304, 422
Mass lumping, 412
Mass matrix, 194, 208, 412
    one-dimensional, 194
Mass transfer, 13, 301
    buoyancy, 305
    channel flow, 304
    concentration, 301
    conservation, 302
    convection, 301, 302
    diffusion, 301
    Dirichlet conditions, 304
    Fick's law, 13, 301
    forced convection, 304–306
    fuel cell, 365
    importance, 1
    mixed convection, 305
    multiple species, 304
    porous media, 321, 347, 348
    Sherwood number, 349
Material property matrix, 134
Mathematical model, 368
Melting, 356
Mesh convergence, 218
Mesh cosmetics, 387, 392
Mesh generation, 379, 406
    advancing front, 381
    automatic, 381, 393
    body fitted, 381
    boundary, 390
    boundary discretization, 381
    Centroidal Voronoi tessellation (CVT),
        389
    conformal, 381
    cosmetics, 382

Mesh generation (*Continued*)
  Delaunay triangulation, 381, 382
  edge contraction, 389
  edge splitting, 389
  edge swapping, 388
  element sizing, 382
  Laplace smoothing, 388
  mesh smoothing, 388
  nonconformal, 381
  nonuniform mesh, 380
  quality, 382
  relaxation index, 389
  structured, 380
  triangulation, 382
  uniform mesh, 379
  unstructured mesh, 380
Mesh generator, 407
Mesh movement, 392
Mesh quality, 387
Mesh smoothing, 388
Mixed convection, 188, 239,
    305
  analytical solution, 242
  channel, 240
  velocity profile, 243
Mixing length, 258
Model heat exchanger, 292
Molar fraction, 370, 371
Molten carbonate fuel cell (MCFC), 367,
    368
Momentum conservation, 255
Momentum equation, 204, 325, 327, 328,
    371
  solidification, 359
Moving body, 8
Multi-dimensional, 169
Multiply connected domain, 412
Mushy region, 358

Natural convection, 238, 313, 337, 401
  adaptive meshing, 401
  axisymmetric, 340
  Nusselt number, 239
  square cavity, 401
  square enclosure, 239

Navier-Stokes equations, 177, 202, 372
  artificial compressibility method, 213
  Characteristic based split, 202
  conservation of mass, 177
  conservation of momentum, 179
  differential approach, 177
  nondimensional form, 184
  post-processing, 215
  Reynolds Averaged, 257
  steady solution, 213
  transient solution, 213
  turbulence, 256
Negative buoyancy, 316
Nernst equation, 369
Neumann boundary, 374
New element size, 394
Newton's law of cooling, 4, 302
Nodal coordinates, 407
Non-conventional energy, 365
Non-Darcy flow, 323, 338
Nondimensional form, 184, 304, 327
  continuity, 360
  continuity equation, 186, 187
  energy, 360
  energy equation, 186, 188
  forced convection, 185
  mixed convection, 188
  momentum, 360
  momentum equation, 186, 188
  natural convection, 186
  one-equation model, 261
  solidification, 360
  Spalart-Allmaras model, 261
  turbulence, 260
Nondimensional scale, 312
Nonisothermal flow, 231, 333
NURBS, 390–392
Nusselt number, 215, 235, 239, 266, 292,
    338, 340, 349

Ohmic loss, 369
Ohmic over-potential, 369
One-dimensional element, 43, 64
One-dimensional, 105
One-dimensional element, 161

One-equation models, 258
  Spalart-Allmaras model, 258
Ordinary differential equation, 158
Outward normal, 411
Overall heat transfer coefficient, 281
Oxygen, 365

p-refinement, 392
Peclet number, 186, 261
Percentage error, 394
Permeability, 322, 323, 359
  fuel cell, 372
  Kozeny–Carman relationship, 359
  solidification, 358
Phase change, 353, 356
Phase change temperature, 354
Phosphoric acid fuel cell (PAFC), 367,
    368
Planar configuration, 375
Plane wall, 105
  composite, 107
  heat source, 112, 115, 117
  homogeneous, 105
Plate, 5
  variable thikness, 145
Point source, 134
Porosity, 322, 374
  constant, 338
  fuel cell, 372
  mushy region, 358
  solidification, 358
  variable, 337
Porous media, 321
  axisymmetric, 340
  Boussenessq approximation, 338
  Brinkman extension, 324
  Darcy number, 328
  Darcy's law, 322
  double-diffusive convection, 347,
      348
  energy equation, 326
  Ergun correlation, 322
  finite element solution, 329
  forced convection, 334
  Forchheimer extension, 322

generalized approach, 324
interface problem, 342, 345
limiting cases, 329
mass balance, 324
momentum equation, 325
natural convection, 337
nondimensional form, 327
nonisothermal flow, 333
Nusselt number, 338
permeability, 322
Poisson equation, 330
porosity, 322
quasi-implicit, 332
reservoir simulation, 322
saturated, 322
semi-implicit, 332
solid matrix drag, 326
spatial discretization, 331
species equation, 327
square enclosure, 338
temporal discretization, 330
Porous medium-fluid interface, 342, 345
Postprocessing, 215, 405, 423
  drag force, 216
  Nusselt number, 215
  stream function, 217
Potential difference, 365
Prandtl number, 186, 188, 261, 286, 328,
    416
Preprocessing, 405, 406
Pressure calculation, 414
Pressure Poisson equation, 330
Property matrix, 143
Proton ceramic fuel cell (PCFC), 367
Proton exchange membrane fuel cell
    (PEMFC), 367

Quadratic element, 46, 55, 69, 74, 115,
    117
  one-dimensional, 115, 117
Quadrilateral element, 58
Quadtree, 386

r-refinement, 392
Radial heat flow, 118

Radiation, 3, 6, 28
  emissivity, 4
  shape factor, 4
  solutal, 306
Rayleigh-Ritz method, 77, 79
Reaction, 365
  anode, 369
  cathode, 369
Rectangular channel, 232, 240
  turbulent flow, 263
  turbulent heat transfer, 265
Rectangular element, 142
  gradient matrix, 143
  load vector, 143
  stiffness matrix, 143
Reference velocity, 176
Refinement strategy, 394
Remeshing, 393
Resistivity, 371
Reynolds Averaged Navier-Stokes
        equations (RANS), 257
  continuity equation, 257
  energy equation, 258
  momentum equation, 257
Reynolds number, 176, 186, 188, 261, 304,
        312, 328
Reynolds stress, 255
Richardson number, 188
Ritz method, 77, 78

Scalar variable, 188, 196
Schmidt number, 304, 312, 328
Second derivative, 398, 399
Second derivatives
  principle direction, 398
  principle values, 398
Semi-implicit, 169
Shape function, 42, 44, 45, 47, 51, 56, 60,
        74, 410
  compatibility, 94
  completeness, 94
  requirement, 94
Shape function derivatives, 410
Shell and tube heat exchanger, 287
Sherwood number, 312, 349

Simultaneous equations, 8, 110, 166
Smoothing operator, 192
Solar, 2, 5
Solid cylinder, 122
  analytical solution, 121
  conduction, 120
Solid oxide fuel cell (SOFC), 367, 368,
        374
  planar, 375
  solution procedure, 375
Solidification, 353
  conduction, 354
  continuity, 359
  convection, 356, 361
  energy equation, 359
  enthalpy formulation, 354
  equations, 358
  interface, 354
  latent heat, 354
  momentum, 359
  mushy region, 358
  one-dimensional, 354
  square cavity, 361
Solutal Rayleigh number, 306
Solution accuracy, 379
Solution update, 420
Spalart-Allmaras model, 258
  eddy viscosity, 259
Spatial discretization, 307
Spatial discretization, 86, 160
Species, 307, 371
Species concentration, 310
Species diffusion, 308
Species equation, 327, 329, 371
Sphere, 233
Spherical coordinates, 12
Square cavity, 401
  solidification, 361
  turbulent natural convection, 272
Square enclosure, 338
Square plate, 135, 140, 144
Stabilization matrix, 195
  one-dimensional, 195
Stability, 169, 201
  von Neumann, 201

Standard electrode potential, 369
Stanton number, 286
Static condensation, 117, 118
Static potential, 373
Steady state, 310, 374
Stefan-Boltzmann law, 4
Stiffness matrix, 93, 108, 110, 119, 121,
          133, 161, 162, 165
  element, 134, 146
Stream function, 215
Stretching ratio, 399
Structured mesh, 380, 406
  multi-block, 381
Subdomain method, 83
Subdomain, 374
Symmetry, 113
System approach, 286

Target error, 394
Taylor expansion, 163, 302, 303
Taylor series, 10
Temperature, 307
Temperature distribution, 133
Temporal discretization, 307, 330,
          412
Tetrahedron, 146
  linear, 72
  quadratic, 74
Thermal conductivity, 134
Thermal diffusivity, 184, 329, 416
Thermal potential, 107
Thermal resistance, 107
Thick wall cylinder, 120
Three-dimensional, 146
Three-dimensional element, 72
  hexahedron, 75
  tetrahedron, 72
Time averaging, 254
  momentum equation, 255
Time constant, 159
Time-dependent, 309
Time discretization, 163
  Crank-Nicolson, 164
  explicit, 164
  finite difference method, 163

finite element method, 166
  first-order, 164
  implicit, 164
  Multi-dimensional, 169
  semi-implicit, 164
  stability, 169
Time-stepping, 416
Tortuosity, 372
Total enthalpy, 359
Transfer coefficient, 370
Transient, 157, 188, 309, 310
  circular cylinder, 228
Transient problem, 28
Triangle, 49, 70
  linear, 49
  quadratic, 55
Triangular element, 132, 405, 408
  linear, 132
  area, 409
  linear, 408
Triangulation, 382
Turbulence
  κ-ε model, 259
  backward facing step, 266
  channel, 263
  circular cylinder, 269
  dissipation, 256
  eddy viscosity, 256, 259
  finite element solution, 262
  friction velocity, 265
  isothermal flow, 263, 266
  large eddy simulation (LES), 275
  mixing length, 258
  natural convection, 272
  nondimensional distance, 265
  nondimensional form, 260
  nonisothermal flow, 265, 272
  Nusselt number, 266
  One-equation models, 258
  Spalart-Allmaras model, 258
  square cavity, 272
  Two-equation models, 259
  unsteady RANS, 269
  wall shear stress, 265
Turbulent eddy viscosity, 255

Turbulent flow, 253
  Boussinesq assumption, 255
  continuity equation, 256
  decomposition, 255
  energy equation, 256
  fluctuating component, 255
  Kolmogorov length scale, 254
  momentum equation, 256
  random variation, 253
  Reynolds stress, 255
  time averaging, 254
  turbulent viscosity, 255
Turbulent heat transfer, 253
  channel, 265
  Prandtl number, 257
  turbulent thermal diffusivity, 256
Turbulent kinematic viscosity, 255, 256
Turbulent mass transport, 317
Turbulent Prandtl number, 257
Turbulent thermal diffusivity, 256
Two-dimensional element, 49
Two-Equation models, 259

Universal gas constant, 369
Unsteady, 157
Unsteady RANS, 269
Unstructured mesh, 308, 380, 401, 406

Variable cross-section, 110
Variable porosity, 337
Variable thickness, 145
Variational method, 90
Vertically divided enclosure, 342
Voroni tessellation, 389

Wall shear stress, 265
Water vapor, 312
Weighted residual method, 77, 82
  collocation, 83
  Galerkin, 84
  least-squares, 85
  subdomain, 83

Zero flux, 307, 312
Zetacomp.com, 405